Textbook on Fundamental of Extension Education

THE AUTHOR

Professor (Dr) Manas Mohan Adhikary, a retired Professor of Agricultural Extension, Faculty of Agriculture, Bidhan Chandra Krishi Viswavidyalaya, West Bengal was attached with the profession of teaching, research and extension for about (40) forty years. His immense potentiality of academic activities and capability to work with the people, sense of research and extension works helped him to serve as Dean, Faculty of Agriculture, Registrar, Bidhan Chandra Krishi Viswavidyalaya, Vice-Chancellor, Bidhan Chandra Krishi Viswavidyalaya, and Director, Directorate of Extension Education (Field Extension Wing) in the same University as well as Head of the Department of Agricultural Extension. Besides that he served as Director, Directorate of Open and Distance Learning, University of Kalyani, Kalyani, Nadia, West Bengal (2019-2021). He has become Course Director for the (8) eight consecutive Winter/Summer School of ICAR sponsored which itself a brilliant achievement in the National Level for any University Professor at his level and class. His hard work and the devotion in the aegis of extension affiliated him as Master Trainers of Experiential learning Cycle in ICAR-USAID programme in 1991. He also contributed at the National Level by becoming Chairman, Site Selection Committee for establishing the Krishi Vigyan Kendra and ushered a new pace in the KVK movement across India. He is the expert of different specialized fields like Group dynamics and leadership, Training of Trainers, Project Evaluation, Participatory Project Management, Value added Agriculture, Media Management *etc.* He has conducted (14) fourteen research and extension projects as Principal Investigator. He has already chaired a series of International Conference in India and Abroad and has already been honoured to Chaired the session on Dynamics of Livelihood changes across the World in IAEUS Congress, Manchester, UK in 2013. To his credit there are 108 numbers of scientific research papers published in National and International Journals, a good score of chapters and 26 numbers of books in related areas. He is widely travelled all through his carrier covering China, Japan, USA, Australia, Turkey, Bangladesh and Thailand *etc.*

Textbook on Fundamental of Extension Education

— *Author* —

Manas Mohan Adhikary

2022

Daya Publishing House®

A Division of

Astral International Pvt. Ltd.

New Delhi – 110 002

ISBN 978-93-5461-214-5

Published by : **Daya Publishing House®**
A Division of
Astral International Pvt. Ltd.
– ISO 9001:2015 Certified Company –
4736/23, Ansari Road, Darya Ganj
New Delhi-110 002
Ph. 011-43549197, 8130496929
E-mail: info@astralint.com
Website: www.astralint.com

Digitally Printed at : **Replika Press Pvt. Ltd.**

Message

Extension Education is the process of teaching rural people how to live better by learning ways to improve their farms, home, and community institutions. Extension Education is a science, which deals with the creation, transmission & application of knowledge designed to bring about planned changes in the behavior-complex of people; to help them to live better by learning the ways of improving their vocations, enterprises & institutions. In our country, every individual does not get a chance to be a part of school education during his/her childhood. Even after school and college education, people need the education to solve many practical problems during their day-to-day life. Today's technology becomes outdated tomorrow. In the modern age of science and technology, innovative inventions are rising day by day. Farming is no exception to this trend. The farmer's need is to supply recent useful and practical information related to agricultural production for better economic returns. Agricultural development will be closely related to the development of the ability of the farmers' understand and adoption of this technology. Many natural and manmade forces interact continuously in this world. This interaction makes our society more complex. It creates many new problems for the people at large. People should know the solution for these problems and extension education helps people at this critical stage. The education received by the people during school life becomes inadequate for an understanding of new emerging problems and their solutions. Therefore, education should be in continuity with the life of a farming community. Such continuity can be made possible through extension education. The major work of agricultural researchers is to develop new agricultural technologies to meet out the requirement of different agro-climatic conditions and other local problems faced by the farming community. Agricultural scientists neither have enough time nor are equipped to convince farmers directly to adopt improved scientific technologies to solve field-related problems and to improve agricultural production. Similarly, it is not easy for all farmers to visit the research stations and obtain knowledge about improved technologies. Thus, there is a need for an agency to interpret the findings of the scientific research to the farmers and also to collect information about the practical problems faced by the farmers to adopt these technologies and pass on the same to the research scientists for the needful solution. This transformation of information has been brought about by extension scientists. Hence, extension scientists bridge the gap between farmers and agricultural scientists.

Extension education is the democratic method of educating the farmers. It respects the individuality of farmers in making their own decision. In this way farmers' status grows, they maintain their self-respect and take balanced decisions for improving their farm income by adopting advanced agricultural practices and technologies. There are many farmer's innovations that they overlook unknowingly. These innovations are of no use unless they are put to practical use. The hidden farmer's technology is refined by the scientists and reaches the people speedily through extension scientists. Extension education is an out-of-school process, which aims to help people themselves in solving their day-to-day problems related to cultivation/cultural practices. The extension workers explain various ways to solve farmers' problems. The extension workers act as a guide and educate them on how to use new knowledge, skills, and attitudes in solving their problems. Extension education helps all classes of people men, women, and youth to solve their present and future economic, social, and cultural problems. There are many types of farmers, some of them are orthodox and rigid and they sometimes do not understand the significance of modern innovations. To motivate such farmers to make use of innovations, the role of extension education is incomparable. Extension education thus contributes to national development. The most important aim of extension education is to develop farmers economically, socially, and culturally as an individual, his family, village, state, and the country as a whole with the help of the implementation of different extension programs throughout the country. It also discusses both the conceptual elements as well as practical aspects that are relevant for applying the extension educational methodology. This publication will be useful to the agricultural extension, education, and training personnel, especially those who are involved in extension strategy development and program planning. This book is expected to equip and arm the agricultural experts in using tools of fundamental extension education and also to promote virtual education in agriculture and rural development. Besides that, the book has accommodated all possible innovative aspects of the basic concept of extension education with its rural developmental aspects followed by different rural-social communication, training, agricultural journalism, and entrepreneurial aspects of classical contents. The book has been written for the entry-level of UG students in full compliance to the basic Extension Educational aspects as per ICAR 5th Dean Committee recommended Under Graduate syllabus, as delineated by the ICAR, New Delhi. I hope the book will be welcomed by a galaxy of scholars and professionals and keep generating valuable feedback in reconstructing it in due course of time.

Lastly, I give my profuse thanks to Sri Anil Mittal 'Ji' the Director of M/S. Daya Publishing House a unit of Astral International Pvt Ltd, New Delhi, inspired me to publish one elementary book on extension education for entry-level UG students. I am indebted to him. I am grateful to my wife Mrs. Shibani Adhikary; son Mayukh Adhikary and daughter Manjeer Adhikary, whose constant efforts will also help me to complete this manuscript in due time without any delay. I also give my plentiful thanks to the Publisher M/S Astral International Private Limited, New Delhi, for their help in bringing out the book in time.

Manas Mohan Adhikary

Preface

The book *"**Textbook on Fundamental of Extension Education**"* has certainly been a selection by the time itself, not by the choice of the author. The book has accommodated all possible innovative aspects of the basic concept of extension education with its rural developmental aspects followed by different rural-social, communication and entrepreneurial aspects of classical contents. The book has been written for the entry level of UG students in full compliance to the basic Extension Educational aspects as per ICAR 5th Dean Committee recommended Under Graduate syllabus, as delineated by the ICAR, New Delhi. The characterization being: **Chapter I – Fundamental of Extension Education: a) Evolution of Extension Education:** Types of education, Meaning, Formal and Informal education, Definition, Concept, importance, Scope, Function, philosophy, process, levels, Objectives, functions and principles; **b)Extension Programme Planning:** Extension programme, meaning, process, objectives, principles, steps in programme planning, model, phases and steps, extension evaluation, nature of evaluation, Steps involved in an extension programme evaluative process; **c) Extension system in India: Glimpses of Pre-Independence era:** Rural Reconstruction at the Sunderbans in Bengal, Gurgaon Project, Shriniketan Project, Seva-gram, Marthandam Project, Baroda Village, Grow More Food Campaign, Indian Village Service, Firka Vikas Yojana; **d) Glimpses of Post-Independence Era:** Etawah Pilot Project, Nilokheri Experiment, Community Development Project, Grow-More-Food Enquiry Committee Report, National Extension Service, Key Village Scheme, Intensive Agricultural District Programme, Operation Flood, Training and Visit (T and V) System, Integrated Rural Development Programme, Intensive Agricultural Area Programmes; **e) First-line extension system of ICAR:** National Demonstration, Operational Research Project, Krishi Vigyan Kendra, Trainers Training Centre, Lab to Land Programme, Technology Assessment and Refinement (TAR) - Institution Village Linkage Programme (IVLP), National Agricultural Technology Project, Agricultural Technology Information Centre, National Agricultural Innovation Project, Rural Landless Employment Guarantee Programme, Jawahar Rozgar Yojana, National Agriculture Extension Project, Watershed Development Programme, National Agricultural Technology Project, Institute Village Linkage Programme, National Agricultural Innovation Project, National Innovations on Climate Resilient Agriculture (NICRA), Attracting and Retaining Youth in Agriculture, Farmers FIRST (Farm, Innovations, Resources, Science and Technology) Programme, Mera Gaon Mera Gaurav, Soil Health Card; **f) New Trends in extension education:** Privatization of Extension, Advantage and disadvantage, Cyber Extension/e-Extension, IT and ICT, Farmer-Led Extension, Market-Led Extension, Expert system, Difference between Conventional Extension and Expert System Extension. **Rural Development: a) Rural Development**: meaning, concept, objectives, functions of extension education in Rural Development, Dimension, aspect, constraints *etc.* List of development programmes in India/Sociology; **b) Community Development Programme**: meaning, genesis, concept, definition, objectives, aspects, principles, philosophy, types and programmes of community development programme; **c) Rural Leadership**: definition, meaning, concepts, principals, types and qualities of a good leader; **d) Extension Administration**: meaning, concept, principals, functions, *etc.*; **e) Monitoring and Evaluation**: definition, scope, meaning, monitoring and evaluation in extension programme; **f) Transfer of Technology**: concept, definition, model, methods, capacity building of extension personnel; **g) Training for Development**: definition, need for training, model, training process, designing extension training programmes, principals, Steps to conduct active extension training programme, build up a training programmes *etc.*; **h) Krishi Vigyan Kendra**: concept, principals, mandate *etc.*; **i) Extension teaching and learning process**: meaning, definition, function, classifications, teaching methods *etc.*; **j) Teaching process in extension**: steps, selection of extension teaching, factors *etc.*; **k) ICT applications in TOT**: role, ICT in extension works, importance, media mix strategies, Factors influencing in selection of extension teaching methods *etc.*; **l) Audio Visual Aids**: definition, meaning, classification, advantages/ disadvantage, importance, use of AV Aids *etc.* n) **Communication**: definition, importance, elements, principals, functions, models, feedback, forms of communication, barrier *etc.* m) **Agricultural Journalism**: definition, meaning, principles of farm journalism, importance, scope, functions *etc.* n) **Diffusion and Adoption of Innovation**: concept, meaning, elements, innovation decision process, model, stages of adoption process, adopter categories *etc.* **Chapter-3: Rural Sociology and Educational Psychology: a) Sociology and Rural Sociology:** Definition, Meaning of Rural Sociology, Nature, Characteristics, Scope, Importance and Significance of Rural Sociology, Origin and Development of Rural Sociology in India; **b) Social Ecology**, Rural Society, Social Group, Social stratification, Structural Concepts. **Social interaction**, Social Institutions; **c) Social Change** and Development, Role of the Individual in Social Change, Importance of Rural Sociology in Extension Education, Interrelationship between Rural Sociology and Extension; **d) Educational Psychology:** Definitions, Meaning, Importance, aims, Role of Educational Psychology in Extension Education, Behaviour, **Domains of learning; e) Personality:** Role of personality in extension education; **f) Perception and Motivation**: Classification of Motives, Theories of motivation, Importance of motivation in agricultural extension; **g) Intelligence**: Types of intelligence. Mental Maturity, Factors affecting Intelligence, Importance of Intelligence in Extension work. **Chapter-4: Communication Skills and Personality Development: a) Communication Skills:** Definitions, Principles, Structural Communication, Elements and their Characteristics of Communication Process, Types of Communication; **b) Listening skills:** Developing listening skills, Note Taking, Make time for note taking activities in class; **c) Writing skills**: Difference between Technical Writing and General Writing; **d) Oral Presentation Skill**: Field Diary, Laboratory Record, Indexing, Footnotes, Bibliographic procedures; **e) Reading:** Comprehension of General and Technical Articles; **f) Writing:** Precise Writing, Summarizing, Precise Writing/Abstracting/Summarizing, Individual and Group Presentations, Impromptu Presentation and Extempore; **g) Speaking**: Public Speaking, Group Discussion, Organizing Seminars and Conferences. **Chapter-5, Entrepreneurship Development and Business Management: a) Entrepreneurship:** Concept of entrepreneur, Concept of entrepreneurship, Functions of entrepreneurs, Meaning of entrepreneurship, Characteristics of Entrepreneurship, Entrepreneurial Process, Importance of Entrepreneurship, Types of Entrepreneurs, Difference between entrepreneur and manager, Scope for entrepreneurship development in Agriculture; **b) SWOT:** Strengths, Weakness, Opportunities and Threats (SWOT) Analysis, Analyzing the Business, Achievement Motivation, Government policy and Programmes, Prime minister's employment generation programme, Schemes for women entrepreneurs, Impact of economic reforms on Agribusiness/Agri-enterprises, Process of Entrepreneurship Development; **c) Skills development:** Business Leadership Skills, Developing Organisational skills, Controlling, Supervising, Problem Solving, Monitoring and Evaluation, Developing Managerial Skills, Business Leadership Skills, Communication, Direction skill, Motivation Skills, Problem Solving Skill; **d) Supply Chain Management**: Total quality management (TQM), Project Planning Formulation and Report Preparation, Financing of Enterprise, Opportunities for Agri-entrepreneurship and rural enterprise. I hope the book will be welcome by a galaxy of scholars and professionals and keep generating valuable feedback in reconstructing it in due course of time.

Manas Mohan Adhikary

Contents

Message v

Preface ix

1. **Fundamental of Extension Education** 1

   a) ***Evolution of Extension Education:*** *Types of education, Meaning, Formal and Informal education, Definition, Concept, importance, Scope, Function, philosophy, process, levels, Objectives, functions and principles.*

   b) ***Extension Programme Planning:*** *Extension programme, meaning, process, objectives, principles, steps in programme planning, model, phases and steps, extension evaluation, nature of evaluation, Steps involved in an extension programme evaluative process.*

   c) ***Extension system in India: Glimpses of Pre-Independence era:*** *rural reconstruction at the sunderbans in Bengal, Gurgaon project, Shriniketan project, Seva-gram, Marthandam project, Baroda village, Grow More Food campaign, Indian village service, Firka vikas yojana.*

   d) ***Glimpses of Post-Independence Era:*** *Etawah pilot project, Nilokheri experiment, Community Development Project, Grow-More-Food enquiry committee report, National Extension Service, Key Village Scheme, Intensive Agricultural District Programme, Operation Flood, Training and Visit (T and V) system, Integrated Rural Development Programme, Intensive Agricultural Area Programmes.*

   e) ***First-line extension system of ICAR:*** *National Demonstration, Operational Research Project, Krishi Vigyan Kendra, Trainers*

Training Centre, Lab to Land Programme, Technology Assessment and Refinement (TAR) - Institution Village Linkage Programme (IVLP), National Agricultural Technology Project, Agricultural Technology Information Centre, National Agricultural Innovation Project, Rural Landless Employment Guarantee Programme, Jawahar Rozgar Yojana, National Agriculture Extension Project, Watershed Development programme, National Agricultural Technology Project, Institute Village Linkage Programme, National Agricultural Innovation Project, National Innovations on Climate Resilient Agriculture (NICRA), Attracting and Retaining Youth in Agriculture, Farmers FIRST (Farm, Innovations, Resources, Science and Technology) Programme, Mera Gaon Mera Gaurav, Soil Health Card.

f) ***New Trends in extension education:*** *Privatization of Extension, Advantage and disadvantage, Cyber Extension/e-Extension, IT and ICT, Farmer-Led Extension, Market-Led Extension, Expert system, Difference between Conventional Extension and Expert System Extension.*

2. Rural Development **63**

a) ***Rural Development****: meaning, concept, objectives, Functions of Extension Education in Rural Development, Dimension, aspect, constraints etc. List of development programmes in India/Sociology.*

b) ***Community Development Programme****: meaning, genesis, concept, definition, objectives, aspects, principles, philosophy, types and programmes of community development programme.*

c) ***Rural Leadership****: definition, meaning, concepts, principals, types and qualities of a good leader.*

d) ***Extension Administration****: meaning, concept, principals, functions, etc.*

e) ***Monitoring and Evaluation****: definition, scope, meaning, monitoring and evaluation in extension programme.*

f) ***Transfer of Technology****: Concept, definition, model, methods, capacity building of extension personnel.*

g) ***Training for Development****: definition, need for training, model, training process, designing extension training programmes, principals, Steps to conduct active extension training programme, build up a training programmes etc.*

h) ***Krishi Vigyan Kendra****: concept, principals, mandate etc.*

i) ***Extension teaching and learning process****: meaning, definition, function, classifications, teaching methods etc.*

j) ***Teaching process in extension****: steps, selection of extension teaching, factors etc.*

k) ***ICT applications in TOT***: *role, ICT in extension works, importance, media mix strategies, Factors influencing in selection of extension teaching methods etc.*

l) ***Audio Visual Aids***: *definition, meaning, classification, advantages/ disadvantage, importance, use of AV Aids etc.*

m) ***Communication***: *definition, importance, elements, principals, functions, models, feedback, forms of communication, barrier etc.*

n) ***Agricultural Journalism***: *definition, meaning, principles of farm journalism, importance, scope, functions etc.*

o) ***Diffusion and Adoption of Innovation***: *concept, meaning, elements, innovation decision process, model, stages of adoption process, adopter categories etc.*

3. Rural Sociology and Educational Psychology **197**

a) ***Sociology and Rural Sociology:*** *Definition, Meaning of Rural Sociology, Nature, Characteristics, Scope, Importance and Significance of Rural Sociology, Origin and Development of Rural Sociology in India,*

b) ***Social Ecology***, *Rural Society, Social Group, Social stratification, Structural Concepts.* ***Social interaction***, *Social Institutions,*

c) ***Social Change and Development***, *Role of the Individual in Social Change, Importance of Rural Sociology in Extension Education, Interrelationship between Rural Sociology and Extension*

d) ***Educational Psychology:*** *Definitions, Meaning, Importance, aims, Role of Educational Psychology in Extension Education, Behaviour,* ***Domains of learning. Personality:*** *Role of personality in extension education,*

e) ***Perception and Motivation***: *Classification of Motives, Theories of motivation, Importance of motivation in agricultural extension.*

f) ***Intelligence***: *Types of intelligence. Mental Maturity, Factors affecting Intelligence, Importance of Intelligence in Extension work*

4. Communication Skills and Personality Development **235**

a) ***Communication Skills:*** *Definitions, Principles, Structural Communication, Elements and their Characteristics of Communication Process, Types of Communication.* ***Listening skills:*** *Developing listening skills, Note Taking, Make time for note taking activities in class.*

b) ***Writing skills***: *Difference between Technical Writing and General Writing.*

c) ***Oral Presentation Skill***: *Field Diary, Laboratory Record, Indexing, Footnotes, Bibliographic procedures.*

d) ***Reading:*** *Comprehension of General and Technical Articles.*

e) **Writing:** Precise Writing, Summarizing, Precise Writing/Abstracting/ Summarizing, Individual and Group Presentations, Impromptu Presentation and Extempore,

f) **Speaking**: Public Speaking, Group Discussion, Organizing Seminars and Conferences.

5. **Entrepreneurship Development and Business Communication** 287

a) **Entrepreneurship:** Concept of entrepreneur, Concept of entrepreneurship, Functions of entrepreneurs, Meaning of entrepreneurship, Characteristics of Entrepreneurship, Entrepreneurial Process, Importance of Entrepreneurship, Types of Entrepreneurs, Difference between entrepreneur and manager, Scope for entrepreneurship development in Agriculture,

b) **SWOT:** Strengths, Weakness, Opportunities and Threats (SWOT) Analysis, Analyzing the Business, Achievement Motivation, Government policy and Programmes, Prime minister's employment generation programme, Schemes for women entrepreneurs, Impact of economic reforms on Agribusiness/Agri-enterprises, Process of Entrepreneurship Development.

c) **Skills development:** Business Leadership Skills, Developing Organisational skills, Controlling, Supervising, Problem Solving, Monitoring and Evaluation, Developing Managerial Skills, Business Leadership Skills, Communication, Direction skill, Motivation Skills, Problem Solving Skill.

d) **Supply Chain Management**, Total quality management (TQM), Project Planning Formulation and Report Preparation, Financing of Enterprise, Opportunities for Agri-entrepreneurship and rural enterprise.

References 335

Index 359

Chapter 1

Fundamentals of Extension Education

EVOLUTION OF EXTENSION EDUCATION

History of agricultural extension is intimately related to the development in technology. The need for extension was felt in Europe and North America during the 18th century in the wake of industrial development. To take advantage of science and technology to the doorsteps of the farmers, agricultural societies were formed. The first society is said to be established in Scotland in 1728 for the spread of knowledge of agriculture through lectures and publications.

History

The roots of extension education, however, can be traced back to 1840 when the term University Extension was first used in Britain to take the knowledge to doorsteps of common people. The first practical steps were taken in 1867-68, when James Stuart, fellow of Trinity College, Cambridge, England gave lectures to women's associations and working men's club. James Stuart is called as the father of University Extension. He approached the authorities of University of Cambridge to organise extension lectures. The system was accepted by the Cambridge as well as Oxford University in 1873. Dr. J. P. Leagans of USA is known as the father of extension, whereas Dr. K. N. Singh is known as father of Extension in the Indian context.

Growth of Agricultural Extension Education as a Discipline in India

The teaching of extension education at undergraduate level started in the year 1950 at the College of Agriculture, Calcutta University. The first post-graduate teaching was launched in 1955, at Bihar Agricultural College, Sabour. Next was the post-graduate programme initiated at the College of Nagpur in 1958. Further, creation of a separate Division of Agricultural Extension at Indian Agricultural Research Institute (IARI), New Delhi with assistance from Ford Foundation under the expert advice of legendary Dr. J. Paul Leagans was a landmark event in the history of extension education.

Conceptual Background

The National Commission on Agriculture (1976) appointed by the Government of India stated the scope and jurisdiction of extension in the following terms:

Extension and Extension education relate to the process of conveying the technologies of scientific agriculture to the farmer in order to enable him to utilise the knowledge for better agriculture and better economy. This consists of provision for non-formal educational facilities through organised extension services, introduction of agricultural education in schools and education through non-degree institutional programme to impart vocational skills to the farmers for improving their productive activities.

Types of Education

There are mainly three types of education, namely, Formal, Informal and Non-formal. Each of these types is briefly described below.

1. Formal Education

Formal education corresponds to a systematic, organized education model, structured and administered according to a given set of laws and norms, presenting a rather rigid curriculum as regards objectives, content and methodology. It is characterized by a contiguous education process named, as "presentable education", which necessarily involves the teacher, the students and the institution. It corresponds to the education process normally adopted by our schools and universities.

- Planned with a particular end in view.
- Limited to a specific period.
- Well-defined and systematic curriculum
- Given by specially qualified teachers.
- Includes activities outside the classroom
- Observes strict discipline.

2. Informal Education

Informal education is quite diverse from formal education and, particularly, from non-formal education, although in certain cases it is capable of maintaining a close relationship with both. It does not correspond to an organized and systematic view of education; informal education life long process in which portionsazure's knowledge, skill, attitude and insight from daily experiences and exposure to the endearment at home.

- ☆ Incidental and spontaneous
- ☆ Not-pre-planned.
- ☆ Not imparted by any specialized agency.
- ☆ No prescribed time-table or curriculum.
- ☆ May be negative also

3. Non-Formal Education

It is organized and systematic education activity carried on outside the frame work of the formal system to provide selected type of learning to the particular group of population.

- ☆ Derived from the expression 'formal education.
- ☆ Outside the realm of formal education.
- ☆ Conscious and deliberate.
- ☆ To be organised for a homogeneous group.
- ☆ Serving the need of the identified group.

Meaning of Extension Education

The word extension is derived from the latin roots 'ex' - meaning 'out' and 'tensio' meaning 'stretching'. Stretching out is the meaning of extension. The word 'extension' came to be used originally in USA during 1914. In other words, the word extension signifies an out of school system of education. Education is an integral part of extension. The basic concept of extension is that it is education. Extension means that type of education which is stretched out to the people in rural areas, beyond the limits of the educational institutions to which the formal type of education is normally confined.

Extension Education is a science, which deals with the creation, transmission and application of knowledge designed to bring about planned changes in the behaviour-complex of people; with a view to help them to live better by learning the ways of improving their vocations, enterprises and institutions. Extension Education is a science that brings about desirable changes in the behaviour of the concerned persons through educational methods, so as to improve their general standard of living with their own efforts. In fact, it deals with the designs and strategies of transfer of technology to the concerned persons. Extension Education is the process of teaching rural people how to live better by learning ways to improve their farm, home and community institutions. Extension as an out of school education and services for the members of the farm family and others directly or indirectly engaged in farm production to enable them to adopt improved practices in production, management, conservation and marketing. Several authors defined extension in various ways emphasizing the importance of one or the other aspect of extension (National Commission on Agriculture, 1976). Extension Education has now developed as a full-fledged discipline, having its own philosophy, objectives, principles, methods and techniques which must be understood by every extension worker and others connected with the rural development. It might be mentioned here that Extension Education, its principles, methods and techniques are applicable not only to agriculture but also to veterinary and animal husbandry, dairying, fisheries, home science, health and family planning. Based upon its application and use, various nomenclatures have been given to it, such as agricultural extension, veterinary and animal husbandry extension, dairy extension, fisheries extension, home science extension, public health extension, and family planning extension.

Differences between Formal Education and Extension Education

Formal Education	*Extension Education*
1. The teacher starts with theory and works up to practical.	1. The teacher (extension professional) starts with practical and may take up theory later on.
2. Students study subjects.	2. Clients study problems.
3. Students must adapt themselves to the fixed curriculum offered.	3. It has no fixed curriculum or course of study and the clients help to formulate the curriculum.
4. Authority rests with the teacher.	4. Authority rests with the clients.
5. Class attendance is compulsory.	5. Participation is voluntary.
6. Teacher instructs the students.	6. Teacher teaches and also learns from the clients.
7. Teaching is only through instructors.	7. Teaching is also through local leaders.
8. Teaching is mainly vertical.	8. Teaching is mainly horizontal.
9. The teacher has more or less homogeneous audience.	9. The teacher has a large and heterogeneous audience.
10. It is rigid.	10. It is flexible.
11. It has all pre-planned and pre-decided programmes.	11. It has freedom to develop programmes locally and they are based on the needs and expressed desires of the people.
12. It is more theoretical.	12. It is more practical and intended for immediate application in the solution of problems.

Origin of Extension Education

- ✰ The word 'Extension' is derived from the Latin roots in which 'tensio' meaning 'stretching and 'ex' meaning 'out'.
- ✰ Thus the term 'Extension Education' means that type of education which is 'stretched out' into the villages and fields beyond the limits of the schools and colleges to which the formal type of education is normally confined.

- The term Extension Education was first coined in the year 1873 by Cambridge University in England.
- The present form of Agricultural extension was primarily stated in USA in 1908, where Rusewelt, then the President of America appointed a Country Life Commission to find out the reasons for migration of people from rural areas to urban area so the their migration can be stopped by engaging them in agriculture and rural development activities.
- In India extension primarily started with **Gurgaon** project which was organized by **F.L. Brayne in 1920.**
- The system of extension has come into existence when a wide gap between research work and farmer's knowledge and practice was observed which ultimately resulted in low productivity and production. In this context an agency to bridge the gap between the research work, and farmers and people at large, to play the dual role of carrying out the results of research to people and problems of people to research station is felt essential.

Definition of Extension Education

- Extension education is an applied science consisting of content derived from research, accumulated field experiences and relevant principles drawn from the behavioural science synthesized with useful technology into a body of philosophy, principles, content and methods focused on the problems of out of school education for adults and youths (**J.P. Leagans (1961**).
- Extension work is an out of school system of education in which adult and young people learn by doing. It is partnership between the government and the people, which provides service and education designed to meet the people. Its fundamental objective is the development of the people (**Kelsey and Harne (1963**).
- Extension is a programme and a process of helping village people to help themselves, increase their production and to raise their general standard of living (**D. Ensminger (1961**)
- We can define extension as the increased dissemination of useful knowledge for improving rural life (**H.W. Butt (1961**).
- Extension is a two – way channel; it brings scientific information to the village people and also takes the problems of the village people to the scientific institutions for solution, it is a continuous educational process, in which both learner and teacher contribute and receive **(B. Rambhai (1958).**
- Agricultural extension is a professional methods of non –formal education aimed at inducing behavioural changes in the farmers for increasing their income through increased production and productivity by establishing firm linkages with research for solving farmer's problems ensuring

adequate and timely supply of inputs and usage proven methods of communication for speeding of the process of diffusion and adoption of innovations (**V.C. Sanoria (1986)**.

- ☆ Extension education is an applied science consisting of content derived from research, accumulated field experiences and relevant principles drawn from the behavioural science synthesized with useful technology into a body of philosophy, principles, content and methods focused on the problems of out of school education for adults and youth.– **J.P. Leagans (1961)**
- ☆ Extension education is the process of teaching rural people how to live better by learning ways that improve their farm, home and community institutions – **J.P. Leagans (1961)**
- ☆ Extension Education is defined as an educational process to provide knowledge to the rural people about the improved practices in a convincing manner and help them to take decision within their specific local conditions. – **O. P. Dahama (1973)**
- ☆ Extension is an education to change the attitudes and practices of the people with whom the work is done – **D. Ensminger (1957)**

From the above definitions, it can be concluded that "Extension education is an education and is bringing a desirable change in behaviour (knowledge, skills and attitudes) of rural people to improve their social, economic and psychological status".

- ☆ The behaviour refers to anything the individual does. According to Leagans (1961), behaviour refers to what an individual knows (knowledge), what he can do (skills-mental and physical), what he thinks (attitudes) and what he actually does (action).
- ☆ Behaviour is, therefore, a function of the person in interaction with the situation. The factors motivating behaviour either in the person or situation are:
- ☆ An environmental determinant
- ☆ The internal urge, wish, feeling, emotion, drive, instinct, need, want, desire, demand, purpose, interest, aspiration or motive which gives rise to the action and
- ☆ The incentive or goal, which attracts or repels an organism.

Why Study Extension Education

To be successful, an extension worker should know not merely 'what' to teach but 'how' to teach people. In other words, it is not enough if he is equipped with the technical knowledge in subject matter field, he should have the ability to successfully communicate his ideas to the people, taking the personnel, social and situational factors into consideration. The captain of a ship has to know not only his ship and the destination he wants to reach. He must also understand ocean currents and tides and the mind systems of the world. All of these are powerful forces which can help him. The extension worker needs to understand not only his own programmes and objectives but also the currents of thoughts in the minds of the people with whom he lives and works. He needs to understand the motives of men; why certain people take to new ideas more rapidly than others, why certain people seek to take the lead and why certain others hesitate. Light is thrown on many such questions by the knowledge accumulated by the social sciences. Knowing that much has been discovered about, currents and winds, no captain will venture out neither without providing himself with that knowledge nor without continuously studying. Similarly, the extension worker, aware of how much is and how very much more is yet to be known about the behavior of people will see that this kind of knowledge is essential to him. He will know that there are dependable currents in village life which can carry people ahead rapidly and he can look for them, confident that they can be found. He will recognize that some of his obstacles are like hidden rocks which he had better to avoid rather than try to drive the ship of his programme straight across them. This kind of knowledge is the field of study of psychology, anthropology and based on them of educational theory, unfortunately, our scientific knowledge of people in India is negligible compared with our scientific knowledge of plants, animals and soils.

The term was first used in a customary in USA. Afterwards it was used by many countries. Extension education is used in various fields, such as Agriculture, Animal Husbandry, Dairy, Veterinary, Health, Home Science, Industry, Cooperative and Forestry. Therefore, it is necessary to understand basic concept of Extension Education for those who are associated with it in some way or the other. The researchers neither have the time nor are they equipped for the job of persuading the villagers to adopt scientific methods and to ascertain from them the rural problems. On the other hand it is impracticable for the millions of farmers to visit the research stations and learn things by themselves. Thus an agency is required to bridge the gap between the research workers and the people at large, to play the dual role of interpreting the results of research to the farmers (in such a way that they accept and adopt the recommendations)as well as of conveying the farmers problems to the research stations for solutions. This agency is termed 'Extension', and the personnel managing this agency/organization are called Extension workers. To equip the prospective extension workers for their job, it is necessary for them to be trained adequately in the formal "teaching institutions". Thus we find that three kinds of inter related services are essential in the process of rural development.

Concepts of Extension

The Basic Concept of Extension as Education

Education is the production of desirable changes in human behaviour. The desired change is in terms of knowledge-things known, attitudes-things felt and skill-things done either in all the three or one or two. Education is a process in which human behaviour is modified so as to be in closer agreement with some model or ideal determined by the values of society. Webster dictionary explained Extension as a branch of university for the students who cannot attend the university. Another popularly used meaning is **out-of-school system of education.** Based on this the following concepts in detail explain extension as education, process and salesmanship.

List of Concepts

1. The Concept of Extension as Educational Process

As an educational process it has five phases:

- ☆ **Analysis of situation is** by collection of facts about people's interests, education, needs, customs, folkways, habits *etc.* and also about their physical situation such as livelihoods, markets, housing conditions, services, communications *etc.* The thorough analysis of these facts will provide the difference between what actually is and what should be.
- ☆ **Objectives and solutions** are set based on the analysis. They help to reach the desirable situation keeping in view the social and economic outcomes, as well as behavioral change.
- ☆ **Teaching plan of work** is deciding what to teach and how to teach, *i.e.*, the content and method of teaching by creating learning situation to achieve the objectives.
- ☆ **Evaluation of teaching** to examine how far the objectives are achieved.
- ☆ **Reconsideration** is review of efforts and results which will reveal new situation. If this new situation reveals need for further action, the entire process has to be initiate

2. The Concept of Salesmanship

- ☆ Extension teaching is sometimes compared with commercial **salesmanship,** as extension agent is primarily involved in selling the ideas.
- ☆ By selling the idea the profit in this context is knowledge, skill and attitudinal change.

Importance of Extension Education

- ☆ Extension uses democratic methods in educating the farmers.
- ☆ Extension Helps in adoption of innovations.
- ☆ Extension helps in studying and solving the rural problems.
- ☆ Extension increases farm yields and improve the standard of living of farmers
- ☆ Extension makes good communities better and progressive.
- ☆ Extension contributes to national development programmes

- In our country, every individual does not get chance to be a part of school education during his/her childhood.
- Even after school and college education people need education to solve many practical problems during their day-to-day life.
- Hence, extension education is required to provide solution to many of the practical problems of people at their door steps.

Need of Extension Education

The need of extension education is realized in many areas such as:

- Today's technology becomes outdated tomorrow.
- In the modern age of science and technology, innovative inventions are rising day by day.
- Farming is no exception to this trend.
- The farmer's need is to supply recent useful and practical information related to agricultural production for better economic returns.
- Agricultural development will be closely related with the development of the ability of the farmers' understanding and adoption of this technology.
- There are many natural and manmade forces which interact continuously in this world.
- This interaction makes our society more complex.
- It creates many new problems for the people at large.
- People should know the solution for these problems and extension education helps people at this critical stage.
- The major work of agricultural researchers is to develop new agricultural technologies to meet out the requirement of different agro-climatic conditions and other local problems faced by the farming community.
- Agricultural scientists neither have enough time nor are equipped to convince farmers directly to adopt improved scientific technologies to solve the field related problems and to improve agricultural production.
- Similarly, it is not easy for all farmers to visit the research stations and obtain knowledge about improved technologies.
- Thus, there is a need of an agency to interpret the findings of the scientific research to the farmers and also to collect information about the practical problems faced by the farmers to adopt these technologies and pass on the same to the research scientists for needful solution.
- Extension education is the democratic method of educating the farmers.
- It respects the individuality of farmers in making their own decision.
- There are many farmers' innovations which they overlook unknowingly.
- These innovations are of no use unless they are put to practical use.

- ☆ The hidden farmer's technology is refined by the scientists and reaches the people speedily through extension scientists.
- ☆ Extension education is an out-of-school process, which aims to help people themselves in solving their day to day problems related to the cultivation/cultural practices.
- ☆ The extension workers explain various ways to solve farmer's problems.
- ☆ The extension workers act as guide and educate them how to use new knowledge, skills and attitudes in solving their problems.
- ☆ The extension education helps all classes of people men, women and youth to solve their present and future economic, social and cultural problems.
- ☆ The extension education, thus contributes to national development.

Scope of Extension Education

Extension education is a science which deals with the creation, transmission and application of knowledge designed to bring about planned changes in the behavior complex of people; with a view to help them live better by learning the ways of improving their vocations, enterprises and institutions.

There is some logic in each of the above concept which may be defined as follows;

The different concepts may be defined as:

- ☆ **Extension service:** An organization and or a programme for agricultural development and rural welfare which employs the extension process as a means of programme implementation.
- ☆ **Extension process:** The extension process is that of working with rural people through out of school education, along those lines of their current interests and needs which are closely related to gaining a livelihood, improving the physical level of living of rural families, and fostering rural community welfare.
- ☆ **Extension Job:** The job of extension in agriculture and home economics is to assist people engaged in farming and home making to utilise more fully their own resources and those available to them, in solving current problems and in meeting changing economic and social conditions. In other words, through the educational and services approach rural people are stimulated to make changes that result in more efficient production and marketing of farm products, conservation of natural resources, more comfortable homes, improved health, and more satisfying family and community life.

The following statements will further amplify the scope of Extension

1. Extension is fundamentally a system of out of school education for adults and youths alike.

2. It is a system where people are motivated through a proper approach to help them by applying science in their daily lives, in farming, home making and community living.
3. Extension is education for all village people.
4. Extension is bringing about desirable changes in the knowledge, attitudes and skills of people.
5. Extension is helping people to help themselves.
6. Extension is working with men and women, boys and girls, to answer their felt needs and wants.
7. Extension is teaching people what to want, (*i.e.*, converting unfelt needs into felt needs) as well as how to work out ways of satisfying these wants and inspiring them to achieve their desires.
8. Extension is teaching through 'learning by doing' and 'seeing is believing'.
9. Extension is working in harmony with the culture of the people.
10. Extension is a two – way channel; it brings scientific information to village people and it also takes the problems of the village people to the scientific institution for solution.
11. Extension is working together (in groups) to expand the welfare and happiness of the people with their own families, their own villages, their own country and the world.
12. Extension is a living relationship between the village workers and the village people. Respect and trust for each other, sharing of joys and sorrows, results in friendship through which village extension work continues.
13. Extension is development of individual in their day to day living, development of their leaders, their society and their world as a whole.
14. Extension is a continuous educational process in which both learner and teacher contribute and receive.

Function of Extension Education

The function of extension is to bring about desirable changes in human behaviour by means of education. Changes may be brought about in their knowledge, skill, attitude, understanding, goals, action and confidence. Change in knowledge means change in what people know. For example, farmers who did not know of a recent HYV crop came to know of it through participation in extension programmes. The Extension Agents (EAs) who did not know of Information Technology (IT) came to know of them after attending a training course. Change in skill is change in the technique of doing things. The farmers learnt the technique of growing the HYV crop which they did not know earlier. The Extension Agents learnt the skill of using IT. Change in attitude involves change in the feeling or reaction towards certain things. The farmers developed a favourable attitude towards the

HYV crop. The Extension Agents developed a favourable feeling about the use of IT in extension programme. Change in understanding means change in comprehension. The farmers realized the importance of the HYV crop in their farming system and the extent to which it was economically profitable and desirable, in comparison to the existing crop variety. Change in goal is the distance in any given direction one is expected to go during a given period of time.

Change in action means change in performance or doing things. The farmers who did not cultivate the HYV crop earlier cultivated it. The Extension Agents who earlier did not use IT in their extension programmes started using them. Change in confidence involves change in self-reliance. Farmers felt sure that they have the ability of raising crop yield. The development of confidence or self-reliance is the solid foundation for making progress. To bring desirable change in behaviour is the crucial function of extension. For this purpose the extension personnel shall continuously seek new information to make extension work more effective.

Philosophy of Extension Education

Philosophy is the pursuit of wisdom, a body of general principles or laws of a field of knowledge. Essentially philosophy is a view of life and its various components. The practical implication is that the philosophy of a particular discipline would furnish the principles or guidelines with which to shape or mould the programmes or activities relating to that discipline. According to Kelsey and Hearne (1967), the basic philosophy of extension education is to teach people how to think, not what to think. Extension's specific job is furnishing the inspiration, supplying specific advice and technical help, and counseling to see that the people as individuals, families, groups and communities work together as a unit in 'blueprinting' their own problems, charting their own courses, and that they launch forth to achieve their objectives.

Basic Philosophy of Extension Education

1. Self help.
2. People are the greatest resource.
3. Programme of change.
4. Cooperative effort.
5. Democratic nature means development of people by the people and for the people.
6. Two channel of knowledge and experience:
   (a) **Problems:** are brought to laboratory.
   (b) **Solutions:** are taken to farmers/villagers.
7. Programme is based on creating interest by seeing and doing.
8. There is voluntarily cooperative participation.
9. Programme is based on attitude and values of the people with whom it works.

Different Philosophy of Extension

Ensminger (1962) has enunciated the philosophy of extension succinctly. He stated:

1. Extension is education for all village people.
2. Extension is changing the attitude, knowledge and skills of the people.
3. Extension is working with men and women, young people, boys and girls, to answer their needs and their wants.
4. Extension is "helping people to help themselves".
5. Extension is "learning by doing" and "seeing is believing".
6. Extension is teaching people what to want, as well as how to work out ways of satisfying these wants and inspiring them to achieve their desires.
7. Extension is the development of individuals in their day-to-day living, development of their leaders, their society and their world as a whole.
8. Extension is working together to expand the welfare and happiness of the people with their own families, their own village, their own country, and the world.
9. Extension is working in harmony with the culture of the people.
10. Extension is a living relationship between the village workers and the village people. Respect and trust for each other, sharing of joys and sorrows, result in friendship through which village extension work continues.
11. Extension is a two way channel-it brings scientific information to the village people and it also takes the problems of the village people to the scientific institutes for solution. Extension is a continuous educational process in which both learner and teacher contribute and receive.

Process of Extension Education

An effective extension educational programme involves five essential and interrelated steps. This concept of the extension educational process is intended only to clarify the steps necessary in carrying out a planned educational effort. It does not imply that these steps are definitely separate from each other. Experience shows that planning, teaching and evaluation take place continuously, in varying degrees, throughout all phases of extension activities

First step: The first step consists of collection of facts and analysis of the situation. Facts about the people and their enterprises are; the economic, social, cultural, physical and technological environment in which they live and work. These may be obtained by appropriate survey and establishing rapport with the people.

The responses obtained are to be analyzed with the local people to identify the problems and resources available in the community. For example, after a survey in a community and analysis of the data, the problem was identified as low income of the farm family from their crop production enterprise.

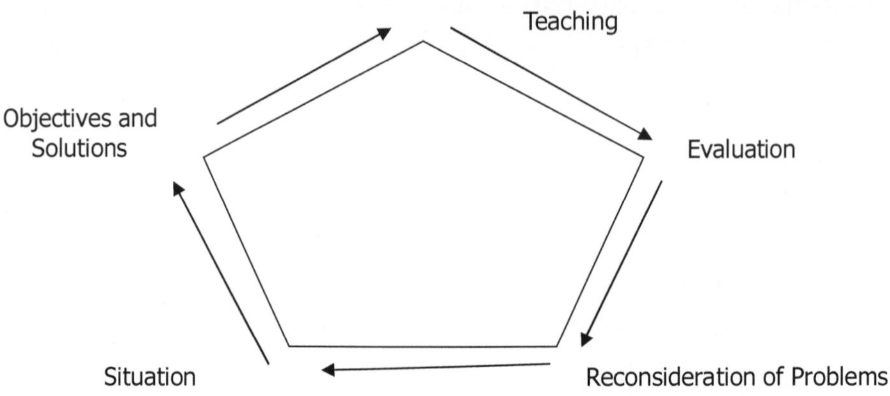

Process of Extension Education.

Second step: The next step is deciding on realistic objectives which may be accomplished by the community. A limited number of objectives should be selected by involving the local people. The objectives should be specific and clearly stated, and on completion should bring satisfaction to the community. Objectives should state the behavioral changes in people as well as economic and social outcomes desired.

In the example, the problem was identified as low income from the crop production enterprise. A deeper probe into the date revealed that low income was due to low yield of crops, which was attributed to the use of local seeds with low yield potential, application of little fertilizer and lack of protection measures.

Third step: The third step is teaching, which involves choosing what should be taught (the content) and how the people should be taught the methods and aids to be used. It requires selecting research findings of economic and practical importance relevant to the community, and selection and combination of appropriate teaching methods and aids.

Fourth step: The fourth step is evaluating the teaching *i.e.*, determining the extent to which the objectives have been reached. To evaluate the results of an educational programme objectively, it is desirable to conduct a re-survey. The evidence of changed behavior should be collected, which shall not only provide a measure of success, but shall also indicate the deficiencies, if any.

In the example, the re-survey after the fixed period of time, indicated that the crop yield had increased by 10 percent. It, therefore, indicated that there was a gap of 10 per cent in crop yield in comparison to the target (objective) of 20 per cent fixed earlier.

Fifth step: The fifth step is re-consideration of the entire extension educational programme on the light of the results of evaluation. The problems identified in the process of evaluation may become the starting point for the next phase of the extension educational programme, unless new problems have developed or new situations have arisen.

Levels of Extension Education

- Extension is generally thought of at two levels, extension education and extension service.
- Extensions at these two levels are interrelated, but at the same time maintain their separate identity.
- The extension education role is generally performed by the higher learning institutions like the Agricultural and other Universities and Colleges, ICAR Institutes, Home Science Colleges and apex level Training and Extension Organizations.
- At the university level, extension is integrated with teaching and research, while at the research institutes, extension is integrated with research.
- At the other apex level organizations, extension is generally integrated with training in extension
- The field extension work of these institutions and organizations is generally limited to the neighboring villages or blocks, which are considered as their extension laboratories.
- Extension Service-It is mainly to provide educational service to the people according to their need, for improving their life through better working.
- The main responsibility of extension service is with the State Government.
- The departments of Agriculture, Horticulture, Animal Husbandry, Veterinary, Forestry, Fishery, Sericulture *etc.* of the State Government carry out extension work with the farmers and rural people over the entire State.
- The departments maintain close contact with the relevant Universities and Research Institutes for obtaining appropriate technology and methodology for extension work, and for providing them with feedback information from the field for research.
- The extension service provided by the departments of the State Government is location-specific, input-intensive and, target and result oriented.
- The extension service works in close coordination with other development departments, input supply agencies, credit institutions, voluntary organizations and Panchayats.
- The extension service has the main responsibility of educating and training the farmers, farm women, and rural youth and village leaders of the State and for this purpose they take the help of the universities, research institutes and, training and extension organizations.

Objectives of Extension Education

Objectives are expression of the ends towards which our efforts are directed.

The fundamental objective of extension is to develop the rural people economically, socially and culturally by means of education.

- ✰ To assist people to discover and analyze their problems and identify the felt needs.
- ✰ To develop leadership among people and help them in organizing groups to solve their problems.
- ✰ To disseminate research information of economic and practical importance in a way people would be able to understand and use.
- ✰ To assist people in mobilizing and utilizing the resources which they have and which they need from outside.
- ✰ To collect and transmit feedback information for solving management problems.

Objective

Objective can be defined as "an expression of the end or direction of movement towards which efforts are directed". The broad objective of extension education is to bring about desirable changes in the behaviour of the people. All the efforts of the extension agents are directed to add new things to the knowledge of the participants who should understand them and use them in solving their problems. Objective is a direction of movement. A well stated objective is always measurable. It is also said to be a goal of growth. Objectives in Extension Education Aims/Goal/Targets

Goal

It is broad and Objectives are narrow. It is an aim of an individual or a group target with a focus on specific subject. Goal is a distance in any given direction, proposed to be covered in a given time. Goal is the tools one uses to reach ones objectives within a time frame. It expresses intended outcomes in general terms and Objectives express them in specific terms. The goal is to hit the bull's eye a destination of the aim. (Example: Improving the socio-economic conditions of livestock owners or empowering the farm woman)

Objectives: These are brief, clear statements that describe the desired learning outcomes of instruction. Attention is focused on the specific types of performances that students are expected to demonstrate at the end of instruction.

Example: To increase the productivity (milk, wool, eggs, meat *etc.* and reproduction status) of livestock, to increase the knowledge level of livestock owner about scientific animal husbandry.

Ambitions: It is a strong desire of one's life time for a bigger achievement without any time frame.

Example: To reach at the top level of management. No time frame. To gets the highest ICAR award.

Target/Aim: It is a specific goal of an individual or a group with a specific time frame of a single or multiple subject. "Aim at the bull's eye". In a broader sense, an amount of uncertainty is involved. You may hit the bull's eye or not. Not all is lost if what is aimed achieved with a reasonable amount of accuracy. You aim to get a 1st prize but ended second. The ambition is not fulfilled but almost achieved with reasonable degree of aberration.

Example: To introduce the crossbred cows in a village (AIM) and crossbred cows in each household of the village (Target). To increase the milk production of buffalo within a month. To be make all the students as doctors or IITans. To join IAS is a goal but to become only IAS/IPS is the target.

Vision: It is an idea, a thinking of an individual or a group or a community for betterment of the group, or world at large. Example: Vision of Rajeev Gandhi .... Introduction of computer in India Vision is often incorporated into an organizational mission (or vision) statement to clarify what the organization hopes to be doing at some point in the future. The vision should act as a guide in choosing courses of action for the organization.

Example: ICAR Vision- 2020, 2030, 2050 likewise IVRI vision – 2020, 2030, 2050 Vision: Food and nutritional security; social equality; right to education.

Objectives: The dictionary meaning of objective is something that one's efforts or actions are intended to attain or accomplish; purpose; goal; target.

Fundamental Objective

Fundamental objective of extension education is the overall development of the rural people. The ultimate objective of dairy extension education is development of dairy stakeholders by improving their living standards. Agricultural extension in our country is primarily concerned with the following main objectives:

1. Dissemination of useful and practical information relating to agriculture, including improved seeds, fertilizers, implements, pesticides, improved cultural practices, dairying, poultry, fisheries, veterinary and animal husbandry, and food and nutrition.
2. Practical application of useful knowledge to farm and home.
3. Improvement in all aspects of the life of the rural people within the framework of the national, economic and social policies involving the population as a whole.

General Objectives of Extension

The following are the general objectives:

1. To assist people to discover and analyze their problems and identify their needs.

2. To develop leadership among people and help them in organizing groups to solve their problems.
3. To disseminate research information of economic and practical importance in a way people would be able to understand and use.
4. To assist people in mobilizing and utilizing the resources which they have or need from outside.
5. To collect and transmit feedback information for solving management problems.

The main objective of dairy extension education is development of dairy stakeholders by improving their Quality of Life.

Functions of Extension Education

Keeping the objectives in view, the function of extension is to bring about desirable changes in human behaviour by means of education. Changes may be brought about in their knowledge, attitude, skill, understanding, goals, action and confidence.

1. Change in Knowledge

Change in knowledge means change in what people know. For example, a farmer who does not know about paneer making technology may know it through participation in extension programme. Similarly, Extension Agent (EA) who does not know about Information Technology (IT), comes to know of it after attending a training programme.

2. Change in Attitude

Change in attitude involves change in the feeling towards any object, person, and thing. For instance, the afore-said farmer developed a favourable attitude towards the paneer making. Similarly, the EA developed a favourable attitude in the use of IT.

3. Change in Skill

Change in skill is change in the technique of doing things. Quoting the afore-said example, the farmer learnt the technique of paneer making which he did not know earlier. Similarly, the EA learnt the skill of using IT.

4. Change in Understanding

Change in understanding means change in comprehension. For example, the afore-said farmer realized the importance of the milk processing (paneer making) and the extent to which it was economically profitable and desirable, in comparison to the selling of raw milk. Similarly, the EA understood the use of IT and the extent to which these would make extension work more effective.

5. Change in Goal

Change in goal is the distance in any given direction one is expected to go during a given period of time. For example, the extent to which the farmer exposed to paneer making technology raised the goal in dairy farming, in terms of increasing milk yield of dairy animals. Similarly, the EA set his goal of getting an improved practice adopted by the clients within a certain period of time by using IT.

6. Change in Action

Change in action means change in the way of doing things. The farmer who did not know paneer making earlier, now has started making it. Similarly, the EA who earlier did not use IT in extension programmes started using them.

7. Change in Confidence

Change in confidence involves change in self-reliance and ability. The afore-said farmer felt sure that he has the ability of making paneer at his own level. Similarly, the EA developed faith on his ability to do better extension work. The development of confidence or self-reliance is the solid foundation for making progress.

Principles of Extension Education

The extension work is based upon some working principles and the knowledge of these principles is necessary for extension personnel. Some of these principles, as related to extension education, are mentioned below:

1. Principle of Peoples Need and Interest

Extension work must be based on the needs and interests of the people. These needs and interests differ from individual to individual, from village to village, from block to block, from state to state; therefore, there cannot be one programme for all people.

2. Principle of Grass Root Level Organisation

A group of rural people in local community should sponsor extension work. The programme should fit in with the local conditions. The aim of organising the local group is to demonstrate the value of the new practices or programmes so that more and more people would participate.

3. Principle of Cultural Difference

Extension work is based on the cultural background of the people with whom the work is done. Improvement can only begin from the level of the people where they are. This means that the Extension professionals has to know the level of the knowledge, and the skills of the people, methods and tools used by them, their customs, traditions, beliefs, values, *etc.* before starting the extension programme.

4. Principle of Cultural Change

Extension education starts with what the learner knows, has and thinks. With this in mind and with an attitude of respect towards clients, the extension professionals must seek to discover and understand the limitations, taboo and the cultural values related to each phase of programme so that an acceptable approach could be selected in the locality

5. Principle of Cooperation and Peoples Participation

Extension is a co-operative venture. It is a joint democratic enterprise in which rural people co-operate with their village, block and state officials to pursue a common cause. Ultimately without the cooperation of people the work cannot be successful and desired result cannot be achieved. The first task of extension education is the cooperation of people and their participation in work. Extension helps people to help themselves. Good extension work is directed towards assisting rural families to work out their own problems rather than giving them ready-made solutions.

6. Principle of Applied Science and Democratic Approach

Extension education is based upon democratic principles. It is based on discussions and suggestions. Discussions are held with the people on actual field conditions, so that they participate in work. Extension professionals provide practical look to the scientific inventions so that farmers can easily adopt them on their farm.

7. Principle of L:earning by Doing

According to this principle, farmers are encouraged to learn by doing the work themselves and by participating in it. When a person does a work, he gains practical knowledge and experiences the difficulties. Extension professionals are able to understand the problems and provide proper guidance to the farmers and thus, they are able to receive proper information/feedback.

8. Principle of Trained Specialists

It is very difficult that extension personnel should be knowledgeable about all problems. Therefore, it is necessary that specialists should impart training to the farmers from time to time.

9. Principle of Adaptability in Use of Extension Teaching Methods

People differ from each other, one group differs from another group and conditions also differ from place to place. An extension programme should be flexible, so that necessary changes can be made whenever needed, to meet the varying conditions. Extension professionals should have knowledge of extension methods so that they can select proper method according to the condition. Teaching methods should be flexible so that they can be properly applied on people according to their age groups, educational background, economic standard and gender. In extension education, two or more methods should be applied according to the principle of adaptability.

10. Principle of Leadership

Extension work is based on the full utilisation of local leadership. The selection and training of local leaders to enable them to help in carrying out extension work is essential to the success of the programme. People have more faith in local leaders and they should be used to put across a new idea so that it is accepted with the least resistance.

11. Principle of Whole Family

Extension work will have a better chance of success if the extension professionals have a whole-family approach instead of piecemeal approach or separate and uninterested approach. Extension work is, therefore, for the whole family, *i.e.* for male, female and children.

12. Principal of Indigenous Knowledge

People everywhere have indigenous knowledge system which they have develop through generation of work experience and problem solving in their own specific situation. The indigenous knowledge systems encompass all aspect of life and people considered it essential for their survival.

13. Principle of Satisfaction

The end-product of the effort of extension teaching is the satisfaction that comes to the farmer and his family members as the result of solving a problem, meeting a need, acquiring a new skill or some other changes in behaviour. Satisfaction is the key to success in extension work. A satisfied stakeholder is the best advertisement.

14. Principle of Evaluation

Extension is based upon the methods of science, and it needs constant evaluation. The effectiveness of the work is measured in terms of the changes brought about in the knowledge, skill, attitude, and adoption behaviour of the people, not merely in terms of achievement of physical targets.

EXTENSION PROGRAMME PLANNING

Extension Programme

The word 'programme' has several distinct meanings in the dictionary. It means a proclamation, a prospectus, a list of events, a plan of procedure, a course of action prepared or announced beforehand, a logical sequence of operations to be performed in solving a problem. When used by an organization, it means a prospectus or a statement issued to promote understanding and interest in an enterprise.

According to Kelsey and Hearne (1949), an "extension programme" is a statement of situation, objectives, problems and solutions'.

According to the USDA (1956), an "extension programme" is arrived at co-operatively by the local people and the extension staff and includes a statement of:

- The situation in which the people are located;
- The problems that are a part of the local situation;
- The objectives and goals of the local people in relation to these problems; and
- The recommendations or solutions to reach these objectives on a long-time basis (may be several years) or on a short-time basis (may be one year or less).

Leagans (1961) says that an "extension programme" is a set of clearly defined, consciously conceived objectives or ends, derived from an adequate analysis of the situation, which are to be achieved through extension teaching activity'.

Lawrence (1962) says that an "extension programme" is the sum total of all the activities and undertakings of a county extension services. It includes: (i) programme planning process; (ii) written programme statement; (iii) plan of work; (iv) programme execution; (v) results; and (vi) evaluation.

From the above definitions, it is clear that an extension programme:

- Is a written statement;
- Is the end product of extension programme planning;
- Includes a statement of situation, objectives, problems and solutions;
- Is relatively permanent but requires constant revision;
- May include long-term as well as short-term programme objectives;
- Forms the basis of extension teaching plans;
- Has been drawn up in advance; and
- Has been built on the basis of content.

So, we can define an **extension programme** as a written statement of situation, objectives, problems and solutions which has been prepared on the basis of an adequate and systematic planning effort and which forms the basis of extension teaching activities in a specific area, for a given period.

Definitions for a set of important terminologies in this regard would provide the needed clarity:

Programme is a written statement containing a more pertinent factual data used in decision-making, the problems agreed upon with priority assignment and the possible solutions to the problems'.

Plan or Plan of work is an outline of activities so arranged as to enable efficient execution of the entire programme. It answers the questions of what, why, how, when, where and by whom the work is to be done.

Project is a single item of the annual plant containing the method of solution of a single selected problem.

Calendar of work is a plan of work arranged chronologically, according to the time when step of work is to be done. It is a time schedule of work.

Aim is a broad objective. It is a generalized statement of direction and may have several objectives. It is also said to be an end in view to give direction to the creative process.

Objective is a direction of movement. A well stated objective is always measurable. It is also said to be a goal of growth.

Goal is a distance in any given direction, proposed to be covered in a given time.

Extension Programme Planning

Having described the concepts of planning and extension programme, now the stage is set to examine the concept of extension programme planning. A few points need to be explicated before attempting a definition.

1. Extension Programme Planning is a Process

The dictionary meaning of 'process' is 'any phenomenon which shows a continuous change in time' or 'any continuous operation or treatment'. If we accept this concept of process, we view events and relationships as dynamic, ongoing, ever-changing and continuous. The basis for the concept of process is the belief that man cannot discover the structure of physically reality; man must create it.

The concept a person has of the extension programme planning process will affect actions and mode of researching the process. Many programme-planning processes take place at any particular time at different levels of the extension organization. For example, programme planning occurs at the national level (five-year plans), at the state level (state plans, annual plans of work) and at the block level. In fact, planning at the block level is taking place when:

- ☆ The long-time plan or projected plan is being developed;
- ☆ The schematic budget is being planned;
- ☆ The annual plan of work is being developed;
- ☆ Detailed plans for individual learning experiences are developed with a major project.

2. Extension Programme Planning is a Decision-Making Process

Planning is basically a decision-making process- and so is extension programme planning. In extension programme planning, scientific facts are put to value judgments of the people through the implementation of a rational planning model in order to decide a programme which will be carried out through the extension teaching activities.

3. Extension Programme Planning Requires Advance Thinking

If we could know 'where we are' and 'where we are to go' we could better judge 'what to do' and 'how to do'. This statement lies at the heart of the nature of planning. Panning does not take place in a vacuum or automatically. It has to be made to happen. It requires knowledge, imagination and reasoning ability. It is a complex exercise as it involves people their needs, their interests, useful technology, educational process, analyzing a situation and making decision about what should be done, determining useful actions, projecting the desired shape of things in future and several other components, which are rarely simple.

4. Extension Programme Planning Requires Kill and Ability on the Part of Planners

Planning effective extension education programmes requires a number of high-level professional skills. Needed abilities include understanding and skill in the following broad areas?

- ☆ Understanding the nature and role of extension education organization.
- ☆ Knowledge and understanding of the technology related to the subject with which the programme is concerned.
- ☆ Ability to clarify the objectives of a programme and to so state them that they are useful in guiding its execution.
- ☆ Skill at seeing the relationship between principles and practice.
- ☆ Skill at inquiry and human relationships.

5. Extension Programme Planning is Built Round Content

A programme regarding any extension activity can only be built on the basis of content. Without some express purpose, there can be no planning. Extension programme planning is build with around available improved technology, the people, their resources, problems, needs and interests.

6. Extension Programme Planning is a Social Action Process

Extension programme planning involves interaction and the decisions so taken in the form of a programme affect others. Interaction assumes some type of communication between two or more people in the planning process. So when the extension staff involving specialists and people's representatives decides on the programme content for extension teaching for the coming year, it is involved in social planning.

7. Extension Programme Planning is a Collaborative Effort

Extension programme planning is a collaborative effort involving identification, assessment, evaluation of needs, problems, resources, priorities and solutions.

8. Extension Programme Planning is a System

Extension programme planning is a system as its procedures and processes are interrelated, ordered and linked progressively to form a collective whole. It includes several sub processes, such as planning, designing, implementing, evaluation *etc.*

9 The End-Product of Extension Programme Planning is an Extension Programme

The first consideration for anyone who is to concern himself with a process or set of procedures for planning is to clearly identify the primary purpose of the planning process to be developed. Many have suggested that the purpose of planning is for educating those who participate.

Objectives of Programme Planning

The general objective of having an extension programme is to influence people to make changes in their way of life and making a living. The assumption is that there is a need for change and if people are not aware, it is necessary to make them aware of this; and to develop their needs. According to Kelsey and Hearne (1967), the following are reasons for having a programme:

- ✰ To ensure careful consideration of what is to be done and why.
- ✰ To furnish a guide against which to judge all new proposals.
- ✰ To establish objectives toward which progress can be measured and evaluated.
- ✰ To have a means of choosing the important (deep rooted) from incidental (less important) problems; and the permanent from the temporary changes.
- ✰ To develop a common understanding about the means and ends between functionaries and organizations.
- ✰ To ensure continuity during changes of personnel.
- ✰ To help develop leadership.
- ✰ To avoid wastage of time and money and promote efficiency.
- ✰ To justify expenditure and to ensure flow of funds.
- ✰ To have a statement in written form for public use.

Principles of Extension Programme Planning

Extension programmes have the definite purpose of improving rural life through individual, group and community action. Extension programme planning has certain principles which holds good irrespective of the nature of the clientele and the enterprises they may be pursuing, *viz.*:

1. Extension programmes should be based on an analysis of the past experiences, present situation and future needs.

2. Extension programmes should have clear and significant objectives which could satisfy important needs of the people.
3. Extension programmes should fix up priority on the basis of available resources and time.
4. Extension programmes should clearly indicate the availability and utilization of resources.
5. Extension programmes should have a general agreement at various levels.
6. Extension programmes should involve people at the local level.
7. Extension programmes should involve relevant institutions and organization.
8. Extension programme should have definite plan of work.
9. Extension programme should provide for evaluation of results and reconsideration of the programme.
10. Extension programmes should provide for equitable distribution of benefits amongst the members of the community.

Programme-Planning Process

Program development involves: analyzing the situation or context; setting objectives/priorities and; designing an action plan; implementing the plan; evaluating inputs, outputs, outcomes, and reconsider the weak and strong points of the programme. The process of extension, as applied to development programmes, involves five essential phases (SOTER) *i.e.* Analyzing the Situation, Objectives or Goals to be Accomplished, Teaching, Evaluating the Teaching, Reconsidering (*J. P. Leagans, 1961*). These steps are intended only to clarify the necessary actions in carrying out a planned extension educational effort. The program development process is on-going and continuous.

Steps in Programme Planning

1. Collection of Facts

Sound plans are based on availability of relevant and reliable facts. This includes facts about the village people, physical conditions, existing farm and home practices, trends and outlook. Besides, other facts about customs, traditions, rural institutions, peoples' organisations operating in the area, *etc.* should be collected. The tools and techniques for collecting data include systematic observations, a questionnaire, interviews and surveys, existing governmental records, census reports, reports of the Planning Commission, Central Bureau of Statistics, and the past experiences of people.

2. Analysis of the Situation

After collecting facts, they are analysed and interpreted to find out the problems and needs of the people.

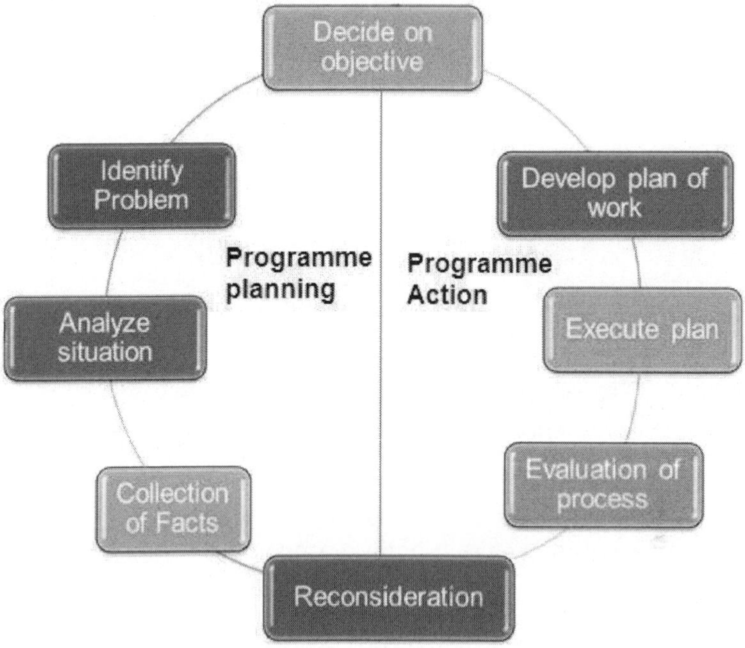

Steps of Programme Planning.

3. Identification of Problems

As a result of the analysis of facts, the important gaps between 'what is' and 'what ought to be' are identified and the problems leading to such a situation are located. These gaps represent the people's needs.

4. Determination of Objectives

Once the needs and problems of the people have been identified, they are stated in terms of objectives and goals. The objectives represent a forecast of the changes in the behaviour of the people and the situation to be brought about. The objectives may be long-term as well as short-term, and must be stated clearly.

5. Developing the Plan of Work

In order to achieve the stated objectives and goals, the means and methods attaining each objective are selected; and the action plan, *i.e.* the calendar of activities is developed. It includes the technical content, who should do what, and the time-limit within the work will be completed. The plan of work may be seasonal, short-term, annual or long-term.

6. Execution of the Plan of Work

Once the action plan has been developed, arrangement for supplying the necessary inputs, credits, teaching aids, extension literature *etc.* has to be made and the specific action has to be initiated. The execution of the plan of work is to be done through extension methods for stimulating individuals and groups to think, act and participate effectively. People should be involved at every step to ensure the success of the programme.

7. Evaluation

It is done to measure the degree of success of the programme in terms of the objectives and goals set forth. This is basically done to determine the changes in the behaviour of the people as a result of the extension programme. The evaluation is done not only of the physical achievements but also of the methods and techniques used and of the other steps in the programme planning process, so that the strong and weak points may be identified and necessary changes may be incorporated accordingly.

8. Reconsideration

The systematic and periodic evaluation of the programme will reveal the weak and strong points of the programme. Based on these points, the programme is reconsidered and the necessary adjustments and changes are made in order to make it more meaningful and sound. Programme planning is not the end-product of extension activities, but it is an educational tool for helping people to identify their own problems and make timely and judicious decisions. From the above mentioned cycle, it is clear that the planning of an extension programme comprises a logical series of consecutive steps. The first 4 steps form the programme-phase. The steps 5-7 form the action-phase. The step 8, *i.e.* reconsideration, joins the 2 phases together, where it leads to the fact-collecting step, thus beginning once more the never ending or continuous process of planning the extension programme.

Extension Programme Planning Model

Sandhu (1965) developed a model for planning extension and rural development programmes. This model has six phases with a number or steps to be followed under each phase.

Planning Process Involves

- ☆ Reaching, Understanding regarding principles, procedures, roles and time schedules
- ☆ Analysis situation
- ☆ Determining programme objective
- ☆ Selection problems
- ☆ Finding solution

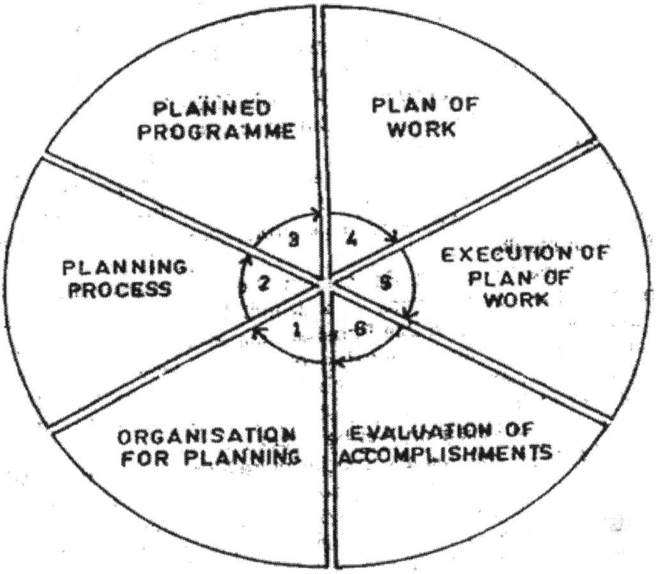

Extension Programme Planning Model.

The various phases and steps involved in this model are:

Organisation for Planning

I. Planning Process

- ☆ Reach understanding regarding principles, procedures, roles and time schedule.
- ☆ Analyze situation.
- ☆ Determine objectives.
- ☆ Select problems with due regard to priorities.
- ☆ Find solutions.

II. Planned Programme

Prepare a written statement of:

- ☆ Situation
- ☆ Objectives;
- ☆ Problems; and
- ☆ Solutions.

III. Plan of work

Prepare a plan of work containing information regarding:

- ☆ People to be reached;

- ✰ Goals, dates and places;
- ✰ Teaching procedures to be followed;
- ✰ Duties, training and recognition of leaders;
- ✰ Roles to be played by extension personnel; and
- ✰ Roles to be played by other agencies.

IV. Execution of Plan of Work

- ✰ Make advance arrangement for inputs and teaching aids.
- ✰ Interpret the approved programme to the staff and people's representatives.
- ✰ Carry out the planned programme, phase by phase, in a co-ordinated manner.

V. Evaluation of Accomplishments

- ✰ Do concurrent evaluation.
- ✰ Do ***ex-post facto*** evaluation.

Extension Evaluation

Extension programmes are mostly funded with public money and are planned and implemented by an organization, which in most cases is a department of a government. In order to justify the appropriation of public funds and continuing support from the people, it is necessary that their management as well as impact be properly and adequately evaluated from time to time. How, to evaluate management, achievements and failures of these programmes has been a challenge to extension workers right from the time when planned extension programmes were introduced. However, 'it was when Tyler's (1950) philosophy of educational evaluation became a part of extension education that educational evaluation became a part of extension education that the pattern of extension educational evaluation took a more usable, understandable form' (Sabrosky, 1966).

The word 'evaluation' has its origin in the Latin word "valere" meaning to be strong or valiant. Its dictionary meanings are the determination of the value, the strength or worth of something, an appraisal, an estimate of the force of or making a judgment of something.

Evaluation as applied to the field of extension education, may be defined as "a process of systematic appraisal by which we determine the value, worth or meaning of an activity or an enterprise". It is a method for determining how far an activity has progressed and how much further it should be carried to accomplish objectives'. Thus to an extension worker evaluation means determining the results of his extension programmes in order to know the extent to which objectives have been achieved and why and what changes would be needed in case the programme is planned again, or in its implementation.

Tyler (1950) developed two basic notions regarding educational evaluation, which equally apply to extension evaluation. These notions are that the process of evaluation (i) is essentially a process of determining behaviour of the people covered under the programme and (ii) the process of determining the degree to which these behavioural changes are actually taking place. Thus extension evaluation may be said to be a process for determining behavioural changes of people resulting from extension programmes. Once evaluation became an integral part of the extension education process, extension managers started applying this process to evaluate programme planning, management and implementation aspects of extension programme.

Definitions of Evaluation

More specific definitions of evaluation are given by persons involved in rural development programmes. While most of these definitions refer specifically to the assessment of the results of programmes of extension education, they can also be applied to the training aspect of such programmes. Some definitions of evaluation are:

- It is a process, which enables the administrator to describe the effects of his programme and thereby make progressive adjustments in order to reach his goal more effectively (Jahoda and Barnit, 1955).
- Programme evaluation is the determination of the extent to which the desired objectives have been attained or the amount of movement that has been made in the desired direction (Boyle and Johns, 1970).
- Programme evaluation is the process of judging the worth or value of a programme. The judgment is formed by comparing the programme should be (Steele, 1970).
- Evaluation is the process of delineating, obtaining and providing useful information for judging decision alternative (Stufflebeam, 1971).
- Evaluation is a coordinated process carried on by the total system and its individual subsystem. It consists of making judgments about a planned programme based on established criteria and known, observable evidence (Boone, 1985).

Nature of Evaluation

1. Evaluation is not Measurement

Evaluation is an integral part of extension education. All aspects of extension work need evaluation. Evaluation does not mean mere measuring of achievements, which is usually done after the programme is executed. Extension being an educational process, it is necessary to evaluate management of the programme and methods used, achievements accomplished in line with the objectives and also to determine the reasons for success or failure.

2. Evaluation is not Exactly Scientific Research

When we think of evaluation as a process of collecting information as a basis for making decisions, forming judgments and drawing conclusions, we realize it has much in common with scientific research. But there is a great difference between our casual everyday evaluation and scientific research. However, the difference is a matter of degree rather than kind. Casual everyday evaluation can be placed at one end of the scale and scientific research at the other end. There are five locations on the scale with no sharp lines of distinction, *i.e.*, casual everyday evaluation, self-checking evaluation, do-it yourself evaluation, extension evaluation studies and scientific research.

Types of Evaluation

- Self-evaluation: This is to be carried out by every worker as a matter of routine. This requires the self-critical attitude which is so essential for extension work. By this self-critical attitude, the chances of an extension worker growing and continuously improving his professional competency become greater.
- Internal evaluation: Evaluation carried to by the agency responsible for the planning and implementation of the programme. Some of the other methods for internal evaluation are: systematic use of diaries and reports of workers, planned visits of staff members to work spot, use of special questionnaires and proforma for observation and inquiry *etc.*
- External evaluation: Evaluation conducted by a person or a committee outside the area of operation. One of the strong features of the Indian Community Development Programme is that simultaneous with its start an independent agency, namely the Programme Evaluation Organization, was established.

Evaluation can also be classified into (i) concurrent and (ii) ex-post facto evaluation.

Evaluate Programme Planning

As a result of experience, theory, research and experimentation, much information has been accumulated about how an extension programme should be planned. Progress in science and technology and the broadening of extension's clientele with the accompanying great variation in needs and interests have made the scientific planning of extension education programmes more important than ever before. There is considerable agreement on certain criteria which, if followed, make for successful extension programme planning at different levels. These criteria represent the ideal with which to compare our practices and procedures or programme planning. Some of the steps needed top evaluates or programming function in view of these criteria includes:

- ☆ Identify the evidence needed to form a judgment about each criterion.
- ☆ Specify the methods that will be used to obtain the evidence, such as personal observation, personal interview or through a systematic survey.
- ☆ On the basis of the evidence gathered, judge whether or not each criterion is being adequately satisfied in the programme planning activities.

Extension Evaluation Process

There are several models of evaluation available in the literature. However, a very simplified version of most of these models may be quite workable for evaluating extension programmes since, as Bhatnagar (1987) has pointed out, any extension evaluation process has to be based on certain assumptions. For example, if some inputs are provided in the form of a programme, specific outputs can be expected and if these outputs happen, then the purpose of the programme can be achieved; if the purpose is achieved, then the development goal is realized. This means that evaluation has to be so designed that the quality types and adequacy of the input measures, outputs and their impact in achieving the programme objectives have to be evaluated systematically.

Steps involved in an extension programme evaluative process may be as follows:

i) Formulate Evaluation Objectives

Specific objectives to be achieved through the evaluative process must be clearly and adequately identified and started. All further efforts should be knit around these objectives.

ii) Classify Programme Objectives

It is assumed that each extension programme, when formulated and implemented, will have specific well-defined objectives. Since evaluation is basically a process of determining the extent to which various extension teaching activities were organized and managed and the extent to which they contributed to achieving the goals, programme objectives must be clearly understood and if necessary, further broke down into measurable terms. This is a crucial step as all further efforts will be directed towards collecting evidence related to these objectives.

iii) Identify Indicators

To identify indicators or the kind of evidence necessary to evaluate achievement in relation to specified programme objectives, it is necessary that specific beneficiaries of the programme be identified, the kind of behavioural changes expected in them be clearly stated, and the kin of learning experiences expected to be provided to them spelled out, together with the level of management to be achieved for provided those learning experiences are specified. Once this is done, identification of specific indicators to measures the achievements will not be difficult.

iv) Decide the kind of Information Needed

Once the indicators for evaluating the management and performance of a programme have been indicated, specific information to be collected may be worked out. Since there is usually more information than an extension worker can collect, he has to be very discriminating about the kind and amount of information that should be collected. Timing for collection of information may also need to be specified.

v) Sampling

The purpose of sampling is to take a relatively small number of units from a population in such a way that the evidence collected from them becomes representative evidence of the entire population. Although there are several sampling methods, perhaps stratified sampling procedures may be most suitable for extension evaluation studies as they allow inclusion of all interested groups and ensure enough heterogeneity in the sample.

vi) Decide the Design of Evaluation

An ideal design of evaluation may be an experimental one. This would allow separating the effect of the programme from other factors, by setting control and treatment groups. Several experimental designs, such as one-group pre-test-post-test design, static groups comparison, pre-test, post-test control group design, Solomon four-group design, longitudinal study design, *etc.* are available in literature and can be used. However, in actual practice, extension progammes are seldom run in a way that allows an experimental design of evaluation.

vii) Collection and Analysis of Evaluation Evidence

There are many methods for collecting information for evaluative purposes, such as the mail questionnaire, personal interview, distributed questionnaires, group interviews, case studies, systematic field observations, systematic study of secondary data *etc.* Selection of the right kind of data collection method will depend on the objectives of the evaluation, kind of information needed, time and resources available and the type of respondents from whom information is to be collected.

However, whatever the method used, a specific questionnaire or interview schedule or data recording sheet must be developed with care. Once the data is collected, it must be tabulated, summarized and analyzed with adequate care. This step should not be rushed. To avoid delay, however, analysis may be done with the help of a computer.

viii) Interpretation of the Results in a Proper Way

It is a very crucial as evaluation results can be missed also. Once tentative generalizations are arrived at, it may be appropriate and they are informally discussed among the interpreters as well as with programme planning and implementation officials, so that the results of evaluation are put in a proper perspective. The evaluation results must clearly state the achievements, failures and future adjustments needed. A written report of the evaluation findings should be prepared and made available to all concerned.

EXTENSION SYSTEM IN INDIA

Rural development is an obligation of the Government to raise the standard of living of the rural people. World Bank defined rural development in terms of an improvement of the economic and social life of the rural poor. Rural development encompasses all sectors of rural life. In its widest sense, it implies development of every aspect of rural life. There are number of rural development programs were started by the Nationalists and Social reformers. Some of these programs gradually disappeared or some were merged with Government sponsored schemes later. 'This is because of various reasons like lack of encouragement of the Government, lack of financial support, inadequate, in experienced and untrained staff. For the clarity, we can divide these Development Programmes in two parts: Pre-Independence Programmes, and Post-Independence Programmes.

Glimpses of Pre-Independence era

Introduction

About 65 percent population of the country is living in the economically undeveloped area. Therefore, many evils such as illiteracy, superstitions, unhealthy practices *etc.* took roots in our country. After Second World War or after independence, it was thought to settle all these problems properly and so, many development programmes were reinvigorated. It is wrong to declare that there was nothing done during slavery period. However, those works, which were executed at that time, were very few in number according to the needs. Looking at a vast country like India, during British rule some selected social workers had started some programmes of rural development. The evolution of extension programme is described in two stages:

- Stage I- Pre-Independence Era (1866-1947)
- Stage II- Post-Independence Era (1947-1953)

Stage I. Pre-Independence Era (1866-1947)

During the pre-independence era, various attempts were made by the individual persons and private agencies *viz.* Gurgaon project in Haryana by Mr. F.L. Brayne; Sriniketan Experiment in Bengal by Sri Rabindra Nath Tagore; Servants of India Society, Pune, in Tamilnadu, Central India (presently Uttar Pradesh and Madhya Pradesh); Sir Daniel Hamilton's Scheme of Rural Reconstruction at the Sunderbans in Bengal; Rural Reconstruction Work by Christian Missions through Allahabad Agricultural Institute; The Christian College, Nagpur; Marthandam project by Dr. Spencer Hatch under the auspices of YMCA; Sevagram experiment in Wardha by Mahatma Gandhi; Adarsh Seva Sangh, Pohri, Gwalior; Indian Village

Service in Lucknow and Etawah in Uttar Pradesh by Mr. A.T. Mosher and Sarvoday Programme in Bombay province by Sri B. N. Gupta.

Scheme of Rural Reconstruction at the Sunderbans in Bengal

In 1903, Sir Daniel Hamilton formed a scheme to create model villages, in an area in Sunderban (Bengal), based on cooperative principles. He organised one village of this type and set on Cooperative Credit Society which functioned up to 1916. In the 1924, he organised a Central Cooperative Bank and Cooperative Marketing Society and established a Rural Reconstruction Institute in 1934. The institute provided training facilities in cottage and subsidiary industries.

Gurgaon Project (1920)

In Gurgaon district, this programme of village development was the first one to be run by the State. It was started by the Mr. F. L. Brayne. In 1920, Mr. Brayne had been appointed on the post of Deputy Commissioner in Gurgaon district and he began this project of rural upliftment in his district, which became famous as Gurgaon Project. The main objectives of this project were:

1. To increase crop production,
2. To control extra expenditure,
3. To improve the health,
4. To develop the feeling of women-education, and
5. Home development work.

Shriniketan Project (1920)

Shriniketan is situated about 170 km. away from Calcutta (Kolkata) in West-Bengal State. This area was backward socially, economically and politically. Sri Ravindra Nath Tagore began this project of village development in this area with the help of sociologist Sri L.M. Hurst. Sri Tagore thought that if some villages were developed, the other villages will get inspiration and the programme of village development will spread all over the country and thus the whole country would be developed. Shriniketan Village Welfare Institute was established in 1920 whose main objectives were the following:

1. To increase the knowledge of rural people.
2. To help the rural people in establishing cottage industry.
3. To inspire the people to follow new technology.
4. Development of dairy farming.
5. To create the feeling of co-operation.
6. To arrange the facility of health and education.
7. To create the feeling of rural leadership.

This Project, in the beginning, was started only in 8 villages but after some time this project was extended to 15 villages. Sri Tagore was the landlord of all these villages. Being centralized over Sri Tagore this Project could not survive after the death of Sri Tagore, and came to an end after his death.

Seva-gram (1920)

For fulfilling this objective, Gandhiji made a programme, which became famous as Gandhian Constructive Programme. The main objectives of this project were:

(1) to use khadi clothes, (2) to introduce Health Programme, (3) the programme of sanitation in the village, (4) the programme of women welfare, (5) the programme of economic help, (6) to uplift the backward classes, (7) primary and adult educational programmes, (8) to improve the conditions of poor people, (9) programme of social harmony, and (10) to popularise the mother tongue and other national dialects.

Marthandam Project (1928)

This project of rural development was initiated in the village Marthandam near Trivendrum of Kerala State by Young Men Christian Association (YMCA) and Christian Church Association under the direction of Dr. Spencer Hatch in 1928. This village was undeveloped economically and the economic condition of the native majority was poor. Here, people used to cultivate only paddy and coconut on some places. The objectives of this project were also made on the basis as to how the public of this area could become Christian. The main objectives of this project were:

a) Spiritual development
b) Mental development
c) Physical development
d) Social development
e) Economic development

Baroda Village Reconstruction Project (1932)

Shree B.T. Krishnamachari in Baroda in the Gujarat State initiated this Project in 1932. This Project Continued for a long period.

a) Objectives

To improve the life style of rural people rapidly. To spread education and industrialization. To develop the necessary factors for the progress of agriculture.

b) Working System

By personal education and contact by the village guides. To use the school teacher of village in the extension of programme. To use the traditional means of extension.

Grow More Food Campaign (1942)

This campaign, started in 1942, was continued after getting the independence. The main object of this campaign was to fulfil the need of food, which had been created due to the Second World War. This campaign was the first one to be organized on a national level. In this campaign, the new seeds and chemical fertilizers were distributed among the farmers. Agricultural departments of state governments organized this campaign.

Indian Village Service (IVS-1945)

In 1945, under the guidance of Dr. W.H. Wisher, this service began in the village Agasoli, District Aligarh in U.P. But, after the partition of the country, the village volunteer M.V. Siddiqi Khan went to Pakistan and this centre was closed.

Firka Bikas Yojana (1946)

The government of Madras (now Tamil Nadu) decided to make efforts for the development of villages at Firka level. The first programme began in 1946. Among Pre-Independence project, this was the biggest project.

GLIMPSES OF POST-INDEPENDENCE ERA

Introduction

The urgent need for stepping up food production was realised even in the pre-Independence era and a Grow-More-Food Campaign was started. Under the campaign, targets for increased agricultural production were laid down for the first time on an all-India basis. But the campaign failed to achieve its targets. Soon after Independence (1947), the Central Government re-defined the objectives of the Grow-More-Food Campaign as the attainment of self-sufficiency in food grains by 1952, and simultaneously increased the targets of production of other crops to meet the shortfall as a result of the partition of the country. At the same time, arrangements were made for integration and co-ordination of the entire campaign for increasing agricultural production.

Etawah Pilot Project (1948)

The idea of intensive all-round development work in a compact area was put into practice as a Pilot Project in Rural Planning and Development in the Etawah District in Uttar Pradesh in 1948, which can be regarded as a forerunner of the Community Development Project in India. **Albert Mayer**, an American Engineer, played the key role in the initiation and implementation of the project. The programme was based on the principle of self-help, democracy, integrated approach, felt needs of the people, rigorous planning and realistic targets, institutional approach, co-operation between governmental and non-governmental organizations, close co-ordination between the extension service and the supply agencies and the collaboration by technical and social scientists.

Nilokheri Experiment (1948)

Sri S. K. Dey, was the founder of the Nilokheri Experiment, when nearly 7000 displaced persons from Pakistan and later integrated with the 100 surrounding villagesin to what came to be rural-cum-urban township. It was built round the vocational training centrer Nilokheri, in July 1948. Sri Dey began this project using 1100 acre of swampy land spreading in the midst of Karnal and Kurukshetra. Sri Dey went on to become the Union Minister of Community Development in 1965. The scheme was also known as "Mazdoor Manzil" because of its principle 'he who will not work, neither shall he eat'.

Community Development Project (1952)

As a result of the Grow-More-Food Enquiry Committee Report and the successful experience of the Etawah Project, 15 Pilot Projects were started in 1952 in selected states with the financial assistance received from the Ford Foundation. Besides helping to increase agriculture production and bettering the overall economic condition of the farmers, these projects were meant to serve as a training ground for the extension personnel. It was soon realised that for the creation of an urge among the rural population to live a better life and to achieve permanent plenitude and economic freedom in the villages, a much bolder and dynamic effort was called for.

Grow-More-Food Enquiry Committee Report (1952)

Though efforts were made to revitalize the Grow-More-Food Campaign, it was observed that the system was not functioning properly and the cultivator's response to the programme was very poor. As a result, the Government of India in 1952 appointed a committee known as the GMF Enquiry Committee to examine the working of the Grow-More-Food Campaign. The findings of this Committee revealed that the problem of food production was much wider than the mere elimination of food imports and that agricultural improvement was a very important part of a much wider problem of raising the level of rural life in the country.

National Extension Service (1953)

The National Extension Service was inaugurated on 2nd October 1953. The scheme of National extension service was designed to provide the essential basic staff and a small fund for the people to start the development work essentially on the basis of self-help. The operational unit of this service was an N.E.S. block comprising about 100 villages and 60,000 to 70,000 people. The N.E.S. blocks were later converted into community development blocks which had higher budget provisions in order to take up more intensive development programmes. The pattern of community development programme was further revised (modified with effect from 1st April 1958). According to this pattern, there were four stages:

- Pre-Extension Stage
- Stage I Blocks
- Stage II Blocks
- Post-Stage II Blocks

Key Village Scheme (1952)

Key Village Scheme launched in August 1952 was the first systematic attempt to improve the quality and productivity of cattle and buffaloes in the country. It was proposed with a view to promote intensive cattle development in compact areas by utilising the limited quantity of high quality breeding stock. Basic objective of the scheme is the rapid multiplication of crossbred cattle for meeting the acute shortage of high quality animals by grading up of the indigenous cattle population.

Intensive Agricultural District Programme (1960)

With a focusing on increasing productivity and production IADP was launched in July 1960 in 7 selected districts of various states in its first phase and later extended to another 9 districts more in second phase during 1963-64. This programme was aimed at integrated and intensive approach to solve the problems of agricultural production through adoption of package of improved practices. The achievements accomplished through this programme were tremendous and it really made a dent into the rural poverty for its eradication through agriculture. This programme was popularly known as Package Programme.

Operation Flood (1970)

Operation flood (also known as the White revolution of India) was a rural development programme started by India's National Dairy Development Board (NDDB) in 1970. One of the largest of its kind, the programme objective was to create a nationwide milk grid. It resulted in making India the largest producer of mil and milk products, and hence is also called the white revolution of India. It also helped reduce malpractices by milk traders and merchants. This revolution followed the Indian green revolution and helped in alleviating poverty and famine levels from dangerous proportions in India during the era. Dr. Varghese Kurien is known as Father of White Revolution.

Training and Visit (T and V) system (1974)

The Training and Visit (T and V) system, the brain child of Dr. Daniel Benor (World Bank Consultant) was introduced in India in 1974 for all round development of agricultural extension system in the country to introduce observation, training and technology transfer to the farmers and extension workers so as to enable them achieving greater productivity and production in the agricultural sector. This system has been introduced at first in 1974 in two states *viz.*, Rajasthan and West Bengal and then spread to another 16 states of the country aimed at building a professional extension service that is capable of assisting farmers in raising production, increasing incomes and of providing appropriate support for agricultural development.

Integrated Rural Development Programme (1978)

The Integrated Rural Development Programme (IRDP) is a rural development program of the Government of India launched in Financial Year 1978 and extended throughout India by 1980. It is a self-employment program intended to raise the income-generation capacity of target groups among the poor. The target group consists largely of small and marginal farmers, agricultural labourers and rural artisans living below the poverty line. The program is implemented in all the blocks in the country as a centrally sponsored scheme funded on 50:50 basis by the Centre and State. The Scheme is merged with another Scheme named *Swarnjayanti Gram Swarozgar Yojana* (SGSY) since 01.04.1999.

Intensive Agricultural Area Programmes (IAAP) 1964-65

Intensive Agriculture Area programme (IAAP) was launched in 1964-65. The core philosophy of the IAAP was that "much greater emphasis should be given to the development of scientific and progressive agriculture in an intensive manner in the areas which have High production potentials". The idea was to cover at least 20 per cent of the cultivated area of the country. The emphasis was on import crops such as Wheat, Rice, Millets, Cotton, Sugarcane, Potato, Pulses *etc.* The Intensive Agriculture Area programme (IAAP) paved the way for Green Revolution in the country.

Even with the short coming of IADP, the spectacular results in improved crop yield obtained in IADP, prompted the Government to think of extending the benefits of improved technology in agriculture in large areas over the country at less cost and with reduced staff strength. This was similar to but a less intensive extension programme in comparison to that of the IADP. 114 districts were selected in the year 1964 and later extended to 150 districts.

Achievements

- ☆ Achieved increased production by exploiting the land resources
- ☆ Package approach were covered in 1410 communities blocks spread over 114 districts in India. In Karnataka 57 C.D. Blocks spread over 14 districts were covered
- ☆ Increased production by 20-25 per cent of the cultivated area was achieved
- ☆ Effective coordination between officials and nonofficial was achieved
- ☆ Multiplication of improved seeds and its distribution to all cultivated areas was possible

Difference between IAAP and IADP

IADP	*IAAP*
The main objective of the IADP was to implement the improved farm practices in the selected potential areas to increase the production	The objective of the IAAP was to extend the concept of IADP to other potential areas to cover at least 20-25 per cent of the cultivated area in the country.
This programme was recommended by a team of Food Ford foundation experts	This programme was recommended by the third Five Year Plan midterm appraisal committee.
The main criteria for selection of the districts was based on areas having assured irrigation facilities minimum natural hazards and well developed cooperatives	The main criteria for selection of districts was based on areas having predominant crops and well developed infrastructure facilities
Twenty eight districts including Mandya and Raichur were covered under this programme ☆ Number of personnel working in each community block were ☆ VLW's -10 ☆ AEO - 1 ☆ SM's - 2 (SM's were responsible for the whole district)	One hundred and fifty districts including fourteen districts of Karnataka were covered under these programmes. ☆ Number of personnel working in each community block were ☆ VLW's -10 ☆ AEO - 1 ☆ SM's - 2 (Similar to IADP)

High Yielding Variety Programme (HYVP) 1966

High Yielding Variety Programme (HYVP) was launched in 1966, which helped the country in attaining self-sufficiency in food. The technological development did not remain confined to the introduction of high yielding crop varieties alone. These were combined with the application of high analysis and balanced fertilizer, irrigation, plant protection, improved implements *etc.* which made a 'Green Revolution' possible in the country.

- ☆ Agricultural scientists found successful in evolving new high yielding varieties in some cereals particularly in wheat, rice and maize, Punjab, Haryana and Western parts of UP were initially selected for this programme
- ☆ The pervasive influence of high yielding technology spread to other area as of farm production such as animal production, fishery, sericulture, social forestry *etc.* The high yielding technologies in these production enterprises also have some common characteristics such as shorter gestation period, good response with better management; higher return fan terms of yield and income, and higher investment in comparison to traditional technologies.

- High yielding technologies by themselves tend to be scale neutral *i.e.* farmers irrespective of the size of their holding can derive economic benefit from them provided they have access to the needed inputs. However, high yielding technologies are not resource neutral. In other words more inputs are required for higher output

Objectives

- To assess the spread of the various high yielding varieties in different parts of the country and also determine the extent of such spread;
- To ascertain the reactions, attitudes of cultivators/participating cultivators; and
- To study the problems of implementation of the programme at different level of administration such as States, district, block and village.
- Besides above, the study also covered certain other important and relevant aspects such as research efforts in evolving of new varieties.

FIRST-LINE EXTENSION SYSTEM OF ICAR

Realizing the scope and importance of integrated working of interrelationship between research, education and extension functions, the ICAR established a section of Extension Education at its headquarters in 1971 which was later on strengthened and renamed as Division of Agricultural Extension. It was intended to enforce this functional relationship down the line in the research institutes, agricultural universities and allied institutions.

There were four main Transfer of Technology (ToT) Projects of the ICAR, namely National Demonstrations (ND), Operational Research Project (ORP), Krishi Vigyan Kendra (KVK) and Lab to Land Project (LLP). All the projects were mobile in nature, except KVKs. The projects were shifted every 5 to 6 years from one location to another, in the larger interest of the farming community and the extension personnel of the area.

The details are given as under:

National Demonstration

A nationwide programme of demonstrations, known as National Demonstration (ND) on major food crops was launched in 1964. It was a nationwide project with a uniform design and pattern. It differed from other demonstrations in the following major aspects:

- There was a specific yield target and there was no separate control plot.
- The area of the demonstration plot was about one hectare.
- The farmers in whose plots the demonstrations were laid out were the actual cultivators with small holdings.
- The agricultural scientists conducted these demonstrations in association with local extension agencies/workers.

To achieve the best results, the most enthusiastic and cooperative farmers were selected the best technologies and techniques used, and a team-approach followed involving scientists from relevant disciplines, including local extension agents. The organization of field days, field visits, and training for the farmers and field extension workers were considered important for rapid spread of technologies. The national demonstrations were intended to show the superiority of the demonstrations to the extension agencies.

Operational Research Project (ORP) 1974-75

The ORPs were initiated in 1974-75. It covered diverse topics like crop farming, mixed farming, integrated pest management, plantation crops, post-harvest technology, improvement of *diara* and *char* lands, land reclamation, arid land management, fisheries *etc.* The ORP demonstrated latest agricultural technologies on the farmers' fields to influence the farmers as well as the State extension agencies. It also studied the socio-economic, technological extension and administrative barriers which were coming in the way of rapid transfer of technologies and pointed out the same to the extension agencies.

The ORPs considered two kinds of problems: first the common agricultural problems affecting the farming community requiring group or community action, *e.g.* plant protection and rodent control; and secondly total resources development of the watershed area. Some of the ORPs attempted to concentrate on the major problematic areas rather than either on a common subject or the total resources development like enhancing agricultural as well as livestock production.

The ORPs were mainly implemented through the agricultural universities, the ICAR institutes and the State departments of agriculture. Selected voluntary organizations with necessary scientific background and support for rural/agricultural development also under took such projects. Linkages of these projects with local and allied institutions were emphasized for mutual interaction and multiplier effect.

Krishi Vigyan Kendra (KVK) (Agricultural Science Centre)

The first KVK was established in 1974 at Pondicherry under Tamil Nadu Agricultural University. The Krishi Vigyan Kendra is designed to impart need-based and skill-oriented vocational training to the practicing farmers, in-service field level extension workers, and to those who wish to go in for self-employment.

The first KVK was established in 1974 at Pondicherry under Tamil Nadu Agricultural University. The priority for establishing KVKs is given to hilly areas, drought prone areas, forest areas, coastal areas, flood prone areas, forest areas, coastal areas, flood prone areas, and areas dominated with tribal farmers, weaker sections, small farmers and landless labourers. The objective is to gradually cover the entire country with one KVK in each district, priority being given to the backward areas. At present there are 722 KVKs in the country.

The basic concepts of a KVK are

1. The center will impart learning through work-experience and, hence, will be concerned with technical literacy, the acquisition of which does not necessarily require as a precondition the ability to read and write.
2. The center will impart training only to those extension agents who are already employed or to practicing farmers and fishermen. In other words, these centers will cater to the needs of those who are already employed, or those who wish to be self-employed.
3. There will be no uniform syllabus for a KVK. The syllabus and programme of each center will be tailored according to the felt needs, natural resources and the potentials for agricultural growth in that particular area

The three fundamental principles of KVK are

- Agricultural production as the prime goal
- Work-experience as the main method of imparting training and Priority to weaker sections of the society.
- The main idea is to influence the productivity to achieve social justice for the neediest and deserving weaker sections of the society like the tribal farmers, small and marginal farmers, agricultural labourers, drought and flood affected farmers, and so on. Need-based training courses are designed for different types of clientele. Courses are based on the information received through family and village survey. No certificate or diploma is awarded irrespective of the duration of the courses. After the training, follow-up extension programmes are organized for converting the acquired skills of the trainees into practice. While designing the courses, the concepts of farming system are taken into account to make the enterprises commercially viable.

Mandate

The mandate of a KVK is unique for it and is determined on the basis of the most important needs of the clientele, their resources and constraints, and nature of the ecosystem. The success of a KVK is judged by the extent to which it fulfils obligations specified in the mandate.

1. On-farm testing on farmers fields of proven technologies in agriculture and allied fields.
2. Organising Vocational Trainings in agriculture and allied areas
3. Conducting frontline demonstrations on major cereal, oilseeds, pulses and other important crops
4. Organinsing in-service training programmes to field/local extension functionaries in emerging advances in agriculture and allied areas.

5. The KVKs are fully funded by the Indian Council of Agricultural Research (ICAR). Initially, one KVK for each district was thought of and now two KVKs are also established in certain districts being the larger ones. Though KVKs are sponsored by ICAR they are working under different administrative controls *viz.*, SAUs, NGOs and ICAR.

Trainers Training Centre (TTC)

Trainers' Training Centres (TTC) was established in 10 specialized subjects, mainly in the ICAR institutes for in-service training of the teachers/trainers of the KVKs and such other non-degree level institutions. The centres offer in-service training courses for the trainers/instructors of the extension training centres, farmers' training centres, agricultural schools, and teachers dealing with work experience and vocational courses. However, these TTC were closed by the ICAR during the year 1996.

Lab to Land Programme (LLP)

The Lab to Land Programme (LLP) was launched by the ICAR in 1979 as a part of its Golden Jubilee celebration. The *overall objective* of the programme was to improve the economic condition of the small and marginal farmers and landless agricultural labourers, particularly scheduled castes and scheduled tribes, by transfer of improved technology developed by the agricultural universities, research institute *etc.*

The programme was initiated with 50000 farm families over the whole country. It was implemented by the agricultural universities, ICAR institutes, state departments of agriculture and some selected voluntary organizations. No separate staff was provided for the programme. Major thrust in the programme was to introduce the most appropriate technologies that would help in the diversification of labour use and introduction of supplementary sources of income. The programme had been in operation in a number of phases. Normally a phase ran for a period of two years with a particular set of farmers.

With effect from April 1992, all first-line transfer of technology projects of the ICAR *viz.* ND, ORP and LLP were integrated into KVKs. Accordingly, the mandates of KVKs were revised and are as follows:

1. Collaborate with the subject matter specialists of the State Agricultural Universities/scientists of the Regional Research Stations (National Agricultural Research Project) and the State extension personnel in 'on-farm testing', refining and documenting technologies for developing region specific sustainable land use systems.
2. Organize training to update the extension personnel with in the area of operation with emerging advances in agricultural research on regular basis.

3. Organize long term vocational training courses in agriculture and allied vocations for the rural youths with emphasis on 'learning by doing' for generating self-employment through institutional financing and on-and off-campus training courses for the farmers with a view to increasing production on their farms.
4. Organize front-line demonstrations in various crops to generate production data and feedback information.

Over the years, the KVKs emerged as a national network of training organization at the base level for imparting improved knowledge and skills to the farmers, farm women, rural youth and extension functionaries.

Technology Assessment and Refinement (TAR) - Institution Village Linkage Programme (IVLP)

In 1995, the ICAR launched this innovative programme and the objectives are to:

1. Introduce technological interventions with emphasis on stability and sustainability along with productivity of small-farm production systems.
2. Introduce and integrate the appropriate technologies to sustain technological interventions and their integration to maintain productivity and profitability taking environmental issues into consideration in a comparatively well-defined farm production system.
3. Introduce and integrate the appropriate technologies to increase the agricultural productivity with marketable surplus in commercial on and off farm production system.
4. Facilitate adoption of appropriate post-harvest technologies for conservation and on-farm value addition of agricultural products, by-products and waste for greater economic dividend and national priorities.
5. Facilitate adoption of appropriate technologies for removal of drudgery, increased efficiency and higher income of farm women.
6. Monitor socio-economic impact of the technological intervention for different farm production systems.
7. Identify extrapolation domains for new technology/technology modules based on environmental characterization at meso and mega level.

National Agricultural Technology Project (NATP)

The National Agricultural Technology Project (NATP) was launched by the Indian Council of Agricultural Research (ICAR) on June 30, 1998, with the support of the World Bank, to strengthen and complement the existing resources and to augment the output of the National Agricultural Research System (NARS).

NATP was the world's biggest World Bank assisted agriculture project worth Rs. 992 crores developed and executed by NARS. The lifespan of NATP was seven years, from 1998 to 2005. NATP was the first project in NARS to shift the focus from discipline oriented research to production system research. NATP was the first project in NARS to involve competitive funding, and have pluralistic approach to involve and fund partners from outside NARS. NATP successfully completed a whopping total of 852 projects.

Agricultural Technology Information Centre (ATIC)

The Agricultural Technology Information Centres (ATIC) were established in the year 2000. These ATICs serve as a single window delivery system for services and products of research for the areas in which the concerned institute is involved. At present, there are 44 ATICs established in ICAR institutes and SAUs.

National Agricultural Innovation Project (NAIP)

The National Agricultural Innovation Project (NAIP) was launched in the year 2006 by ICAR. The overall objective of NAIP is to facilitate the accelerated and sustainable transformation of Indian agriculture in support of poverty alleviation and income generation through collaborative development and application of agricultural innovations by the public organizations in partnership with farmers groups, the private sector and other stakeholders.

Rural Landless Employment Guarantee Programme (1983)

Generating gainful employment for Landless rural poor. Creating proactive assets in rural areas. Improving the overall quality of rural life.

Jawahar Rozgar Yojana (NREP + RLEGP) (1983)

Rural Poor Families. Generation of additional gainful employment for the employed and underemployed persons in the rural areas

National Agriculture Extension Project (1982-83)

To strengthen the reorganized agricultural extension system. To conduct district level bimonthly technical workshops. To conduct district level diagnostic survey. To develop literature on crops and agricultural enterprises suitable for the district. To organize farmers interaction sessions. Farmers

Watershed Development Programme (1994-95)

To develop wastelands or degraded lands, drought-prone and desert areas on watershed basis, keeping in view the capability of land, site conditions and local needs. To promote the overall economic development and improving the socioeconomic condition of the resource poor and disadvantaged sections inhabiting the programme areas. To mitigate the adverse effects of extreme climatic conditions such as drought and desertification on crops, human and livestock population for their overall improvement.

National Agricultural Technology Project (1998)

The overall objective of NATP was to revitalize the agricultural technology generation assessment, refinement and dissemination systems. It had three components. These were: ICAR organization and management reforms Expansion of location specific research on production system and Testing of innovations in technology dissemination.

Institute Village Linkage Programme (1999)

Introduce technological intervention with emphasis on stability and sustainability along with productivity of small production system. To introduce and integrate the appropriate technologies to sustain technological interventions and their integration to maintain productivity and profitability taking environmental issues into considerations in comparatively well defined production systems. To introduce and integrate the appropriate technologies to increase the agriculture productivity with marketable surplus in commercial farm production systems. To facilitate adoption of appropriate post - harvest technologies for conservation and on - farm value addition of agriculture products, by products and wastes for greater economic dividend and national priorities. To facilitate adoption of appropriate technologies for removal of drudgery increased efficiency and higher income of farm woman. To monitor socio-economic impact of the technologies intervention for different production systems. To identity extrapolation domains for new technology modules based on environmental characterization at meso and mega levels.

National Agricultural Innovation Project (2006)

To alleviate the accelerated and sustainable transformation of Indian agriculture in support of poverty easing and income generation via cooperative development and application of agricultural innovations by the public organizations in partnership with farmers groups, the private sector and other stake holders. All stakeholder of agriculture.

National Innovations on Climate Resilient Agriculture (NICRA) (2011)

To enhance the resilience of Indian agriculture covering crops, livestock and fisheries to climatic variability and climate change through development and application of improved production and risk management technologies. To demonstrate site specific technology packages on farmers' fields for adapting to current climate risks. To enhance the capacity of scientists and other stakeholders in climate resilient agricultural research and its application farmers.

Attracting and Retaining Youth in Agriculture (2015)

To attract and empower the Youth in Rural Areas to take up various Agriculture, allied and service sector enterprises for sustainable income and gainful employment in selected districts. To enable the Farm Youth to establish network groups to take up resource and capital intensive activities like processing, value addition and marketing. To demonstrate functional linkage with different institutions and stakeholders for convergence of opportunities available under various schemes/ program for sustainable development of youth.

Farmers FIRST (Farm, Innovations, Resources, Science and Technology) Programme (2016)

To enhance farmer-scientist interface, enrich knowledge and facilitate continued feedback. To identify and integrate economically viable and socially compatible technological options as adoptable models for different agro-ecological situations. To develop modules for farm women to address drudgery reduction, income enhancement and livelihood security. To study performance of technologies and perception of the farmers about agriculture as a profession in the rural settings. To build network of linkages of organizations around the farm households for improving access to information, technology, input and market To institutionalize Farmer FIRST process.

Mera Gaon Mera Gaurav (2015)

To promote the direct interface of scientists with the farmers to accelerate the lab to land process. To ensure that farmers benefit from best farm practices by providing required information, knowledge and advisories on regular basis by adopting villages. To create awareness among farmers about the organisations and their programmes and schemes as well as government policies related to the farming sector. To imbibe a sense of ownership among the agricultural scientists of the country.

Soil Health Card (2015)

To provide a basis to address nutrient deficiencies in fertilization practices. To strengthen functioning of Soil Testing Laboratories (STLs) through capacity building. To diagnose soil fertility related constraints with standardized procedures. To develop and promote soil test based nutrient management. To disseminate soil testing results through SMSs.

NEW TRENDS IN EXTENSION EDUCATION

Privatization of Extension

Privatization broadly refers to a process by which the government reduces its role in an activity and encourages private sector to take up these roles.

Concept and Definition

Chandra Shekara, 2001: The Process of funding and delivering the extension services by private individual or organisation is called Private Extension.

Bloom indicated that private sector extension involves personnel in the private sector? extension involves personnel in the private sector.

Vanden Ban and Hawkins (1996) state that in private sector extension, farmers are expected to share the responsibility for the service and pay all or part of the cost.

Saravanan and Shivalinge Gowda (1999) operationalized: Privatization as follows: "Privatization of extension service refers to the services rendered in the area of agriculture and allied aspects by extension personnel working in private agencies or organisations for which farmers are expected to pay a fee (or free) and it can be viewed as supplementary or alternative to public extension services".

These concepts about the privatization emphasize three aspects:

It involves extension personnel from private agency/organization.

- ☆ Clients are expected to pay the service fee.
- ☆ Act as supplementary or alternative to public extension service.
- ☆ Through the process of privatization, extension effectiveness is expected to improve by: reorienting public sector extension with limited and well focused functions,
- ☆ More number of extension providers (institutional pluralism) resulting from active? encouragement by the public sector to initiate, operate and expand. more private participation leading to the availability of specialised services hitherto not
- ☆ Available from the public system user contributions to extension leading to improved financial sustainability
- ☆ Support and control by clients leading to client orientation

Privatisation can take place broadly in two ways:

- ☆ Dismantling of Public sector
- ☆ Controlled Privatisation

Reasons Behind Privatization of Extension Services

Following are some of the reasons behind privatization in agricultural extension services:

1. Fiscal crisis: Many less developed country's governments have found it difficult to make adequate resources available for extension. Financial pressures have, in turn led to the search for ways of reduction in public sector costs.
2. Disappointing performance of public extension system: Impact of public extension system in agricultural development is disappointing. Extension work is ineffective, does not match with farmer's needs, there is little consideration for cost-effectiveness and less competent extension personnel have no accountability to farmers. Moreover, public extension alone will never answer to entire demand of farming community.

3. Changing contexts and opportunities: In recent past Indian agriculture is shifting from mere subsistence level to commercialized agribusiness to meet all challenges and opportunities of globalization and liberalization, effective alternative extension approach is required. Increasing commercial and specialized nature of agriculture will demand quickly and technically sound advice, based on scientific analysis with appropriate marketing information. Further India is facing population explosion which had resulted into reduction in land holding size. Due to improved transport network and better communication facilities private agencies and communication media are also reaching to remote villages. Therefore, there is an urgent need for an efficient technological guidance along with recent market information, which is only possible through privatization of extension services.
4. The extension worker: the extension worker and farmers ratio is very wide in India *i.e.* 1:1000 and it is further widens due to engagement of extension workers in administrative work, official correspondence, report *etc.*

Advantages of Private Extension System (Sulaiman and Sadamate, 2000):

1. Reorienting public sector extension with limited and well focussed functions. More number of extension providers (institutional pluralism) resulting from active encouragement by the public sector to initiate, operate and expand.
2. More private participation leading to the availability of specialized services hitherto not available from the public system.
3. Support and control by clients leading to client orientation.

Disadvantages of Private Extension System (Vanden Ban and Hawkins, 1998):

1. Privatization may hamper free flow of information.
2. Farmers may be less interested to disseminate technologies to other farmers what they have learnt from private extension.
3. Only commercial farmers will be benefited through privatization and subsistence farmers will seldom be able to pay the fee.
4. The commercial interest of the private agencies may jeopardise the efforts of research and extension of eco-friendly and sustainable agriculture.
5. Contact between farmers and extension agents get declined.

Cyber Extension/e-Extension

Cyber

According to Oxford dictionary the word Cyber means "relating to Information technology, the Internet and virtual reality, the Cyber space. The word has its origin from cybernetics.

Cyber Space

The cyber space is the imaginary or virtual space of computer connected with each other on networks, across the globe. These computers can access information in form of text, graphics, audio, video and animation files. Software tools on network provide facilities to interactively access the information from connected services. The cyber space thus can be defined as the imaginary space behind the interconnected telecommunications and computer networks, the virtual world.

Extension

Extension stands for "the action or process of enlarging or extending something". It could be extension of area, time or space.

E-extension and cyber extension are more or less synonymous and can be used interchangeably. Cyber extension and e-extension are extension approaches; whereas ICT is the tools by which various services are delivered to the clientele fulfilling the objectives of e-extension/cyber extension (Mondal, 2019).

Cyber Extension

According to Sharma (2005) Cyber extension can be defined as the "Extension over cyber space". Cyber extension means "using the power of online networks, computer communications and digital interactive multimedia to facilitate dissemination of agricultural technology". Cyber extension includes effective use of information and communication technology, national and international information networks, internet, expert system, multimedia learning systems and computer based training systems to improve information access to the farmers, extension workers, research scientists and extension managers.

Cyber agricultural extension is an agricultural information exchange mechanism over cyber space, the imaginary space behind the interconnected computer networks through telecommunication means. It utilizes the power of networks, computer communications and interactive multimedia to facilitate information sharing mechanism (Wijekoon, 2003).

IT and ICT

IT (Information Technology) refers to computer based technology and telecommunications. IT refers to an entire industry that uses computers, networking, software and other equipment to manage information. Generally, IT departments are responsible for storing, processing, transmitting, retrieving and protecting digital information of the company.

ICT (Information Communications Technology) can be seen as an integration of IT with mediation broadcasting technologies, audio/video processing and transmission and telephony. Therefore, ICT can be seen as an extended acronym for IT.

ICTs are those technologies that can be used to interlink information technology devices such as personal computers with communication technologies such as telephones and their telecommunication networks. The PC, laptop and tablet with e-mail and internet provides the best example (Mondal, 2019).

Information Provided through Cyber Extension

Cyber extension can provide large amount of information to farmers. Some of those as mentioned by Sharma are:

1. Weather: Daily information of maximum and minimum temperature, day length, direction and speed of wind, rain fall, relative humidity and weather forecasting.
2. Alert: Information regarding type of disaster, sudden change in weather, outbreak of diseases and insects *etc.*
3. Insurance: Detail information about insurance of farmers, his crop, animal and agricultural assets like tractors, implements and tools *etc.*
4. General awareness: Information about vaccination, cleanliness, health, family planning and nutrition.
5. Rural Finance: Information about micro finance and subsidy *etc.*
6. Education: Detail information about farmers training and educational knowledge of his children about professional and non-professional institution.
7. Animal Science: The detail regarding the local animal and well known diseases, better breeds, balanced feed, shed management, health *etc.*

Advantages of Cyber Extension

1. Information can be provided quickly to farmers.
2. Round the clock service to farmers.
3. Information can be accessed from any place on the earth.
4. Helpful in enhancing the communication efficiency.
5. Preserves the quality of message.
6. Experts can be contacted directly.
7. Per unit cost is less as compared to traditional system of extension.

Limitation of Cyber Extension

1. Lack of reliable telecom infrastructure in rural areas.
2. Erratic or no power supply.
3. Lack of ICT trained manpower (willing to serve) in rural areas.
4. Lack of content (locally relevant and in local language).
5. Low purchasing power of rural people.

6. Lack of holistic approach.
7. Issues of sustainability.

Farmer-Led Extension

Concept

The farmer led extension approach gives farmers the opportunity to share their experiences and practices through a method demo with fellow farmers in the area. It was noted that farmers who were successful in their farming venture have established credibility among their peers. In selecting the farmer extensionist, the primary consideration is- he should be an innovative farmer, active and hardworking, honest and credible interested in learning, accepted and committed to the community, and most importantly interested to share his knowledge and skills. Likewise a farmer extensionist must have conducted a techno demo trial and field day, preferably attended the FFS. He must also have the capacity and willingness to finance the cost of technology.

Farmer-to-Farmer extension is defined here as "the provision of training by farmers to farmers, often through the creation of a structure of farmer promoters and farmer trainers.

There are 5 types of farmers-led extension:

1. Farmers to farmer
2. Farmer field school
3. The problem census
4. Problem solving approach
5. NGO government collaboration

Paradigm Shift from Production-Led to Farmers-Led Extension System (Kokate *et al.*, 2009)

Components	*Production-led*	*Farmers-led*
Purpose/ Objective	Transfer of production technologies	Capacity building (especially farmers extensionist), create para professional technologies extension workers, creating or strengthening local institutions
Goal	Food self-sufficiency	Livelihood security including food, nutrition, employment to alleviate poverty. Sustainability and conserving biodiversity
Approach	Top-down, commodity and supply driven	Participatory, bottom-up and demand driven
Actors	Mostly public institutions	Pluralistic with public, private, nongovernment and farmers organizations as a partner rather than competitors

Components	*Production-led*	*Farmers-led*
Mode	Mostly interpersonal/ individual approach	Integration of clients oriented on-farm participatory/experiential learning methods supported by ICTs and media
Role of extension agents	Limited to delivery mode and feedback to research system	Facilitation of learning, building overall capacity of farmers and encouraging farmers experimentation
Linkages/liaison	Research-Extension-Farmers	Research-Extension-Farmers Organizations (FIGs, CIGs, SHGs)
Emphasis	Information management, Production "Seed to Seed"	Knowledge management and sharing
Nature of technology	Input intensive, crop based and general recommendations as per agro-climatic zone, fixed package of information	Knowledge intensive, broad based, farming system perspective and blending with ITKs
Critical areas	Improvement, production and protection	Decision support system, integrated farming system approach, natural resource management, client group formation and community empowerment
Critical inputs	Money and material	Access to information, building human and social capital
Accountability	Mostly government	To farmers rather than donors

Market-Led Extension

Concept

With globalization of the market, farmers need to transform themselves from mere producers' sellers in the domestic market to producers cum sellers in a wider market sense to best realize the returns on their investments, risks and efforts.

Agriculture without economy is nothing. So the market linkage is a must for agriculture. Market is congregation of prospective buyers and sellers with a common motive of trading a particular commodity. Extension is the spreading/reaching out to the mass. Market-led extension is the agriculture and economics coupled with extension is the perfect blend for reaching at the door steps of common man with the help of technology. Market led extension is the market ward orientation of agriculture through extension.

Basics of Market Led Extension

- Market analysis
- Market intelligence
- Use of technology

Aspects of Market Led Extension

- Community mobilization
- Forming CIGs
- Forward and backward linkages
- Improvement of effectiveness of existing markets
- Improvement of efficiencies of marketing channels

Paradigm Shift from Production-Led Extension to Market-Led Extension (Adopted from MANAGE, 2008)

Aspects	*Production -led extension*	*Market-led extension*
Purpose/ Objective	Transfer of production technologies	Enabling farmers to get optimum returns out of the enterprise
Expected end results	Delivery of messagesAdoption of package of practices by most of the farmers	High returns
Farmers seen as	Progressive farmer High producer	Farmer as an entrepreneur "Agripreneur"
Focus	Production/yields "Seed to seed"	Whole process as an enterprise/High returns "Rupee to Rupee"
Technology	Fixed package recommended for an agro-climatic zone covering very huge area irrespective of different farming situations	Diverse baskets of package of practices suitable to local situations/ farming systems
Extensionists' interactions	Messages Training Motivating Recommendations	Joint analysis of the issues Varied choices for adoption Consultancy
Linkages/liaison	Research-Extension-Farmer	Research-Extension-Farmer extended by market linkages
Extensionists' role	Limited to delivery mode and feedback to research system	Enriched with market intelligence besides the TOT function Establishment of marketing and agro-processing linkages between farmer groups, markets and processors
Contact with farmers	Individual	Farmers' Interest Groups Commodity Interest Groups/SHG's
Maintenance of Records	Not much importance as the focus was on production	Very important as agriculture viewed as an enterprise to understand the cost benefit ratio and the profits generated
Information Technology support	Emphasis on production technologies	Market intelligence including likely price trends, demand position, current prices, market practices, communication net work, *etc.* besides production technologies

Challenges to Market Led Extension

- Gigantic size of Public Extension System of the public extension in the country.
- Generation of data on market intelligence would be a huge task by itself.
- The present extension system suffers from several limitations of stationery, mobility,
- allowances, personal development of cadres *etc.* Infrastructure challenges.
- Storage and preservation of produce
- Minimization post harvest losses.
- Capacity building of extension personnel and farmers.

Expert system

Concept

An Expert System is a software application that attempts to reproduce the performance of one or two more human experts. The concept of expert system development came from the subject domain of Artificial Intelligence (AI). The expert system is used to behave like a human expert to solve the problem with the help of pre-set conditions in the software application. The computer programme uses knowledge and inference procedures to solve problems.

Definition

Daniel Hunt (1986) – Expert system is an intelligent computer program that uses knowledge and inferences procedures to solve problems.

According to Prasad and Babu (2008) Expert system can be defined as a tool for information generation from knowledge. It is a computer programme designed to stimulate the problem-solving behaviour of an expert in a narrow domain or discipline.

An expert system is typically composed of at least three primary components. These are the inference engine, the knowledge base, and the User interface.

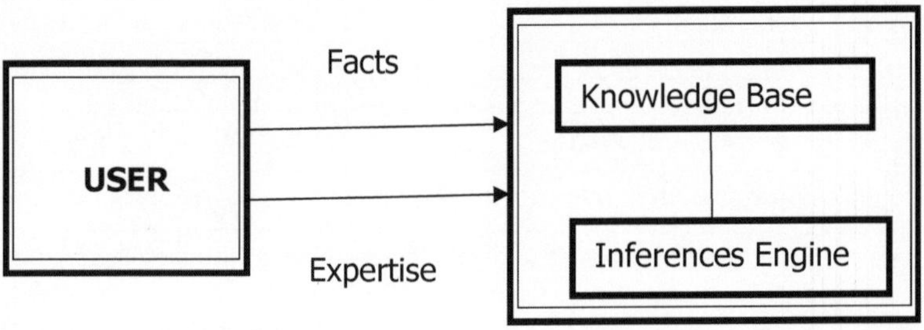

Expert System

Knowledge base - This component consists of data, facts and rules for a certain topic, industry or skill, usually equivalent to that of a human expert.

Interference engine - This component uses the facts and rules in the knowledge base to find and learn new knowledge or patterns. It is responsible for gathering the information from the user, by asking various questions and applying it wherever necessary. It seeks information and relationships from the knowledge base and to provide answers, predictions and suggestions the way a human expert would.

User interface - A user interface is the method by which the expert system interacts with a user.

Expert systems use information technology to gain and use human expertise. Obviously, this can be very beneficial to organizations. Expert Systems can:

1. Provide answers for decisions, processes and tasks that are repetitive
2. Hold huge amounts of information
3. Minimize employee training costs
4. Centralize the decision making process
5. Make things more efficient by reducing the time needed to solve problems
6. Combine various human expert intelligences
7. Reduce the number of human errors
8. Provide strategic and comparative advantages that may create problems for competitors
9. Look over transactions that human experts may not think of

Difference between Conventional Extension and Expert System Extension (Bahal *et al.*, 2004)

Sl.No.	*Conventional Extension*	*Expert System of Extension*
1	Universal approachability of same information is a problem	Universal approachability of same information is possible
2	Information is given whatever is available without considering needs and resources	Information is chosen based on their needs and resources
3	No Cost benefit analysis	Cost benefit analysis
4	Information flow depends on availability of agent	Information through Cyber Cafe at any place at any time
5	Require users to draw their own conclusion from facts	Conclusion is drawn based on the decision given by the expert

Objectives (Mondal, 2019)

The main objectives of developing expert system in agriculture are:

1. To enhance the performance of agricultural extension personnel and farmers.

2. To make farming more efficient and profitable.
3. To reduce the time required in solving the problems and performing the routine tasks of extension workers.
4. To maintain the expert system by continuously upgrading the database.

Some Examples of Expert System in India

Rice-Crop: The MANAGE has developed an expert system to diagnose pests and diseases for rice crop and suggest preventive/curative measures. The rice crop doctor illustrates the use of expert-systems broadly in the area of agriculture and more specifically in the area of rice production through development of a prototype, taking into consideration a few major pests and diseases and some deficiency problems limiting rice yield.

1. Expert System on Wheat Crop Management - EXOWHEM is an Expert system on wheat crop management is an expert system which includes all aspects of wheat production on India. The main goal of the system is to provide the users with all kinds of suggestions and advices regarding the wheat crop production. The system is designed as web based application by IASRI New Delhi that covers agricultural operations, fertilizer application, variety selection, as well as the economic benefits. It is purely a rule based expert system.
2. AMRAPALIKA is an Expert System for the diagnosis of pests, diseases, and disorders in Indian Mango. The system makes diagnosis on the basis of response/responses of the user made against queries related to particular disease symptoms. A rule-based expert system is developed using Expert System Shell for Text Animation (ESTA). The knowledge base of the system contains knowledge about symptoms and remedies of 14 diseases of Indian mango tree appearing during fruiting season and non-fruiting season.
3. e-Sagu is a tool for IT–based personalized Agro-Advisory system. ("Sagu" means cultivation in Telgu language). It aims to improve farm productivity by delivering high quality personalized (farm–specific) agro-expert advice in a timely manner to each farm at the farmer's door steps without farmer asking a question. In e-Sagu, the development of IT such as (Database, Internet, and Digital Photography) is extended to improve the performance of agricultural extension services. The e-Sagu system was implemented during 2004 by delivering advisory to 1051 cotton farms for the farmer of three villages in Warangal district in Andhra Pradesh.

Advantages

The advantages of expert system are:

1. Expert Systems are useful in many aspects and ready to use by end user as advisory system.

2. Provides consistent answers for repetitive decisions, processes and tasks.
3. Holds and maintains significant levels of information.
4. Encourages human expert to clarify and finalize the logic of their decision-making.
5. Never "forgets" to ask a question, as a human might.
6. They can respond at great speed due to the inherent advantages of computers over humans.
7. Unlike humans, they do not get tense, fatigue or panic and work steadily during emergency situations.

Disadvantages

However, there are also disadvantages to expert systems, such as:

1. Lacks common sense needed in some decision making.
2. Cannot make creative responses as human expert would in unusual circumstances.
3. Not capable of explaining the logic and reasoning behind a decision
4. Cannot adopt to changing environments, unless knowledge base is changed.

Chapter 2

Rural Development

RURAL DEVELOPMENT

Rural Development has become buzz word in the arena of human development, especially with the setting of Millennium Development Goals to be achieved globally by 2020. India should be one among the top list of the countries walking towards rural development, as life in India is mostly in rural areas. Rural geographically refers to the villages and demographically to the people living in villages. Simply defining development is growth. Development is the process of change in terms of expansion, progress and advancement in quality of life. Rural development is a multidimensional process of attaining progress and advancement in living standards both geographically and demographically. Rural development is a complex process of change in rural sub-systems and their interaction leading to desired improvements in rural employment opportunities, income distribution, rural welfare and other aspects of rural life. It is a package of action which covers a series of mutually supporting projects and activities concern with different sectors of rural communities.

Meaning

Simply expressing, rural development is developing rural people and areas. There is no universally accepted definition and the term is used in different ways and in vastly divergent contexts. As a concept, it indicates overall development of rural areas. In this sense it is comprehensive and multidimensional concept and includes the development of agriculture and allied activities like village and cottage industries, crafts, socio-economic infrastructure, community services and facilities and above all the human resources in rural areas. As a phenomenon, it is the result of interactions between various, physical, technological, economic, socio-cultural;

and institutional factors. As a strategy, it is designed to improve the economic and social well being of a specific group of people *i.e.*, the rural poor. As a discipline, it is multidisciplinary in nature representing an intersection of agricultural, social, behavioral engineering and management of sciences.

Concept

A rural development is a strategy to enable a specific group of people *i.e.*, poor rural women and men, to gain for themselves and their children more of what they want and need. It is empowering the poorest of the poor *i.e.*, small farmers, tenants and the landless, who seek livelihood in the rural areas to demand and control more of the benefits of rural development. It is a process of enhancing the standard of living rural poor through increased production, productivity, employment, and mobilizing the land, labor and water. It is establishment of required rural infrastructure with the participation of people.

Objectives

The objectives of rural development, according to the World Bank, are not restricted to any single department but spread over several, and the resultant mix serves to raise agricultural output, create new employment, improve health and education, expand communications, provide housing *etc.* According to Singh (1999), the main objectives of rural development in all societies, irrespective of their economic, political and socio-cultural systems are:

- to increase the availability and improve the distribution of the life-sustaining goods, such as food, clothes, shelter, health and security;
- to raise per capita purchasing power and improve its distribution by providing better education, productive and remunerative jobs and cultural amenities
- to expand the range of economic and social choices to individuals by freeing them from servitude and dependence.
- Therefore, a measure of rural development should provide, at the minimum, an indication of per capita availability of life-sustaining goods or per capita income in rural areas, as well as some idea of the distribution of income, assets and other means of socio-economic welfare.

Aims

There are three basic elements behind any development programme-Life sustenance, Self-respect and Freedom.

Life-sustenance: People have certain basic needs food, shelter, clothing, health care and security, without which it is difficult to survive. They are also considered as life sustaining requirements. When there is short supply of these facilities, the situation is called as underdevelopment. When there is adequate economic growth they can be adequately supplied. Hence it can be concluded as economic growth is a necessary condition for development.

Self-respect: Every person seeks some form of self respect, dignity and honours. Absence of self-esteem is an indication of underdevelopment. Child labour, atrocities, child marriages *etc.* are included in the indicators for ranking the nation on human development index.

Freedom: Freedom here refers to political and social opportunities. If the individual is bound by the servitude of men to nature, ignorance, other men, institutions and dogmatic beliefs, it cannot claimed to have achieved servitude the goal of development. Servitude in any form reflects a state of underdevelopment.

Functions of Extension Education in Rural Development

Extension Education is the process of developing capabilities of the individuals through education which is stretched out to people to meet their needs. Rural development aims at enhancement in the standard of living of the poor through various programmes designed and planned to meet the developmental needs. It can be concluded that Rural Development is the programme and Extension Education is the process of taking the programme to the right people in a right way to meet their developmental needs, thus enhances the standard of living.

The nine functions of Extension Education process in rural development programme

1. **Planning** – It identifies the needs of the people and sketch out a programme to meet the identified needs.
2. **Convergence** – It identifies the suitable rural development programme in operation, which is relevant to meet the needs and decides the extent, situation and action of utilization.
3. **Coordination** – It brings the people with needs, the institutions with programmes and personnel with functions on one plat form to meet the common goal.
4. **Cooperation** – It motivates planners, executers and users to work as a team with a combined force to meet the goals.
5. **Collaboration** – It pools all the essential resources from all directions to make them accessible to achieve the goal.
6. **Facilitation** – It plays a role of catalyst in the implementation of development programme and smoothens the progress
7. **Monitoring** - It keeps an eye on the progress of the programme for the purpose of overcoming the obstacles
8. **Evaluation** – It quantitatively assess the quality and quantity of progress at periodic intervals to reconsider any of the issues that are not met.
9. **Network**- The process of extension education finally set the functional relationships between institutions, which works as input and output device in the field of human development.

Dimension of Rural Development

Although in rural development programmes priority should be given to the marginalized, poor and deprived section of society, it should encompass all aspects of rural life, including nature and its elements. The important dimensions of rural development works are presented below:

Dimension 1: Development of Agriculture and Allied Sectors

In rural development programmes, emphasis should be given for development of modern agriculture and allied sectors. This includes high yielding and hybrid seeds, fertilizers and chemicals, knowledge of Integrated Pest Management (IPM), and Integrated Nutrient Management (INM), availability of improved agricultural implements, preservation of biodiversity, collection and maintenance of Indigenous Technical Knowledge (ITK) and germplasm. Besides introduction of new and value added crops, food processing, post-harvest technology, promotion of agri-business and small scale industries, dairy, poultry, forestry, fish-culture, sericulture *etc.* are important areas which need to be developed.

Dimension 2: Socio-economic and Political Development

Rural development also encompasses improvement of socioeconomic and political conditions of the rural people. Rural development programmes may be meaningful and significant when the rural people have freedom to express their opinion, free and fair franchise in democracy, increased purchasing power, uplifted from below the poverty line (BPL), freedom of indebtedness from private money leaders *etc.*

Dimension 3: Rural Disaster Risk Mitigation

For sustainable rural development, disaster risk mitigation is an important aspect which the policy makers and project administrators should take into account. For this purpose, community risk perception, risk identification, risk mitigation and risk transfer through insurance like crop insurance, Kisan Credit Card *etc.* should be taken into account. Prevention of river bank erosion, flood and drought control measures, community capacity building is also included under this dimension.

Dimension 4: Improving Quality of Rural Life

Improving quality of rural life is also in important dimension of rural development. In this dimension, activities may include construction of rural durable assets and infrastructure like roads, bridges, power supply, educational facilities, improving health and sanitation, adopting of Information Communication Technology (ICT) in rural areas, new product development for rural market *viz.* fast moving consumer goods (for example, cosmetics, toiletries, foods and beverages *etc.*), durable agricultural goods and services like insurance, banking, telephone *etc.*

Livelihood Scenario and Evolution of Rural Development Programmes

The rural development programmes in India are surrounded around the major livelihoods, namely Agriculture, Animal Husbandry and related subsidiary activities.

The following is the livelihood scenario around which the rural development programmes are planned.

Sl. No.	*Major Occupations*	*Resources*	*Inputs*	*Stakeholders*	*Output*
1	Agriculture	Land, water, Knowledge	Seed, pesticides, fertilizers	Producers	Products, value added products, by products
2	Livestock Management	Livestock, land, water	Feed, fodder, Technology	Service providers Support institutions	By products, bio-waste
3	Other Occupations	Skill, landwater	Technology Machinery, materials	Technologists	Knowledge and skill, experience

Outcomes of Rural Development Programmes

- ✰ Improved Resource Management
- ✰ Increased production
- ✰ Increased well being
- ✰ Decreased vulnerability
- ✰ The rural development programmes are planned around every element that contributes to the development

The Five Assets in Rural Development Programme

The five capitals in rural development programmes

- ✰ Human capital- energy, indigenous knowledge, skill, experience, interest *etc.* of human beings, who are the partners of development.
- ✰ Natural capital-natural resources
- ✰ Financial capital- money, materials, support *etc.*
- ✰ Physical capital- infrastructure, implements, machinery *etc.*
- ✰ Social capital- culture, social systems, groups, social relation *etc.*, existing among the human beings

Constraints in Rural Development

A constraint is something that comes in the way of any function, activity, policy, strategy and limits the progress or development or growth. In the context of rural development, the constraint is the one that is obstructing the process of development *i.e.*, right from planning, implementation, monitoring and evaluation.

General Constraints

In general the following are the constraints commonly encountered in any development programme,

- ☆ Lack of commitment by any or all i.e the participants, extension workers and organization, lead to the failure of a programme.
- ☆ Lack of meaningful objectives and goals, make the programme in effective and in-actionable.
- ☆ Failure to develop and implement strategies, results in astray of objectives and action plan.
- ☆ Lack of rationality in the programme, leads planning of unreasonable steps or leaving crucial steps.
- ☆ Excessive reliance on experience, by pass the current scientific information
- ☆ Failure to identify the limiting factors, may lead to adversity
- ☆ Lack of organizational support, weakens the extension agent in decision making process
- ☆ Resistance to change, on the part of the target group

Process Constraints

- ☆ Migration to urban areas
- ☆ Occupational transition
- ☆ Dependency on middleman marketing
- ☆ Increasing health hazards.
- ☆ Cultural practices
- ☆ Inadequate and unequal distribution of resources
- ☆ Untapped indigenous knowledge
- ☆ Gender disparity
- ☆ Lack of policy advocacy

Product Constraints

- ☆ Lack of skill and knowledge
- ☆ Communal conflicts and civil wars
- ☆ Malnutrition and health hazards
- ☆ Population explosion
- ☆ Untimely and inadequate rainfall
- ☆ Infertile soils

List of Development Programmes in India/Sociology

Here is a list of top forty seven development programmes or schemes adopted in India for its rural development.

1. Pradhan Mantri Jan Dhan Yojana (PMJDY)

Financial inclusion and access to financial services for all households in the country is the main objective of PMJDY. Pradhan Mantri Jan Dhan Yojana is a national mission to bring comprehensive financial inclusion of all the households in the country. Under the PMJDY, any individual above the age of 10 years and does not have a bank account can open a bank account without depositing any money. The scheme was to ensure the access to financial services such as banking/savings and deposit Accounts, remittance, credit, debit cards, insurance and pension in affordable manner. The scheme was mostly targeted to the people belonging to the Below Poverty Line but is beneficial to everyone who does not have a bank account. Jan Dhan Yojana has seen a great success, about 21 crore accounts have been opened in just about one and half year under the scheme. Out of the total 12.87 crore in rural area and 8.13 crore accounts have been opened in urban areas. Despite of zero minimum balance, there is 33074.89 crore rupees balance in these accounts with 28.88 per cent accounts opened with zero balance.

2. Sukanya Samriddhi Yojana

The main aim of this scheme is to encourage parents to build a fund for the future education and marriage expenses for their female child. Sukanya Samriddhi Yojana is an ambitious small deposit savings scheme for a girl child. Under the scheme, a saving account can be opened in the name of girl child and deposits can be made for 14 years. After the girl reaches 18 years of age, she can withdraw 50 per cent of the amount for marriage or higher study purposes. After the girl completes 21 years of age, the maturity amount can be withdrawn including the interest at rates decided by Government every investments and returns are exempt from section 80C of Indian income tax act. The maximum investment of Rs. 1.5 Lakh per year can be made while minimum deposit is Rs. 1000/- per year. In case of more than one girl child, parents can open another account on the different name but only for two girl child. Only exception is that the parents have twins and another girl child.

3. Pradhan Mantri Mudra Yojana (PMMY)

The main objective is to provide financial support for growth of micro enterprises sector. Pradhan Mantri MUDRA (Micro Units Development and Refinance Agency) Yojana was launched with the purpose to provide funding to the non-corporate small business sector. Pradhan Mantri Mudra Yojana (PMMY) is open and is available from all Bank branches across the country. The small businesses/startups or entrepreneurs can avail loans from Rs. 50 thousand to 10 Lakh to start/grow their business under the three, Shishu, Kishore and Tarun categories of the scheme. As per the official website of PMMY, 27,344053 numbers of loans have been sanctioned under the scheme till 26 February, 2016. The amount sanctioned has reached more than Rs. 1 Lakh Crore.

4. Pradhan Mantri Suraksha Bima Yojana (PMSBY)

The main objective is to provide accidental insurance cover to all Indian citizens Pradhan Mantri Suraksha Bima Yojana which is also a government backed accident insurance scheme in India aimed at increasing the penetration of accidental insurance cover in India. The scheme is open and available to all Indian citizens between the ages of 18 to 70 years. Under the scheme, the policy holder can get a life insurance cover of Rs. 2 Lakh with an annual premium of just Rs. 12 excluding service tax. All the Indian citizens between 18-70 years of age with a saving bank account are eligible to avail the scheme.

5. Atal Pension Yojana (APY)

The main objective is to Increase the number of people covered under any kind of pension scheme. Atal Pension Yojana is one of the three Jan Suraksha schemes launched by PM Narendra Modi. APY is aimed at increasing the number of pension scheme beneficiaries across the country. The scheme is especially targeted to the private unorganized sector and is open to all Indian citizens between the age of 18 to 40 years.

6. Pradhan Mantri Awas Yojana (PMAY)

Achieve housing for all by the year 2022, 2 crore in Urban and 3 crore homes in Rural areas is the main objective. Pradhan Mantri Awas Yojana is an ambitious scheme of Narendra Modi Government. Under the PMAY, the government aims to provide about 5 Crore affordable homes to the people belonging to EWS and LIG categories by the year 2022. There is a target of building 2 crore homes in urban area and 3 crore in rural areas across the country. Under the scheme, the government will provide financial assistance to the poor home buyers, interest subsidy on home loan and direct subsidy on homes bought under the scheme.

7. Sansad Adarsh Gram Yojana (SAGY)

Social, cultural, economic, infrastructure developments in the villages *i.e.*, development of model villages called "Adarsh Gram" Under the Yojana; Members of Parliament (MPs) will be responsible for developing the socio-economic and physical infrastructure of three villages each by 2019 and a total of eight villages each by 2024.

Funding

a) Funds from existing schemes, such as the Indira Awas Yojana, Pradhan Mantri Gram Sadak Yojana, Mahatma Gandhi National Rural Employment Guarantee Scheme, and Backward Regions Grant Fund, *etc.*,

b) The Member of Parliament Local Area Development Scheme (MPLADS),

c) The gram panchayat's own revenue,

d) Central and State Finance Commission Grants, and
e) Corporate Social Responsibility funds.

8. Pradhan Mantri Fasal Bima Yojana (PMFBY)

The main aim is to provide insurance cover to Rabi and Kharif crops and financial support to farmers in case of damage of crops. In order to make crop insurance simpler and cheaper for the farmers and to provide them with better insurance services, a Central Sector Scheme of Pradhan Mantri Fasal Bima Yojana (PMFSY) was launched by the Government of India replacing NAIS and MNAIS.

9. Pradhan Mantri Gram Sinchai Yojana (PMGSY)

The main objective is irrigating the field of even farmer and improving water use efficiency to provide 'Per Drop More Crop'. The scheme is aimed to attract investments in irrigation system at field level, develop and expand cultivable land in the country, enhance ranch water use in order to minimize wastage of water, enhance crop per drop by implementing water-saving technologies and precision irrigation. All the States and Union Territories including North Eastern States are covered under the programme. The government has approved Rs.50, 000 crore for the implementation of Pradhan Mantri Krishi Sinchai Yojana for next 5 years, *i.e.* up to 2020.

10. Pradhan Mantri Garib Kalyan Yojanaye (PMGKY)

The main objective is to implement the pre-poor welfare schemes in more effective way and reaches out to more poor population across the country. Garib Kalyan Yojana is a Poverty Alleviation Scheme, which is primarily a work shop that you can pay and attend. The effort of the campaign and workshop is to motive and apprise the member of parliaments to help them effectively implement the government run schemes for the welfare of poor in the country.

11. Pradhan Mantri Jan Aushadhi Yojana (PMJAY)

Providing drugs/medicines at affordable cost across the country is the main objective. Under the scheme, over 500 medicines will be sold through Jan Aushadhi stores at price less than the market price. Private hospitals, NGO's, and other social groups are eligible to open the Jan Aushadhi stores with a onetime assistance of Rs. 2.5 Lakh from the central Government.

12. Make in India

The main objective is to encourage multi-national, as well as domestic companies to manufacture their products in India and create jobs and skill enhancement in 25 sectors. The major objective behind the initiative is to focus on job creation and skill enhancement in 25 sectors of the economy. The initiative also aims at high quality standards and minimising the impact on the environment. The initiative hopes to attract capital and technological investment in India.

13. Swachh Bharat Abhiyan

The main objective is to fulfil Mahatma Gandhi's dream of a clean and hygienic India. Swachh Bharat Mission is being implemented by the Ministry of Urban Development (UD) and by the Ministry of Drinking Water and Sanitation (DWS) for urban and rural areas respectively.

14. Kisan Vikas Patra

The main objective is to provide safe and secure investment avenues to the small investors. Unlike in PPF the Kisan Vikas Patra certificates would be available in the denominations of Rs 1,000, 5,000, 10,000 and 50,000 and there is no upper limit on investment in KVPs.

15. Soil Health Card Scheme

The main objective is to help farmers to improve productivity from their farms by letting them know about nutrient/fertilizer requirements for their farms. The soil health card studies and reviews the health of soil or rather we can say a complete evaluation of the quality of soil right from its functional characteristics, to water and nutrients content and other biological properties. It will also contain corrective measures that a farmer should adopt to obtain a better yield.

16. Digital India

The main objective is to deliver Government services to citizens electronically by improving online infrastructure and by increasing Internet connectivity. The Digital India programme is a flagship programme of the Government of India with a vision to transform India into a digitally empowered society and knowledge economy. The vision of Digital India programme is to transform India into a digitally empowered society and knowledge economy.

17. Skill India

Train over 40 crore people in India in different skills by 2022 is the main objective. The main goal of Skill India Program is to create opportunities, space and scope for the development of talents of the Indian youth. The scheme also targeted to identify new sectors for skill development and develop more of those sectors.

18. Beti Bachao, Beti Padhao Yojana

To generate awareness and improving the efficiency of welfare services meant for women are the main objectives. The scheme is to have as focused intervention and multi-section action in almost 100 districts with low Child Sex Ratio (CSR).

19. Mission Indradhanush

The main objective is to immunize all children as well as pregnant women against seven vaccine preventable diseases namely diphtheria, whooping cough (Pertussis), tetanus, polio, tuberculosis, measles and hepatitis 'B' by 2020. The aim of Mission Indradhanush is to achieve full immunization in 352 districts which includes 279 mid priority districts, 33 districts from the North East states and 40 districts from phase one where huge number of missed out children were detected.

20. Deen Dayal Upadhyaya Gram Jyoti Yojana (DDUGJY)

Electric supply feeder separation (rural households and agricultural) and strengthening of sub-transmission and distribution infrastructure including metering at all levels in rural areas. DDUGJY will help in providing round the clock power to rural households and adequate power to agricultural consumers. The earlier scheme for rural electrification *viz.* Rajiv Gandhi Grameen Vidyutikaran Yojana (RGGVY) has been subsumed in the new scheme as its rural electrification component.

21. Deen Dayal Upadhyaya Grameen Kaushalya Yojana (DDUGKY)

Achieving inclusive growth, by developing skills and productive capacity of the rural youth from poor families are the main objectives. DDUGKY aims to train rural youth who are poor and provide them with jobs is having regular monthly wages. It is one of the cluster initiatives of the Ministry of Rural Development that seeks to promote rural livelihoods. It is a part of the National Rural Livelihood Mission (NRLM) – the Mission for poverty reduction called Aajeevika.

22. Pandit Deendayal Upadhyay Shramev Jayate Yojana (PDUSJY)

The main objective is to consolidate information of Labour Inspection and its enforcement through a unified web portal, this will lead to transparency and accountability in inspections. A Unified Labour Portal, known as the Shram Suvidha portal was launched under the scheme as a platform to facilitate the implementation of a transparent system for information and data base management.

23. Atal Mission for Rejuvenation and Urban Transformation (AMRUT)

The main aim is to provide basic services (*e.g.* water supply, sewerage, urban transport) to households and build amenities in cities which will improve the quality of life for all, especially the poor and the disadvantaged. The purpose of Atal Mission for Rejuvenation and Urban Transformation (AMRUT) is to-

a) Ensure that every household has access to a tap with assured supply of water and a sewerage connection;

b) Increase the amenity value of cities by developing greenery and well maintained open spaces (*e.g.* parks); and

c) Reduce pollution by switching to public transport or constructing facilities for non-motorized transport (*e.g.* walking and cycling).

24. Swadesh Darshan Yojana

The main aim is to develop world class tourism infrastructure. As part of the Swadesh Darshan Scheme, Theme Based Tourism Circuits (TBCT) around specific themes such as religion, culture, ethnicity, niche, *etc.* are identified for infrastructure development across the country.

25. Pilgrimage Rejuvenation and Spiritual Augmentation Drive (PRASAD)

The main aim is to develop world class tourism infrastructure in Amritsar, Ajmer, Amaravati, Dwaraka, Gaya, Kanchipuram, Kedarnath, Kamakhya, Mathura, Puri, Varanasi and Vellankani *etc.*

26. National Heritage City Development and Augmentation Yojana (HRIDAY)

The main objective is bringing together urban planning, economic growth and heritage conservation in an inclusive manner to preserve the heritage character of each Heritage City. With duration of 27 months (completing in March 2017) and a total outlay of INR 500 crores, the Scheme is being implemented in 12 identified Cities namely, Ajmer, Amaravati, Amritsar, Badami, Dwarka, Gava, Kanchipuram, Mathura, Puri, Varanasi, Velankanni and Warangal.

27. Udaan Scheme

The main objective is encouraging girls for higher technical education and aims to provide a platform that empowers girl students and provides them with better learning opportunities. It is a mentoring and scholarship scheme to enable meritorious girl students to transit from schools to technical education without much difficulty and also aims to enrich and enhance teaching and learning of mathematics and science at senior secondary school level by providing free online resources for all.

28. National Bal Swachhta Mission

The main objective is to provide hygienic and clean environment, food, drinking water, toilets, schools and other surroundings to the children. The Bal Swachhta Mission is a part of the nationwide sanitation initiative of 'Swachh Bharat Mission' launched by the Prime Minister on 2nd October, 2014.

29. One Rank One Pension (OROP) Scheme

The main objective is to provide same pension, for same rank, for same length of service, irrespective of the date of retirement. This is not a Modi Government scheme; however, the government is making its efforts to implement the long pending scheme.

30. Smart City Mission

Under the mission, the NDA Government aims to develop smart cities equipped with basic infrastructure and offer a good quality of life through smart solutions. Assured water and power supply, sanitation and solid waste management, efficient urban mobility and public transport, robust IT connectivity, e-governance and citizen participation along with safety of its citizens are some of the likely attributes of these smart cities.

31. Gold Monetisation Schemes

The main objective is to reduce the reliance on gold imports over time. The main objective is to develop 100 cities all over the country making them citizen friendly and sustainable. The programme is to lure tonnes of gold from households into the banking system. Under the scheme, people can deposit gold into the banks and earn interest based on the value of the gold.

32. Start-up India, Stand-up India

The main objective is to provide support to all start-up businesses in all aspects of doing business in India. Under the scheme, the start-ups will adopt self-certification to reduce the regulatory liabilities. An online portal, in the shape of a mobile application, will be launched to help start-up founders to easily register. The app is scheduled to be launched on April 1.

33. Digi Locker

The main objective is to provide a secure dedicated personal electronic space for storing the documents of resident Indian citizens. It is a part of Digital India programme, Digital Locker has been designed to reduce the administrative overhead of government departments and agencies created due to paper work. It will also make it easy for the residents to receive services by saving time and effort as their documents will now be available anytime, anywhere and can be shared electronically. The Government of India will provide financial support of Rs. 45,800 crore over the entire implementation period of IPDS under which strengthening of sub-transmission network, metering, IT application, customer care Services, provisioning of solar panels will be implemented.

34. Shyama Prasad Mukherji Urban Mission

The main objective is to create 300 rural clusters across the country and strengthen financial, job, and life style facilities in rural areas. Urban Mission is a solution for both villages and cities in the country- that would promote growth of villages and its residents at the place where they are. Under the mission, the government will identify and develop 300 rural clusters with urban like facilities in the next 3 years.

35. Sagarmala Project

The main objective is to transform the existing ports into modern world class ports. The prime objective of the Sagarmala project is to promote port-led direct and indirect development and to provide infrastructure to transport goods to and from ports quickly, efficiently and cost effectively. The Sagarmala Project, aimed at port-led development in coastal areas, is bound to boost the country's economy and the government has lined up about Rs 70,000 crore for its 12 major ports only.

36. 'Prakash Path' – 'Way to Light' – The National LED Programme

The main objective is to distribute LED bulbs and decrease the power consumption. This is one of the many schemes launched by Narendra Modi government India. The programme has been launched to distribute and encourage the use of LED light bulbs to save both cost and consumption.

37. UJWAL Discom Assurance Yojana (UDAY)

The main objective is to obtain operational and financial turnaround of State owned Power Distribution Companies (DISCOMs). The Scheme aims to reduce the interest burden, reduce the cost of power, reduce power losses in distribution sector, and improve operational efficiency of DISCOMs.

38. Vikalp Scheme

The main objective is to provide confirmed accommodation in next alternative train for the wait listed passengers. Vikalp scheme is available only for the tickets booked through internet for six months and option will be limited to mail and express trains running on Delhi-Lucknow and Delhi-Jammu sectors.

39. National Sports Talent Search Scheme (NSTSS)

The main objective is to identify sporting talent among students in the age group of 8-12 years. The scheme is being implemented by the Sports Authority of India (SAI), under the Ministry of Youth Affairs and Sports for spotting talented young children in the age group of 8-14 years from schools and nurturing them by providing scientific training.

40. Rashtriya Gokul Mission

The main objective is to conserve and develop indigenous bovine breeds. Rashtriya Gokul Mission aims to conserve and develop indigenous breeds in a focused and scientific manner. It is a focused project under National Programme for Bovine Breeding and Dairy Development, with an outlay of Rs 500 crore during the 12th Five Year Plan.

41. PAHAL-Direct Benefits Transfer for LPG (DBTL) Consumers Scheme

The main objective is to send the subsidy money of LPG cylinders directly into the bank accounts of the consumers and increase efficiency and transparency in the whole system. Under the scheme, the LPG consumer can now receive subsidy in his bank account by two methods. Such a consumer will be called CTC (Cash Transfer Compliant) once he joins the scheme and is ready to receive subsidy in the bank account.

42. The National Institution for Transforming India (NITI AAYOG)

The main objective is to foster involvement and participation in the economic policy-making process by the State Governments of India. The National Institution for Transforming India (NITI) which replaced 65 year old Planning Commission will act more like a think tank or forum, in contrast with the Commission which imposed five- year-plans and allocated resources to hit set economic targets.

43. Pradhan Mantri Khanij Kshetra Kalyan Yojana (PMKKKY)

The main objective is to Safeguard Health, Environment and Economic Conditions of the Tribal's. Pradhan Mantri Khanij Kshetra Kalyan Yojana (PMKKKY) is meant to provide for the welfare of areas and people affected by mining related operations, using the funds generated by District Mineral Foundations (DMFs).

44. Namami Gange Project

The main objective is to integrate the efforts to clean and protect the Ganga River in a comprehensive manner. Namami Gange approaches Ganga Rejuvenation by consolidating the existing ongoing efforts and planning for a concrete action plan for future. The interventions at Ghats and River fronts will facilitate better citizen connect and set the tone for river centric urban planning process.

45. Setu Bharatam Project

The main objective is to free all national highways from railway level crossings and renovate the old bridges on national highways by 2019. Setu Bharatam is an ambitious programme with an investment of Rs. 50,000 crore to build bridges for safe and seamless travel on National Highways. 208 new "road over bridges/road under bridges" are envisaged for construction, while 1500 bridges will be widened, rehabilitated or replaced.

46. Pradhan Mantri Ujjwala Yojana

The main objective is to distribute free LPG connections to the women belonging to 5 Crore BPL families across the country. According the estimates, about 1.50 Crore BPL families will be benefitted under the scheme in the year 2016-17. The scheme will cover 3.5 Crore more BPL families in the next two years. The scheme provides a financial support of Rs 1600 for each LPG connection to the BPL households. This is the first ever welfare scheme by the Ministry of Petroleum and Natural Gas which would benefit crores of women belonging to the poorest households.

47. Rashtriya Gram Swaraj Abhiyan

The main objective is to help Panchayat Raj Institutions deliver Sustainable Development Goals. It is a new proposed scheme announced in the union financial budget 2016-17 by the Finance Minister Mr. Arun Jaitley. Rashtriya Gram Swaraj Abhiyan has been proposed with allocation of Rs. 655 crore.

COMMUNITY DEVELOPMENT PROGRAMME

Definition

Community: is a group of people who live in a geographical area and have an interest in each other for development.

Development: connotes growth or maturation. It implies gradual and sequential phases of change. It refers to the upward or increasing differentiation.

Community Development: It is a movement designed to promote better living for the community with the active participation and/or the initiative at the community. It is a method by which people of villages are involved in helping to improve their own economic and social conditions and thereby they become more effective groups in programmes of their national development. It is a process of change by which people's efforts are united with those of group authorities to improve their economic, social and cultural conditions of communities, into the life of the nation and to enable them to contribute fully to national programmes.

History

Review of the pre and post-independence era extension programmes recommended a much bolder and dynamic effort for creation of an urge among the rural population to live a better life and to achieve development. It depends upon whole hearted co-operation of the beneficiaries, government officials and non-officials at every stage. Consequently, the Government of India entered into an operational agreement with the Government of the U.S.A under the Technical Co-operation Programme Agreement. Under this Agreement, 55 Community Development Projects were started in different parts of the country on 2 October, 1952 for three years.

Aims of CDP

The main aims of these projects were:

- To increase agricultural production by all possible means,
- To tackle the problems of unemployment,
- To improve village communications,
- To foster primary education, public health and recreation,
- To improve housing,
- To promote indigenous handicrafts and small-scale industries,
- To improve the villager's lot through their own primary effort.

In short, the programme aimed at achieving all-round socio-economic transformation of the rural people.

Genesis of CDP

Genesis of Community Development Programme (CDP - 1952). Based on the experiences within the country and abroad, the recommendations of the Fiscal Commission (1949) and the Grow More Food Enquiry Committee (1952) the CDP was launched on October 2, 1952, the birth-day of Mahatma Gandhi. At the initial stage in 1952 there were 55 community projects in 3 blocks. Each of the community development projects covered an area of about 450-500 sq.miles with about 300 villages and a population of about 2 lakhs. Each project area has been divided into three development blocks. A development block consisted of about 100 villages with about 150-170 sq. Miles and a population of about 60-70 thousands. Each block was further divided into groups of 5-10 villages each. Each such group formed the area of operation for a village level worker (Gram Sevak) who was the basic level extension functionary in the community development programme. The need for rapid extension of the programme to other parts of the country led to the National Extension Service (NES) alongside the CDP on October 2, 1953 covering the entire country within a period of 10 years. In 1957, the Balvantroy Mehta team recommended the establishment of statutory elective of local leaders *i.e.*, Panchayat Raj Institutions (Village Panchayat, Panchayat Union and District Development Council). This programme laid emphasis on all round development of the whole community with special emphasis on weaker and under privileged sections through the use of area development, self-help and integrated approaches. The CDP, however, could not bring expected change in improving the conditions of rural masses or there was not much development in the area of agricultural development.

Meaning of CDP

Community is a group of people live in a geographical area and have an interest in each other for the purpose of making a living. Development indicates gradual growth of maturation which is sequential phase. It can also be expressed as increasing differentiation in upward direction. It can be concluded that community development is bringing forth the potential elements, abilities and qualities of group of people who live in a common territory and who have an interdependent relationship with each other.

Fields of Coverage

The community development programme was started for overall development of rural people. It consisted of agriculture, animal husbandry, irrigation, cooperation, public health, education, social education, communication, village industries *etc.* There are officials for each activity from district to village level to plan, execute and evaluate the programme.

Organizational Concept

Concept

Initially this project was launched in 55 Community Development Blocks with an operational area of 400 to 500 square miles comprising of 300 villages and 2 lakh population in each block. The project area was divided into three development blocks with 100 each villages and 60 to 70 thousand population, groups with 5-10 villages. The project was headed by a Project Officers and Subject Matter Extension Officers in the disciplines of Agriculture, animal husbandry, industries, rural engineering, social education *etc.*, and 60 multipurpose Village Level Workers (VLWs) one for each group villages. Due to positive impact of community development programme, a need for rapid expansion was urgently felt. In 1953 the NES programme was launched with the idea of having wider coverage at a lower cost and more of people's participation. NES block was headed by Block Development Officer (BDO) with Extension officers (Eos) and VLWs limited to 10 in 100 villages and 65,000 population and funds were drastically reduced compared to CD Blocks. NES was thought of as the agency and CD as the method to bring about socio-economic transformation of people. This was the major development in the sphere of rural reconstruction in India.

Objectives of Community Development Programme

1. Help in Planning

The Community Development Projects are aimed at helping the villages in planning and developmental activities. Through these projects, schemes for improving the agricultural production are undertaken.

2. Involving Villages in National Reconstruction

An important objective of Community Development Project is to make village people self-dependent and encourage them to take part in the activities of national reconstruction. This is done through the following methods:

a) Changing the traditional and conservative outlook of the village people.

b) Organizing youth and women bodies to take part in the developmental and welfare activities.

c) Making arrangements for recreation of the village youth and women.

3. Providing Educational Facilities in Villages

Through these projects, an attempt is being made to provide educational facilities for the village people. This is done by the following methods:

a) Arranging centres of social education.

b) Arranging recreational programmes.

c) Training village people in the activities of planning and development.

4. Improving the Standard of Living of Villagers

The main objective of Community Development Programmes is to improve the standard of living of the village people. They have been provided with various employment facilities and opportunities to set up industries and the training to improve their agricultural production. Cooperative societies and other such bodies have been set up for them to undertake various activities of development planning.

5. Political Training

Through the Community Development Projects, village people are also given training in administrative activities. Through the working of Community Development Projects, people are trained in the practical working of democracy. The village people are trained in the area of civic affairs.

6. Other Objectives of Community Development

The above mentioned objectives are the general objectives of the community development. There are also specific objectives intended basically at 'all-round Development of the Rural Society'.

The Specific Objectives of Community Development Programme

1. Agriculture and improvement of agricultural production
2. Setting up of cooperative societies in each village
3. Animal husbandry
4. Public health
5. Rural education
6. Improving the means of communication and transport in rural areas
7. Setting up village level small-scale cottage industries
8. Organizing and strengthening the village panchayats

Philosophy and Principles of Community Development Programme

Community Development Project is a new experiment in the development and planning activities of the world. Such projects have been specially launched only in Asian Countries. It is an experiment intended at fulfilling the various need of the village society and also making it self-dependent. It is a part of the Indian Constitution aimed at establishing a socialist society in this country.

Principles of Community Development Programme

1. Drawing programmes for fulfilling various needs of the community.
2. Involving the people in planning and developmental activities.
3. Bringing about material as well as psychological betterment.
4. Teaching rural people the political set up in democracy.
5. Creating the local leadership.
6. Drawing up of national policy for the development of the country.
7. Setting up of cooperative societies for carrying developmental works.
8. Development based on socialism.

These Community Development Projects are based on the decentralization of economy, administration and political power. They are the means of training the people in the art of 'self-administration'. These programmes employ all the resources for the development of nation. They focus their attention at training people to become self-dependent and self-reliant. These programmes are basically aimed at changing the face of rural society.

Features of Community Development Programme

This project in India is an integrated programme for rural development, which has the following features:

1. The basic characteristic feature of community development is self-dependence. It means that through this programme people are trained to become self-reliant economically.

2. To bring about a change in the administration is another important feature of community development.
3. The most important feature of this programme is the involvement of the people in the system.
4. All-round development of blocks is another important feature of community development.

Community Development Projects and their Implementation

As already stated, these programmes are socio-economic programmes undertaken for the reconstruction of rural society; implementation is aimed at the involvement and participation of rural people, which is carried out by the institutions of local self-government and local administration.

Institutions for Implementation of Programmes

The programmes of community development are carried out by the following three basic institutions at the village level:

1. Village Panchayat
2. Cooperative Societies
3. Village School

In rural development and programmes of community development, the village school and the village teacher play a very vital role. The village teacher is the centre of activity and encourages people and involves them in the programme of village development and reconstruction.

Community Centres

In villages, there are various community centres that look after the various aspects of life of village people.

These community centres perform the following functions

1. Establishing reading rooms and libraries.
2. Encouraging village people to improve their standard of living.
3. Taking up various cultural programmes.
4. Establishing organizations for the welfare of boys and girls.
5. Establishing schools for the village children.
6. Undertaking the programmes of national integration and also observing national festivals to strengthen national unity.

Budgetary Planning

As part of community project, it has been planned to set up blocks covering 100 villages with about 80,000 people. On the basis of this, all the states have been divided into blocks, regions and zones.

The budget is allotted to carry out the following stages of the programme

1. Pre-extension Block Development
2. First stage of Block Development
3. Second stage of Block Development

Aspects of Community Development

1. Community Development as a Process

Community development is intact a process and a method to bring about proper development and upliftment of the village society. This is a process to bring socio-economic changes at the social and psychological level.

2. Community Development as a Method

Community development is not only a process of social change but also a method for achieving this objective. In a democratic country like India bringing about a change in the society can be possible only with the cooperation of the people.

3. Community Development as a Programme

Community development is also a programme on the basis of which the process of change is carried out. Without proper programme, neither the method nor the process shall have any utility. This programme implies development of the society through stringent methods and process.

4. Community Development as a Movement

Community development is a movement, which is undertaken for eradication of illiteracy, unemployment and poverty and improving village life.

Programmes of Community Development

The Community Development Project has a large number of programmes intended to bring about the changes in the rural community.

The programmes included in the project are as follows:

1. Agriculture

Development of agriculture sector is the first aim of community development. The facilities include providing improved means of implements to the agriculturists, providing them with the improved seeds and better irrigation facilities and so on. The real task is to bring about the improvement in agriculture so that the condition of the rural society may improve.

2. Development of Communication

Rural conditions improve only when the means of communication and transport in rural areas are developed. The project aims at construction of new roads and also improving transport and communication facilities.

3. Development of Education

The Constitution of India envisages compulsory primary education. It stresses on adult education. Under Community Development Project, centres of social education, night schools and other institutions for imparting education have been planned. Apart from these libraries, reading rooms and centres of social education have to be set up. These institutions aim at expansion of education and extension of educational facilities.

4. Development of Health Services

To improve the standards of rural life, one has to change facilities regarding health need improvement. It is done by providing facilities of medical and health services, maternity and child welfare services, *etc.* Provision of health services is one of the developmental measures taken up by Community Development Project.

5. Employment Facilities

Rural societies are characterized by seasonal employment, as agriculture is the only means of livelihood. It also suffers from unemployment and underemployment. Community Development Programme aims at setting up small-scale cottage industries in rural areas, through which employment could be provided for rural people.

6. Housing Facilities

Rural areas in India suffer from lack of proper houses. The houses in rural areas do not have the basic facilities and are unhygienic. The development programmes include providing better housing facilities.

7. Social Welfare Activities

Social welfare activities such as providing better facilities for recreation, women's education have been taken up in Community Development Project.

Essentials of Community Development

The projects should be initiated based on the expressed needs of people. There should be intensive action with multipurpose programmes. Equal to material achievement, change in the attitudes of people is important. There should be participation of people at every stage and activity of the programme. Development of local leadership is essential. Participation of women and youth should be there. Self-help projects should receive both internal and external resources. Adequate administrative and policy support with involvement of Non-Governmental organizations.

Steps in CDP

Identification of felt needs of people with their participation. Need → Programme planning → Institution. Needs are addressed through a programme, which is to be implemented by an institution.

Stages in CDP

The Pattern of Community Development Programme was further revised (modified with effect from 1 April 1958). According to this pattern, there were four stages:

Pre-Extension Stage: At this stage, there was a budget provision of Rs.18, 800 for one year.

The staff consisted of a block development officer, one agricultural extension officer and five village-level workers whose main duties were to survey the whole block area, to prepare the ground work for the intensive stage and conduct agricultural demonstrations.

1. **Stage I Blocks:** After one year of the Pre-Extension stage, the block stepped into Stage I, with a budget provision of Rs.12 lakhs for five years for intensive development. During this period, provision was made to have one block development officer, eight extension officers, ten village level workers(VLWs), two gram sevikas, three stockmen (veterinary), one physician and 3-4 midwives, one compounder, a sanitary inspector and the necessary office staff.
2. **Stage II Blocks:** In this stage, the staffing pattern was more or less the same, but there was a difference in the budget provision. The budget in this stage was five lakhs of rupees for a period of five years, *i.e.*, one lakh per year.
3. **Post-stage II Blocks:** This was the permanent stage. It was meant for the follow up work. The budget was one lakh of rupees per year. At this stage, all the developmental activities as in Stage II, under the various heads, such as agriculture, animal husbandry and co-operation, were to continue. Needs are successfully satisfied only when, needs are identified, but not imagined; programme is planned for the need but not for the institution; institution is relevant enough to implement the programme.

RURAL LEADERSHIP

Leadership is the art of influencing people to attain group objectives willingly. What a minister does in his State, a captain does on the playground; the manager has to do in his organisation. Leaders in all walks of life should have some basic qualities. They should be able to establish contact with their equals, deal with their subordinates and guide them, mediate in conflicts, resolve issues by weighing various alternatives, allocate scarce resources properly and take risks and initiatives.

The environment in which a leader is placed is important. The organisational culture, the economic and social set-up, the extent of unionisation and other factors may demand different types of leaders in different situations. A task-oriented leader, for instance, may be more successful in situations which are either very favourable or very unfavourable to him, while a relations- oriented leader may be more effective in intermediate situations.

Definition

According to Livingston – 'Leadership is the ability to awaken the desire to follow a common objective'.

According to C.I. Bernard – 'Leadership is the quality of behaviour of the individuals whereby they guide people or their activities in organised efforts'.

According to Bernard Keys and Thomas – 'Leadership is the process of influencing and supporting others to work enthusiastically towards achieving objectives'.

Leadership is essentially a continuous process of influencing behaviour. It may be considered in context of mutual relations between a leader and his followers. The leader tries to influence the behaviour of individuals or group of individuals around him to achieve desired goals.

Meaning

Leadership in rural areas requires particular traits. It is an interesting social phenomenon. In villages, the degree of interaction between the leader and villagers is high. Various groups in villages determine the pattern of leadership in villages. Leadership in rural areas functions amidst of small groups. The human relations approach is also important, as most of the villagers are ignorant, innocent, needy and sensitive. Various writers have tried to define the term 'leadership'. However, still a comprehensive definition for the term does not exist.

According to Linderman, 'a leader is an individual whose relations, judgments and feelings are accepted (responded to) by the group, as the bases of belief and action.' In the view of Allport, 'leadership according to our present usage means the direct, face-to-face contact between leader and followers; it is personal social control.' A more recent definition has been given by Robert Tannenbaum and Fred Massarik. According to them, 'leadership is an inter-personal influence exercised in situation and directed through the communication process towards the attainment of specified goal or goals.' From the above definitions, leadership may be described as the interaction between leaders and followers in a particular situation. It involves attempts on the part of the leader (influence) to affect the behaviour of followers in a situation.

Characteristics

Some of the essential characteristics of leadership are as follows:

1. Leadership traits – though there is no general agreement on the qualities of leaders in all situations, still there are certain personality traits, which will be helpful in all situations. Dwight Sanderson emphasizes on some qualities such as good physique, strength and stature, self-assertiveness

and self-confidence, superior mental ability, amiability, friendliness and sympathy. Tead also mentions ten qualities of a leader – physical and nervous energy, a sense of purpose and direction, enthusiasm, friendliness and affection, integrity, technical mastery, decisiveness, intelligence, teaching skill and faith.

2. Leadership is related to situation. It functions only in accordance with the prevailing conditions.
3. Leadership involves group mechanism. All our behaviour, ideals, and thoughts are molded in terms of the group, which we belong to. Particularly, the rural leaders are essentially a part of group mechanism.

Concept of Leadership

Leadership is a dynamic process, which deserves study. It is a relational process involving interactions among leaders, members and sometimes outside constituencies. Good leaders are made not born. If you have the desire and willpower, you can become an effective leader.

Good leaders develop through a never-ending process of self-study, education, training, and experience. To inspire your workers into higher levels of teamwork, there are certain things you must be, know, and, do. These do not come naturally, but are acquired through continual work and study. Good leaders are continually working and studying to improve their leadership skills; they are NOT resting on their laurels.

Leadership is a process by which a person influences others to accomplish an objective and directs the organization in a way that makes it more cohesive and coherent. Leaders carry out this process by applying their leadership attributes, such as – beliefs, values, ethics, character, knowledge and skills.

Concept of Leadership

- BE a professional. Examples – Be loyal to the organization, perform selfless service, and take personal responsibility.
- BE a professional who possess good character traits. Examples – Honesty, competence, candor, commitment, integrity, courage, straightforwardness, imagination.
- KNOW the four elements of leadership – follower, leader, communication, and situation.
- KNOW yourself. Examples – strengths and weakness of your character, knowledge, and skills.
- KNOW human nature. Examples – Human needs, emotions, and how people respond to stress.
- KNOW your job. Examples – be proficient and be able to train others in their tasks.

- ✰ KNOW your organization. Examples – where to go for help its climate and culture, who the unofficial leaders are.
- ✰ DO provide direction. Examples – goal setting, problem solving, decision making, planning.
- ✰ DO implement. Examples – communicating, coordinating, supervising, evaluating.
- ✰ DO motivate. Examples – develop moral and spirit in the organization, train, coach, counsel.

Good Qualities of a Leader

Some of these qualities may be enumerated as follows:

1. Height, weight, skin colour - though there are exceptions to this. A short person may be more effective leader than a tall one, a light-weight can be more effective leader than a heavy one, and a dark-skinned can surpass the fair-skinned in providing leadership. India's Lai Bahadur Shastri, K. Kamraj and Jagjivan Ram can be cited as ready examples of this exception.
2. Energy, both nervous and physical; how long he can work energetically.
3. Mental ability - Being well-read and well-informed.
4. Personality - A captivating, magnetic personality.
5. Initiative - Inventiveness and boldness to implement new plans.
6. Imagination - Creativeness and original thinking.
7. Emotional stability - Mental stability in adverse situations.
8. Desire to accept responsibility - Courage to own responsibility.
9. Flexibility - Ability to adjust and adapt to changing situations.
10. Honesty - Truthfulness and openness.
11. Sincerity - Earnestness and authenticity.
12. Determination - Resolve and will-power.
13. Persistence - Doggedness and perseverance.
14. Endurance - Staying power, stamina.
15. Integrity - Reliability, uprightness.
16. Judgment - Decisive, conclusive.
17. Courage - Guts, bravery.
18. Good looks - both physical and sartorial.

Principals of Leadership

1. **Know yourself and seek self-improvement** - In order to know yourself, you have to understand your *be, know,* and *do,* attributes. Seeking self-improvement means continually strengthening your attributes. This

can be accomplished through self-study, formal classes, reflection, and interacting with others.

2. **Be technically proficient** - As a leader, you must know your job and have a solid familiarity with your employees' tasks.
3. **Seek responsibility and take responsibility for your actions** - Search for ways to guide your organization to new heights. And when things go wrong, as they often tend to do sooner or later - do not blame others. Analyze the situation, take corrective action, and move on to the next challenge.
4. **Make sound and timely decisions** - Use good problem solving, decision making, and planning tools.
5. **Set the example** - Be a good role model for your employees. They must not only hear what they are expected to do, but also see. *"We must become the change we want to see."* - Mahatma Gandhi
6. **Know your people and look out for their well-being** - Know human nature and the importance of sincerely caring for your workers.
7. **Keep your workers informed** - Know how to communicate with not only them, but also seniors and other key people.
8. **Develop a sense of responsibility in your workers** - Help to develop good character traits that will help them carry out their professional responsibilities.
9. **Ensure that tasks are understood, supervised, and accomplished** - Communication is the key to this responsibility.
10. **Train as a team** - Although many so called leaders call their organization, department, section, *etc.* a team; they are not really teams. they are just a group of people doing their jobs.
11. **Use the full capabilities of your organization** - By developing a team spirit, you will be able to employ your organization, department, section, *etc.* to its fullest capabilities.

Different Types of Leaders in Rural Context

(i) Traditional Leaders

- ☆ Emerge out of tradition and stick to tradition.
- ☆ Static type and do not accept change.
- ☆ Have vested interests.

  e.g. Tribal chiefs

(ii) Caste leaders

- ☆ Followers belong to the same caste as that of leaders.
- ☆ Provide leadership in matters related to caste.

(iii) Religious Leaders

- ✰ Become leader by preaching religious doctrines or performing religious acts or rituals.
- ✰ Advise followers on religious matters.

(iv) Political Leaders

- ✰ Arise out of the political system.

(v) Functional Leaders

- ✰ Recognized because of their specialized knowledge and function in the society.

  e.g. Professionals like teachers, doctors *etc.*,

(vi) Opinion Leaders

- ✰ These are the people to whom other people go for opinion or advice on certain issues.
- ✰ Also act as legitimizes and influence decision making of the followers.
- ✰ In comparison to opinion seekers, opinion leaders have more social participation, extension agency contact and exposure to mass media. Opinion leaders are more cosmopolite and more innovative. They amplify the dissemination.

EXTENSION ADMINISTRATION

Definition and Meaning

Administration is the guidance, leadership and control of the efforts of a group of persons towards some common goals. Administration implies conscientious efforts directed towards organizing and controlling human activities in order to achieve agreed or desired goals. It involves, therefore, the organization, direction and control of persons and facilities in order to accomplish specified ends. Administration involves essential activities of the people charged with ordering, forwarding and facilitating the efforts of individuals or groups brought together for desirable purposes. It involves efforts such as decision-making, programming, communicating relevant information, controlling and evaluating various actions. Administration is used during policy formulation for developing staff, understanding the planning process, job description, training and evaluation processes.

Management versus Administration

Management is a part of Administration. Management is an administrative technique in the conduct of public affairs. The traditional framework within which a civil service exercises its responsibilities is described as an administration. Administration lays emphasis on proper procedure, regulation and control. It employs ease of communication to achieve an un-ambiguity – interpreting words and issues to ease the operation of a system. Administration takes place in the health sector, universities, army, church, agriculture, industries, business and social organizations such as the agricultural extension service.

The Concept of Extension Administration

Agricultural extension involves various activities directed to improve food production and living standards of the people. Extension, therefore, requires direction and control of these activities in order to achieve the desired goals. Extension administration depicts the effort of the administrator to direct, guide and integrates the activities of members of his staff in order to attain the goal of the extension service and help farmers realize their objectives. Thus, extension administration refers to the art of directing, guiding and controlling human and material resources in an extension system. It involves articulate joining of all aspects of agriculture such as animal and crops husbandry, soil management, technology design and supply services in order to increase food production and income levels of farmers. Administration in extension is concerned mainly with bringing human beings together for the purpose of executing extension functions.

Members of an extension service are categorized into directive or executive cadre. Extension administration recognizes the existence of people with diverse interests, aptitudes, attitudes and social background. Therefore, extension is designed to serve farmers with differences in orientation. Extension has the task of limiting various interests and discouraging the uninteresting nature of farmers in order to salvage the whole agricultural system and achieve overall improvement in national economy.

Principals of Extension Administration

It is not possible to draw up a blueprint of exactly how to serve most effectively as an extension administrator. Most principles of management are pragmatic rules of thumb and outgrowths of experience. The principles are at best limited tools but, if wisely and carefully used, they can be very valuable. They are a guide to action.

Principles are a guide to action in an organization.

1. The structure of an administrative organization is established in a subordinate – super-ordinate hierarchy. Each worker should know his supervisor. Each supervisor is expected to supervise a person. Each member of staff within the extension agency should report to one and only one leader. This structure should be made flexible and modifiable.
2. The authority delegated to an individual in an organization should be commensurate with the responsibility assigned to him. Responsibility implies accountability for the performance of the duty. Responsibility cannot exist without authority. Responsibility cannot be delegated but authority can be delegated. Authority is the supreme coordinating power within an organisation.

3. Effective administration occurs when authority and responsibilities assigned to an individual or group are clearly defined and understood. A job description which clearly sets out the duties of each person and the degree of authority given is very crucial. What is job description? Job description is a statement describing the duties of a worker. Job specification is a statement that explains the minimum requirements of the staff qualification.
4. A two-way channel of communication must exist for a meaningful administration of an extension system to be achieved. Vertical and horizontal communication should be maintained to develop desirable staff attitude and interest in an organization. Communication is a process whereby decisions are transmitted from one member of extension organization to another through agreed channels.
5. Effective use of specialist knowledge in planning and implementing programmes is very necessary and greatly required in perform dance-oriented extension administration. One of the serious problems facing administration the world over is inadequate use of specialist knowledge in policy formulation, planning and execution of programmes. Extension administration requires specialist knowledge in various aspects of agriculture and extension services.

Conditions Necessary for Successful Extension Administration

The success of administration in an extension service depends on the following conditions:

1. The extension service should be seen as an institution established for the purpose of assisting rural people through educational process to make maximum use of their available resources.
2. Administrators should realize the need for the identification of sources of technological information to be used for the solution of problems of rural people.
3. Extension administration should regard extension education as the sharing of technical knowledge useful for solving recognized problems of rural people.
4. Extension education should be available to all rural people irrespective of race, creed, social class, or sex.
5. Extension administrators and extension policy makers should develop an extension service adapted to the social realities and social organizations of the society being served.
6. There is need to establish institutions for the training of personnel required for the effective running of the extension service.
7. There is the need for organizing the educational and training methods to suit the skills, attitudes, abilities, knowledge and resources of the people.

8. The principle of local involvement or participation in the planning of extension programmes should be ensured at all times.
9. Efforts must be made for auxiliary services such as transport, veterinary services and credit.

Functions of Extension Administrations

Functions of an Extension Administrator (Extension Director)

The Director of Extension is the organization's Chief Executive. The function of the Director is to plan, organize, staff, direct and controls the activities of the organization. The position includes responsibility for supervision, personnel management, salary administration, facilities management, fiscal management, programme development and co-ordination; programme execution and programme evaluation. The Director is ultimately responsible for managing the organization and for giving it leadership and direction.

The above functions must be delegated to some extent to others in the organization. There is, also, a great need for effective communication within the organization.

The Extension Director should have the following minimum qualifications:

a. University degree in a relevant field and, usually, a postgraduate degree, and
b. Several years of professional experience, preferably at different levels in the extension organization.

Other qualities expected of the Extension Director are:

a. the ability to make decisions and follow through on those decisions
b. a positive attitude in dealing with superiors, subordinates, and the general public
c. a highly developed sense of co-operation
d. the ability to inspire in others a sense of responsibility and duty to the people being served
e. a basic understanding of the educational philosophy and methods of an effective extension organization
f. the technical skills of management, and
g. an understanding of agriculture and the ability to relate to rural people.

Other Responsibilities and Functions of an Extension Administrator

A. Responsibilities

a. Delineation of the objectives of extension service - since the extension service exists to fulfill set objectives, there is a central source from which activities are initiated.

b. Drawing broad plans of action - he is responsible for drawing broad plans for restructuring activities in the extension service. The administrator must be fully conversant with the functions in other areas of agriculture.

c. Assembling of various categories of staff - provision of support staff at all levels is a major responsibility of top administrative staff. The top administrator should assemble an adequate working team of experts whose activities are complementary to each other.

d. Delegation of function - it is the usual practice for the chief executive to delegate functions. He should be able to assign specific functions to individual members of staff. In all cases, responsibilities are clearly written down and not merely stated verbally. Periodic review of job description is made by the administrator to reflect changing situations.

e. Directing (supervising) delegated functions - this responsibility assumes that the administrator has a clear understanding of his authority and how it is exercised. Directing and overseeing requires that an officer to whom function is assigned should report back to the administrator. The report could be oral or written.

f. Definition and standardization of performance - it is the responsibility of the administrator to clearly define and set standards of performance. This is needed to ensure that the quantity and quality of the performance of the organization and staff is established.

g. Provision for evaluation of performance - evaluation of the overall result of extension effort is the responsibility of top administrators. Evaluation refers to the appraisal of the degree to which goals set for various programmes have been attained.

h. Provision for stimulation and energization of personnel - when large numbers of people are brought together in an organization, it is not feasible to have common interest spontaneously. They need guidance and stimulation from their superior officers.

B. Functions

The functions of an administrator in the administrative process are:

a. Planning
b. Organizing
c. Staffing
d. Directing
e. Coordinating
f. Reporting
g. Budgeting

MONITORING AND EVALUATION

"Monitoring and Evaluation are central functions of development management. These concepts are familiar in every project these days. Monitoring and evaluation (M and E) are used for performance assessment as they may feed back into the implementation process, aiding management to improve the quality of its decision-making, and aiding it to develop a learning orientation. The debate about M and E has often centred on the search for appropriate methodologies.

Meaning and Scope of Monitoring and Evaluation

Meaning

Monitoring is a continuous assessment both of:

(i) the functioning of the project activities in the context of implementation schedules, and

(ii) the use of project inputs by beneficiaries in the context of design expectations (Murphy, 1993). It is a day-to-day management activity. Monitoring is the gathering of information on the utilization of project inputs, on unfolding of project activities, on timely generation of project outputs, and on circumstances that are critical to the effective implementation of the project.

Scope

In extension, managers make decisions at four levels and require information in four categories: According to Murphy (1993), the decision levels are:

a. What type of services do the people need and how should they be delivered?
b. Are these services being delivered as planned?
c. Are the farmers aware of the availability of services being provided and are they using them? Reasons should be sought for their use or lack of use.
d. What differences have occurred due to the services provided?

Extension managers require four categories of information: diagnostic information, implementation information, utilization information and impact information.

a. **Diagnostic:** Information involves description of the current farming situation, and analysis, which makes it possible to understand why a situation is as it is and what it means in terms of extension programme design or redesign. Issues here include current technology in use, types of farming households identified or felt needs and constraints.
b. **Implementation:** Information (Physical and financial information or inputs) includes a description of the implementation of the programme.

This involves staff availability, training sessions held, number of visits, cash flow situation, number of demonstration plots planted and so on. These will help to establish if implementation is going on as planned.

c. **Utilization:** Information (also known as on-going evaluation, evaluation or beneficiary contacts) includes key indicators such as rates of participation in extension demonstrations by different categories of farmers, adoption rates of major technological changes in relation to farmers reached by the extension service, and in relation to all possible users of the technology in that area (to see if there is a spread of information from farmers to farmers). Have the farmers actually modified their behaviour in response to the extension advice?

d. **Impact:** Information measures the difference resulting from the shift in technology. Expectations include increase in productivity, change in cropping pattern, as well as increased income and well-being of the farmers. Impact information to collect, also, includes data on activities outside the extension programme itself, useful to managers in the planning and finance ministries, and donors (Murphy, 1993).

Steps in Conducting on Evaluation

a. Develop an evaluation plan – determine all resources required. Determine the personnel and financial resources available and needed for carrying out the evaluation. Determine, also, the amount of time needed to complete the evaluation.
b. Select and define the part of the programme to study – training methods, results of accomplishments, and so on.
c. Determine the need for the evaluation – define the objectives of the evaluation, stating what evidence is needed to show that the programme is reaching the people, *e.g.* accomplishments, changed behaviour *etc.*
d. Decide the method of collecting information – for example, questionnaires, observations *etc.*
e. Determine the potential audiences (the population) to be sampled and what sample is necessary and feasible.
f. Determine the criteria for evaluating the programme - *e.g.* number of beneficiaries *etc.*
g. Collect information, after training interviewers - Employ field supervisors.
h. Collate field data collected and analyze data.
i. Interpret the data and prepare a report of major findings.
j. Plan for use of the findings (workshops, seminars) and determine their implication for future work. The findings should be applied and utilized to improve the on-going programme and/or in modifying a future one.

Definitions of Evaluation

More specific definitions of evaluation are given by persons involved in rural development programmes. While most of these definitions refer specifically to the assessment of the results of programmes of extension education, they can also be applied to the training aspect of such programmes. Some definitions of evaluation are:

- It is a process, which enables the administrator to describe the effects of his programme and thereby make progressive adjustments in order to reach his goal more effectively (Jahoda and Barnit, 1955).
- Programme evaluation is the determination of the extent to which the desired objectives have been attained or the amount of movement that has been made in the desired direction (Boyle and Johns, 1970).
- Programme evaluation is the process of judging the worth or value of a programme. The judgment is formed by comparing the programme should be (Steele, 1970).
- Evaluation is the process of delineating, obtaining and providing useful information for judging decision alternative (Stufflebeam, 1971).
- Evaluation is a coordinated process carried on by the total system and its individual subsystem. It consists of making judgments about a planned programme based on established criteria and known, observable evidence (Boone, 1985).

Nature of Evaluation

1. **Evaluation is not measurement:** Evaluation is an integral part of extension education. All aspects of extension work need evaluation. Evaluation does not mean mere measuring of achievements, which is usually done after the programme is executed. Extension being an educational process, it is necessary to evaluate management of the programme and methods used, achievements accomplished in line with the objectives and also to determine the reasons for success or failure.
2. **Evaluation is not exactly scientific research:** When we think of evaluation as a process of collecting information as a basis for making decisions, forming judgements and drawing conclusions, we realize it has much in common with scientific research. But there is a great difference between our casual everyday evaluation and scientific research. However, the difference is a matter of degree rather than kind. Casual every day evaluation can be placed at one end of the scale and scientific research at the other end. There are five locations on the scale with no sharp lines of distinction, *i.e.*, casual everyday evaluation, self-checking evaluation, do-it yourself evaluation, extension evaluation studies and scientific research.

Types of Evaluation

- **Self-evaluation**: This is to be carried out by every worker as a matter of routine. This requires the self-critical attitude which is so essential for extension work. By this self-critical attitude, the chances of an extension worker growing and continuously improving his professional competency become greater.
- **Internal evaluation**: Evaluation carried to by the agency responsible for the planning and implementation of the programme. Some of the other methods for internal evaluation are: systematic use of diaries and reports of workers, planned visits of staff members to work spot, use of special questionnaires and proforma for observation and inquiry *etc.*
- **External evaluation**: Evaluation conducted by a person or a committee outside the area of operation. One of the strong features of the Indian Community Development Programme is that simultaneous with its start an independent agency, namely the Programme Evaluation Organization, was established.

Evaluation can also be classified into (i) concurrent and (ii) ex-post facto evaluation.

Evaluate Programme Planning

As a result of experience, theory, research and experimentation, much information has been accumulated about how an extension programme should be planned. Progress in science and technology and the broadening of extension's clientele with the accompanying great variation in needs and interests have made the scientific planning of extension education programmes more important than ever before. There is considerable agreement on certain criteria which, if followed, make for successful extension programme planning at different levels. These criteria represent the ideal with which to compare our practices and procedures or programme planning. Some of the steps needed top evaluates or programming function in view of these criteria includes:

a. Identify the evidence needed to form a judgement about each criterion.
b. Specify the methods that will be used to obtain the evidence, such as personal observation, personal interview or through a systematic survey.
c. On the basis of the evidence gathered, judge whether or not each criterion is being adequately satisfied in the programme planning activities.

Extension Evaluation Process

There are several models of evaluation available in the literature. However, a very simplified version of most of these models may be quite workable for evaluating extension programmes since, as Bhatnagar (1987) has pointed out, any extension evaluation process has to be based on certain assumptions. For example, if some inputs are provided in the form of a programme, specific outputs can be expected and if these outputs happen, then the purpose of the programme can be achieved; if the purpose is achieved, then the development goal is realized. This means that evaluation has to be so designed that the quality types and adequacy of the input measures, outputs and their impact in achieving the programme objectives have to be evaluated systematically.

Steps involved in an extension programme evaluative process may be as follows:

i) Formulate Evaluation Objectives

Specific objectives to be achieved through the evaluative process must be clearly and adequately identified and started. All further efforts should be knit around these objectives.

ii) Classify Programme Objectives

It is assumed that each extension programme, when formulated and implemented, will have specific well-defined objectives. Since evaluation is basically a process of determining the extent to which various extension teaching activities were organized and managed and the extent to which they contributed to achieving the goals, programme objectives must be clearly understood and if necessary, further broke down into measurable terms. This is a crucial step as all further efforts will be directed towards collecting evidence related to these objectives.

iii) Identify Indicators

To identify indicators or the kind of evidence necessary to evaluate achievement in relation to specified programme objectives, it is necessary that specific beneficiaries of the programme be identified, the kind of behavioural changes expected in them be clearly stated, and the kin of learning experiences expected to be provided to them spelled out, together with the level of management to be achieved for provided those learning experiences are specified. Once this is done, identification of specific indicators to measures the achievements will not be difficult.

iv) Decide the Kind of Information Needed

Once the indicators for evaluating the management and performance of a programme have been indicated, specific information to be collected may be worked out. Since there is usually more information than an extension worker can collect, he has to be very discriminating about the kind and amount of information that should be collected. Timing for collection of information may also need to be specified.

v) Sampling

The purpose of sampling is to take a relatively small number of units from a population in such a way that the evidence collected from them becomes representative evidence of the entire population. Although there are several sampling methods, perhaps stratified sampling procedures may be most suitable for extension evaluation studies as they allow inclusion of all interested groups and ensure enough heterogeneity in the sample.

vi) Decide the Design of Evaluation

An ideal design of evaluation may be an experimental one. This would allow separating the effect of the programme from other factors, by setting control and treatment groups. Several experimental designs, such as one-group pre-test-post-test design, static groups comparison, pre-test, post-test control group design, Solomon four-group design, longitudinal study design, *etc.* are available in literature and can be used.

vii) Collection and Analysis of Evaluation Evidence

There are many methods for collecting information for evaluative purposes, such as the mail questionnaire, personal interview, distributed questionnaires, group interviews, case studies, systematic field observations, systematic study of secondary data *etc.* Selection of the right kind of data collection method will depend on the objectives of the evaluation, kind of information needed, time and resources available and the type of respondents from whom information is to be collected.

viii) Interpretation of the Results in a Proper Way

It is a very crucial as evaluation results can be missed also. Once tentative generalizations are arrived at, it may be appropriate and they are informally discussed among the interpreters as well as with programme planning and implementation officials, so that the results of evaluation are put in a proper perspective. The evaluation results must clearly state the achievements, failures and future adjustments needed. A written report of the evaluation findings should be prepared and made available to all concerned.

TRANSFER OF TECHNOLOGY (ToT)

Technology transfer is the movement of data, designs, inventions, materials, software, technical knowledge or trade secrets from one organization to another or from one purpose to another. The technology transfer process is guided by the policies, procedures and values of each organization involved in the process.

Also known as transfer of technology (TOT), technology transfer can take place between universities, businesses and governments, either formally or informally, to share skills, knowledge, technologies, manufacturing methods, and more. These forms of knowledge transfer helps ensure that scientific and technological developments are available to a wider range of users who can then help develop or exploit it. This transfer can occur horizontally across different areas or vertically by moving technologies, for example, from research centres to research and development teams.

Tech transfer is promoted at conferences organised by groups like the Association of University Technology Managers, so that investors can assess the prospect of commercialization for a ground breaking new product or service.

This commercialization can involve the creation of joint ventures, licensing agreements and partnerships to share the risks and rewards. This can also be coupled with the raising of venture capital, which is generally more common in the United States than in Europe, for example. Research institutions, governments and businesses may also use the services of technology transfer offices to help with the process. These offices may include economists, engineers, lawyers, marketing experts and scientists.

An important part of tech transfer is the protection of intellectual property (IP) associated with innovations developed at research institutions. This can mean licensing patented intellectual property to outside businesses or the creation of start-up companies to license the IP.

However, before innovations can be brought to market they need to be developed through technology readiness levels (TRL). TRLs 1-3 focus on research while levels 6-7 and higher sees a product move towards production. Bridging the gap between these different levels can be complex and time-consuming, as it requires the development of research into prototypes and then to fully tested and reliable finished products.

Technology transfer, also called transfer of technology (TOT), is the process of transferring (disseminating) technology from the person or organization that owns or holds it to another person or organization. These transfers may occur between universities, businesses (of any size, ranging from small, medium, to large), governments, across geopolitical borders, both formally and informally, and both openly and secretly. Often it occurs by concerted effort to share skills, knowledge, technologies, manufacturing methods, samples, and facilities among the participants. To ensure that scientific and technological developments are accessible to a wider range of users who can then further develop and exploit the technology into new products, processes, applications, materials, or services. It is closely related to (and may arguably be considered a subset of) knowledge transfer. Horizontal transfer is the movement of technologies from one area to another. At present transfer of technology is primarily horizontal. Vertical transfer occurs when technologies are moved from applied research centers to research and development departments.

Meaning of Technology Transfer

The process of disseminating knowledge, skills, and other know-how that manifests in the form of technology from its owner (individual or an organization) to another person or organization is known as technology transfer. It is also popularly known as Transfer of Technology. Various stakeholders amongst whom technology transfer takes place include universities, business organizations, research and innovation societies, and others.

Such transfer takes place with the motive to share skills, knowledge, technologies, methods of manufacturing, and other related profit motives. The transfer is further done with an intention to provide improved accessibility to a wide range of users who can then further develop and exploit the technology to develop new products, processes, applications, materials, or services.

Types of Technology Transfer

Technology transfer can be broadly classified into vertical and horizontal technology transfer.

a. **Vertical Technology Transfer**– This chain of transfer includes basic research to applied research, applied research to development, and from development to production. It is also known as internal technology transfer. This type of transfer is mostly carried out between research associations, universities, and governments, among others.

b. **Horizontal Technology Transfer**– When the technology that has already been put in place or use within one organization is further transferred and used in another place, the transfer is known as horizontal technology transfer. It is also known as external technology transfer. This type of transfer takes place between private companies, small and large business organizations, among others.

Methods of Technology Transfer

Technology transfer can take place using the following instruments.

1. **Licensing**: An agreement between the owner of the technology (Licensor) and the receiver (Licensee) which gives the right to use the technology developed or owned by the transferring individual or company for a specified time period is known as licensing.

   The two broad categories of licensing include the one which grants exclusive rights to use the technology and another which grants non-exclusive rights wherein the owner reserves the right to further transfer the technology to other company apart from the receiver. It may also include the right to sub-license, permitting the licensee to grant someone else the right to use the technology.

2. **Joint Venture Agreement**: The Company executes a joint venture agreement with respect to technology transfer for a particular business with a vision to incorporate long-term cooperation between the parties, the motivation of all participants in the successful transfer, and to incur lower costs as compared to working independently.

3. **Franchising**: It is one of the most preferred methods of transferring technology. The companies generally transfer technical know-how or skill involved under this type of agreement.

4. **Original Equipment Manufacturer**: It is a kind of sub-contracting agreement wherein a foreign company transfers a relevant portion of its technologies and a local company manufactures according to the specifications in the agreement. Such agreement enables local companies and firms to absorb technologies and restructure their production mechanism.
5. **Buy-Back Contracts**: It is a form of agreement between stakeholders from developing countries and large foreign companies, wherein a foreign company supplies industrial equipment in exchange for profits derived from the sale of raw materials or goods produced. This kind of technology transfer is often used in the construction of new plants and other related businesses.

Model of Transfer of Technology

Technology transfer involves the implementation of scientific or technological information in a context different from that for which it was developed. Successful transfer proceeds in a certain sequence of activities, with evaluation and information feedback at each of 4 decision stages:

1. The search stage, in which the recipient and the donor determine their respective needs and capabilities, the compatibility of their policies and priorities, and their personal compatibility;
2. The adaptation stage, in which the project's social and economic desirability and feasibility are studied;
3. The implementation stage, in which the necessary capital and human resources are marshalled; and
4. The maintenance stage, in which the project is put into full-scale operation. If at any stage a project is found to be ineffective, it should be altered or stopped. This paper emphasizes technology transfer to developing countries, although the model presented is equally applicable to transfer within a country or between developed countries.

Technology transfer models rely on a progression of studies to bridge the gap between the controlled, artificial laboratory conditions and field applications. They are closely related to the idea of analog models and purposive sampling in those earlier studies serve as analogs to justify inferences with regard to yet-unstudied organisms or domains.

Through a programmatic series of research studies, the technology-transfer model can support inferences with regard to target universes (*e.g.*, target populations, target effect, side effects; *i.e.*, external validity question

1. Robustness and moderating conditions (*e.g.*, drug indications and contraindications; external validity question
2. Novel universes (*e.g.*, off-label use; *i.e.*, external validity question

3. Capacity Building can be defined as "activities which strengthen the knowledge, abilities, skills and behaviour of individuals and improve institutional structures and processes such that the organization can efficiently meet its mission and goals in a sustainable way. Training is one of the essential components of capacity building.

TRAINING FOR DEVELOPMENT

Training is teaching, or developing in oneself or others, any skills and knowledge or fitness that relate to specific useful competencies. Training has specific goals of improving one's capability, capacity, productivity, and performance. It forms the core of apprenticeships and provides the backbone of content at institutes of technology (also known as technical colleges or polytechnics). In addition to the basic training required for a trade, occupation, or profession, training may continue beyond initial competence to maintain, upgrade and update skills throughout working life. People within some professions and occupations may refer to this sort of training as professional development. Training also refers to the development of physical fitness related to a specific competence, such as sport, martial arts, military applications, and some other occupations.

Capacity Building can be defined as "activities which strengthen the knowledge, abilities, skills, and behavior of individuals and improve institutional structures and processes such that the organization can efficiently meet its mission and goals in a sustainable way. Training is one of the essential components of capacity building.

Training has become an inseparable part of HRD. It has become one of the components, which enables any institution to churn out its employees as the most productive and most suitable ones.

Definition and Meaning of Training

Training is the art of increasing the knowledge and skill of an employee for doing a particular job. Training is a learning process, which seeks a relatively permanent change in behavior that occurs as a result of experience. Some important definitions of training are as under.

1. According to **Garry Dessler**, "Training is the process of teaching new employees the basic skills they need to perform their jobs."
2. According to **Jack Halloran**, "Training is the process of transmitting and receiving information related to problem-solving."
3. Edwin **B. Flippo** Said, "Training is the act of increasing the knowledge and skills of an employee for doing a particular job."
4. In the words of **Dale S. Beach**, "Training is the organized procedure by which people learn knowledge and improve skill for a definite purpose."
5. In the words of **Michael J. Julius**, "Training is a process by which the aptitudes, skills, and abilities of employees perform specific jobs are increasing."

6. According to **Edwin B. Flippo**, "Training is the act of increasing the knowledge and skill of an employee for doing a particular job."
7. Training is the process of aiding employees to gain effectiveness in their present or future work through the development of appropriate habits of thought, action, skills, knowledge, and attitude **(Milton Mall, 1980)**.
8. Training is the process of changing employee behaviour, attitudes, or opinion through some type of guided experience **(Krietner, 1989).**
9. Training is a systematic process of changing the behavior, knowledge, and or motivation of present employees to improve the match between employee characteristics and employment requirements **(Milkovich and Boudreau, 1998)**

In simple words, providing the employee's ability to perform a specific job is called training. Thus, the art, knowledge, and skill to accomplish a specific job in a specific way are called training. In simple words, providing the employee's ability to perform a specific job is called training. Thus, the art, knowledge, and skill to accomplish a specific job in a specific way are called training. At all levels of the organization, training and development programs are needed to make qualitative improvements in the employees' work.

Need for Training

The process of training has caught up mainly in industries. This can be attributed to the sudden and competitive change that is occurring in the world. However, the need for training can be fixed down to the following:

- Rapid changes in technologies and jobs people do.
- Immediate and long term skill shortage
- Changes in the expectation and composition of the workforce
- Competition and market pressure for improvement in quality of products and services.

Types of Training given to Extension Personnel

This is of broadly two types:

1. Pre-service Training

It is a process through which the individuals are made ready to enter a certain kind of professional job, as in agriculture, medicine or engineering. It is professional training prior to any appointment, oriented to make an individual prepared to enter into a new profession. Swanson (1984) defines it as a programme of training activities that prepares an individual for a career in extension and usually leads to some type of diploma, certificate, degree, or other qualification in one or more of the following agriculture, fisheries, forestry, animal and/or veterinary science or home science. The state departments of Agriculture now prefer University graduates for entry into their extension services and similarly, the Veterinary department prefers to take only Veterinary graduates released from the Universities.

2. In-Service Training

It is meant for in-service candidates who are on the job. In-service training is a process of staff development for the purpose of improving the performance of an incumbent holding a position with assigned job responsibilities. It promotes the professional growth of individuals. In-service training is a problem-cantered, learner-oriented, and time-bound series of activities, which provide the opportunity to develop a sense of purpose. Broaden the perception of the participants and increase their capacity to gain knowledge and mastery of techniques.

According to Arnon (1987), even for the University graduate, learning cannot cease on completion of formal studies. He said that the in-service training is given with the following objectives:

1. To keep up with research by regular meetings between researchers and extension workers, joint colloquia, *etc.*
2. To impart basic knowledge not only in the fields directly related to agriculture but also in sociology, economics, psychology, *etc.*
3. To improve extension methods, by constant evaluation of methods, the joint study of research findings and extension methods, exchange of experiences.

In-Service training are of different types, some of them are as follows:

i. Orientation Training

This training is given usually to newly appointed extension personnel. It provides an introduction to public employment and provides answers to questions which a newly recruited person is likely to ask. This term is also used for training in-service extension personnel in a new responsibility likes a new operational programme so that personnel are appropriately oriented towards meeting the requirements of the new situation.

ii. Induction/Portal/Vestibule Training

Induction training is given to new extension personnel immediately after they have been employed and before they are assigned to work in a particular area usually as an Assistant Agriculture Officer or Agriculture Officer, or Extension Officer.

iii. Maintenance or Refresher Training

This training is originally started for trainers of the training institutes and Universities for refreshing their knowledge and skills for imparting them to trainees. The term indicates any new training for updating professional competence of extension personnel notably in the subject matter area of specialization. This training is usually imparted in the later career of extension personnel. This training is having considerable importance to extension personnel as it relates to updating to technical knowledge and competence of extension personnel. This deals with new information and new methods and review of older materials. This type of training is given to the employees to keep them at their peak performance level and also prevent them from getting into a rut.

iv. Retraining

It refers to the efforts designed to prepare an individual for a new assignment or a broadened aspect of the old specialty.

v. Career or Development Training/Training for Professional Qualification

This type of training is designed to upgrade the knowledge, skills, and ability of employees to help them assume greater responsibility in higher positions. This training may lead to the acquisition of higher degrees (undergraduate or postgraduate) or diplomas by the employees, to motivate them to move up to higher levels of administrative hierarchy (promotions) The Directorate of Extension is operating such a scheme on a yearly basis under which, in addition to salary and allowances which personnel get from their own employing organizations, it pays fixed monthly stipends to extension personnel to cover their cost of boarding, lodging and tuition fees. Only meritorious extension personnel and that too below the age of 45 years are eligible for such courses.

Training to Farmers

There is a regular farmer training programme in all agricultural universities. There are training centers for young farmers. In some states, they also arrange short courses for the farmers. The training includes crop raising, animal feeding, and management, plant protection. For such training, the following points should be considered.

1. **Time of holding the training:** It should be at the convenience of the farmers *i.e.*, when they are comparatively free from such of agricultural operations. This will differ according to the seasons and climate. In case T.N March to May for Kharif crop and August to September for rabi crop is the ideal time for conducting training courses in Agriculture.
2. **Duration of course:** For farmers who are engaged in farming, a one-week course is sufficient for special topics such as the use of irrigation facilities and water management, operation of implements and plant protection, *etc*, it may be of two or three days duration.
3. **Venue of course:** Besides physical facilities, the appropriate environment under which the course is to be conducted i.e, where the farmer's car see the actual crop, method demonstrations, operations with some machines and implements or some treatments such as fertilizer application, the venue has to be given due considerations.

4. **Production cum demonstration camps and discussion groups of the farmers:** These should be arranged in the villages because the farmers cannot afford to remain away from their farms and homes. These should be organized before each main crop. The duration should be 1-2 days only, and the trainees or participants should be from the same village or groups of nearby villages so that the farmers can walk back to their homes the same evening. This will provide technical knowledge to the farmers' rights in their villages, and the topics can be related to their local problems.

Model of Training

Most training design models contain five steps. One of the most commonly used models is the ADDIE model, which stands for analysis, design, development, implementation, and evaluation. Analysis stands for needs analysis, where the need for the training is studied. Design is the phase where the training program is outlined and planned. Development is where the training is rolled out to the field in whatever form the design phase is stipulated. The evaluation concludes the process and measures how effective the training program was at achieving its goals.

Analysis

Analysis is the first phase of the training model. During this phase, trainers analyze all aspects of a training problem and begin seeking answers while proposing a solution. Timelines are established, training objectives are created, and the first outlines of the training program begin to take shape. Potential causes and possible solutions are explored, and initial budgets are proposed. Barriers to success are investigated, and the target audience is analyzed. Great training solutions must begin with analysis.

Design

Design is the phase of the training model where learning objectives and outcomes are determined. The core of the potential training solution is created and explored. Storyboards and initial prototypes of the training solution are proposed and reviewed with the client. Feedback is received, and initial training solutions begin to take shape. Types of training solutions, classroom, web-based and blended learning programs are discussed and explored. Much like an architect's renderings, the blueprint for your training solution begins to take shape.

Development

Development is the phase of the training design model where the training program is created and written. Whether the program is classroom-based or designed to be taken online, materials are created and produced in this phase. The design phase produced the outline or blueprint, but it is in this portion of the training model where everything comes together in production. Supporting materials are produced, trainers are trained, and the target audience is notified of the training dates.

Implementation

Your training program is delivered to your employees in the implementation phase of the training model. Classes are taught or taken online. Students receive their training and practice how to use their new skills. Materials and training products are distributed to participants, and classes begin. Initial results are measured, and the program begins to take shape in your company. If the preceding phases are conducted properly, implementation runs smoothly and the training is taken and received as it was intended.

Evaluation

Evaluation completes the training model. Measuring the results of your training program begins during the implementation phase. Learning is measured after each class, and results are analyzed. Evaluation of the entire program is conducted after all the training is completed. Measurements and feedback determine whether adjustments to the initial design are needed, and results are reviewed with the client. Students are contacted and instructors, designers, developers, and anyone involved with the program meet for a "lessons learned" review. The model then begins again.

Steps in Training Process/Phases of Training

1. Decide If Training is needed.
2. Determine What Type of Training is needed.
3. Identifying Goals and Objectives.
4. Implementing Training.
5. Evaluation of the Training Program.

Step 1: Decide if Training is needed

To compete effectively, firms must keep their employees well trained. The first step in the training process is a basic one, to determine whether a problem can be solved by training. The first step in the training process is to determine Training needs. The overall purpose of the assessment phase is to determine if training is needed and, if so, to provide the information required to design the training program. Training is conducted for one or more of these reasons:

- required legally or by order or regulation,
- to improve job skills or move into a different position,
- for an organization to remain competitive and profitable.

If employees are not performing their jobs properly, it is often assumed that training will bring them up to standard. This may not always be the case. Ideally, training should be provided before problems or accidents occur and should be maintained as part of quality control.

The assessment consists of three levels of analysis: organizational, task, and person.

1. **Organizational Analysis:** It examines the kinds of problems that an organization is experiencing and where they are located within an organization.
2. **Task/Operational Analysis:** An operational analysis identifies the kinds of skills and behaviors required of the incumbents for a given job and the standards of performance that must be met.
3. **Personnel Analysis:** The objective of the personnel analysis is to examine how well individual employees are performing their jobs.

Training should be given to those who need it. Assigning all employees to a training program, regardless of their skill levels, is a waste of organizational resources and creates an unpleasant situation for employees who do not need training. The training objectives must be clarified, related to the areas identified in the task analysis, and should be challenging, precise, achievable, and understood by all.

Step 2: Determine What Type of Training is Needed

The employees themselves can provide valuable information on the training they need. They know what they need/want to make them better at their jobs.

Just ask them! Also, regulatory considerations may require certain training in certain industries and/or job classifications. Once the kind of training needed has been determined, it is equally important to determine what kind of training is not needed. Training should focus on those steps on which improved performance is needed. This avoids unnecessary time lost and focuses the training to meet the needs of the employees.

Step 3: Identifying Goals and Objectives

Once the employees' training needs have been identified, employers can then prepare for the training. Clearly stated training objectives would help employers communicate what they want their employees to do better or stop doing! Learning objectives do not necessarily have to be written, but they should be clear and thought out before the training begins for the training to be as successful as possible.

Step 4: Implementing Training

Professionals should conduct training with knowledge and expertise in the given subject area.

Nothing is worse than being in a classroom with an instructor who does not know what they are supposed to be teaching! Use in-house, experienced talent, or an outside professional best option.

The training should be presented so that its organization and meaning are clear to employees. An effective training program allows employees to participate in the training process and practice their skills and/or knowledge. Employees should be encouraged to become involved in the training process by participating in discussions, asking questions, contributing their knowledge and expertise, learning through hands-on experiences, and even though role-playing exercises. Actually, for making the training program effective, the targeted group of employees and the use of methods such as On-the-job or Off-the-job training should select first. The capacity and knowledge of trainers and their acceptance by the participants are of secondary importance.

1. **On the job:** Training is administered at the actual work site using the actual work equipment
2. **Off the job:** Training is administered away from the actual work site. It may be any prominent hall room or auditorium, but the required training environment equipment and materials should be available or arranged.

The training program that results from the assessment should directly respond to an organizational problem or need. Approaches vary by location, presentation, and type.

Step 5: Evaluation of the Training Program

One way to make sure that the training program is accomplishing its goals is by evaluating the training by both the trainees and the instructors Training should have, as one of its critical components, a method of measuring the effectiveness of the training.

Training Process

In the case of training, the focus will be on a person-on-the job-in organization. Whereas in the case of the training process, the focus will be both at the starting point and at the end with difference. The application of what a person has learned during the training process is called the effectiveness of training. The training process has three phases as follows:

- ✰ Pre-training
- ✰ Training
- ✰ Post-training

Pre-training Phase

- ✰ The pre-training process starts with understanding the situation, which calls for behavior that is more effective.
- ✰ The key aspect of the process is an analysis of the situation and job on which improved performance is to be achieved.
- ✰ Pre-training begins with a description of the job to be changed by it.
- ✰ The technical requirement of the job is not enough but also knowledge on the operational description of the job is required so that the training programme can be designed to meet out those requirements.

Training Process – Models

There are several models for training processes, of which there are three important models.

1. Simple model of training process
2. Elaborated model of training process
3. Spiral model of training process

Simple Model of the Training Process

More effective behaviour of people-on-the-job-in-the-organization is the primary objective of the training process as well a whole. In the simple training process, improvement is dependent variable and participants and organizations independent variables. A model of training in its simplest form is presented in Figure 1.

But training is actually a more complex process than Figure 1 suggests. In the first place, the training system itself needs to be included. It may be temporary system, such as an occasional program, or a permanent institution, such as training department. In either case, the trainers-in-the-system also learns through

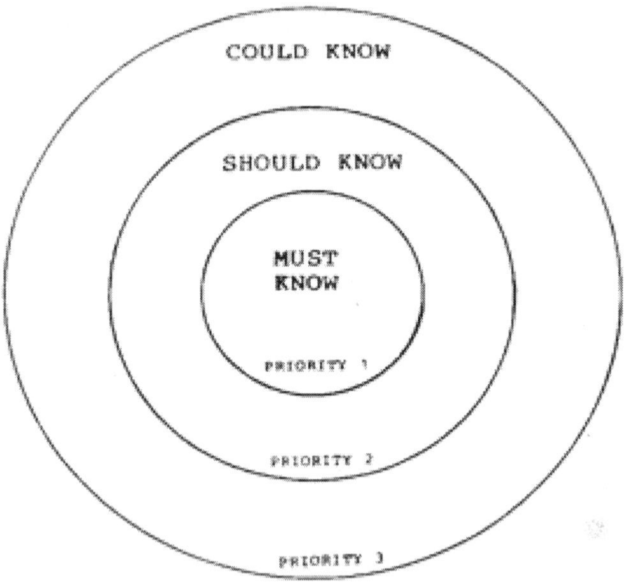

Figure 1: Simple Model of the Training Process.

the various opportunities available for checking their effectiveness, *i.e.* through feedback. Thus the independent and intervening variables become dependent variables. This elaboration is shown in Figure 2.

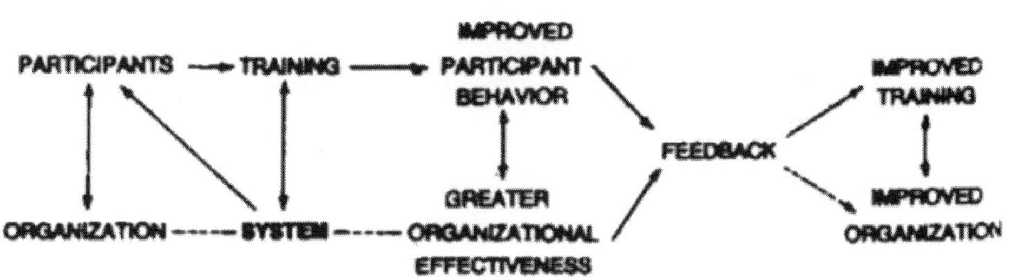

Figure 2: Elaborated Model of the Training Process.

The process as a whole, with its three partners and three stages, is depicted in figure 3. It is a spiral model, overlapping the people-on-the-job-in-the-organization. This spiral model is useful in visualizing a training program as an entity, as well as each event and series of events that make up the program. For instance, the phase of actual acquisition and development of new knowledge and skills.

Spiral Model

The spiral model in the Figure 3 shows the phases through which participants pass as they learn, then return (with enhanced capabilities, we hope) to their jobs. At various stages in the process the other two partners contribute "inputs" to assist the participants. These inputs are shown as arrows: arrows originating inside the spiral depict inputs of the work organization; arrows originating outside the spiral, inputs of the training institutions. The arrows are merely visual conveniences. The large spiral itself comprises a spiral feedback system. We will use this spiral model to explicate the training process, further focussing turn on participants, work organizations.

Figure 3: Spiral Model of the Training Process.

Training Strategy

Effective training calls, in the first place, for clarity of objectives and means. In that respect it is just like a manufacturing process, both the ends and means must be appropriate to the purpose. Relating them demands clear specifications for each part of the training task, including the resources of time, skill, and facilities required for it. Ensuring this is a responsibility of the first order.

It is necessary to consider four strategic questions, which shall be examined in turn. The first and second are of external strategies,

- Establishing training goals
- Defining training specifications
- The next two are outcomes of internal strategy
- Organizing the training inputs and
- Improving the training institution
- The second aspect is the organization's receptivity to the more effective behavior of the trained people. How far the organization would respond to the change in knowledge and skill of the trainee after he returns from training is also to be considered.
- Who feels the need for this new behaviour? Is it his immediate superior of the organization itself or does the trainee himself have to be taken into account for an effective training programme.

Training Phase

- Most of the training programmes would be for a session or an evening course or a residential program.

- In the training program, the trainee is exposed to a new subject matter, new people, new atmosphere and the participant would be at unease for a while, later when the subject which would be useful and stimulating is taught the participant would focus his attention on the subject of his interest and would be in line with other participants.
- There would be several questions in his mind, such that he is lacking, the skill required for his job or is it an opportunity given for his sincere work in the organization or is it a plan of the organization to keep him away from the organization so that it would implement the programme which he had strongly opposed.
- With all such questions in his mind, there will be no guarantee that the trainee will learn what he has chosen to learn. His mind would deviate and he would learn something of his interest from the training program provided. This error in selection would be due to the lack of necessary capabilities of the trainee or irrelevant training design and methodology followed by that training institution *etc.*,
- Finally, after overcoming all the hurdles in the initial stage of the training programme, the participant would explore in the training situation what interests him the more. After exploring, if he finds it useful he tries it again and checks for its effectiveness and satisfaction. There would be several trials repeatedly.
- If he is satisfied with the results, he decides to incorporate it in his organization, but if he finds it to be not useful he discards it and tries some other variant, in some cases, he may discontinue his learning.

Post-training Phase

- Here the situation changes, the participant goes back to his work place, meets his colleagues, family members *etc.* He goes prepared with some anticipation; as he had been away from them for a while and also had come back learning some new ideas.
- Newly learned skills undergo modifications to fit in with the work situation. If the organization were encouraging and helping, the participant would use his training for the betterment of his organization. Some organization would offer support to the participants to have contact with the training institution even after the training program.
- On the other hand, if the organization resented his absence and if his table is loaded with work, he would feel extra burden and would work to make up for lost time. He would loose his interest to make use of his training and the contact with the training institution is also broken off.

Principles of Training

When a company puts its employees through training programs, it must ensure that they are efficient and relevant to the employees' tasks in the organization as it is estimated that only 20-30 per cent of training given to employees are used in the month later. To help mitigate this issue, some general principles should be followed to increase employees' desire to take part in the program. These include:

- **Self-efficacy:** This means to increase the learner's belief that they can fully comprehend the teachings.
- **Attitude:** An uncooperative attitude towards learning could hinder the individual's capability to grasp the knowledge being provided.
- **Competence**: This is the skill an individual develops that enables them to make good decisions in an efficient manner.
- **External motivators**: These are the behaviours individuals present when a reward or extrinsic goal is given to them.

Motivation is an internal process that leads to an employee's behaviour and willingness to achieve organizational goals. Creating a motivational environment within an organization can help ensure employees achieve their highest level of productivity. Motivation can create an engaged workforce that enhances individual and organizational performance. The model for motivation is represented at the most basic level by motivators separated into two different categories:

- **Intrinsic factors**: These represent the internal factors to an individual, such as the difficulty of the work, achievement recognition, responsibility, opportunity for meaningful work, involvement in decision making, and importance within the organization.
- **Extrinsic factors:** These are external factors to the individual, such as job security, salary, benefits, work conditions, and vacations.

Conducting an Active Training Programme

A training program can be totally successful only if effective exercises have been worked out previously. Attention should be paid on the physical setup, rapport building and content of the program. Programs that look gorgeous on paper are worthless if the trainer doesn't have delivery skills to carry out the design requirements.

Steps To Conduct an Active Extension Training Programme

1. Preparing Yourself Mentally

- Feeling comfortable with the course content
- Thorough preparation well in advance
- Preparation of material activities for the training program
- Course materials, manuals, rooms, audio-visual equipment's, *etc.* and get connected with the participants

- ☆ If a question is asked and you do not know the answer give it as a group exercise; another way is to write them down and promise to find the answer.
- ☆ Repeating a course may be a bored one to the trainer but not to a new trainee - Focus your attention on the participants and not on oneself and make opportunities to learn from their experience through discussions.

2. Arranging the Physical Environment

- ☆ The physical set up at first the participants receive will create permanent impressions of the program.
- ☆ The seating arrangements should depend on many factors like number of participants, method of speech, and the like.
- ☆ In case the program has little of writing work, the participants can get rid of tables and they can arrange the chairs to their comfort.
- ☆ In other case, if small subgroups should be formed, care should be taken to leave enough space so that one group does not disturb another group.
- ☆ A well-known arrangement is horseshoe type. This can be modified into a square or a circle. All these arrangements favour group discussions with face to face contact with each participant.
- ☆ These arrangements can be formed with the help of the participants themselves for their own comfort.

3. Greeting Participants and Establishing Program

- ☆ The best and most desired start is a good welcome address. The trainer should be able to reach the mind of each person, make each feel good in the new situation, allow his feelings to flow without any hesitation. Hence a trainer should ensure that his program should have a good greet and hence to build rapport with the participants.
- ☆ A short refreshment before the actual training enables participants to mix well among themselves. Trainer himself can build relationship by knowing their names and making them feel comfortable. During the opening session he/she should introduce the participants to each of them and he himself should be introduced with a touch of un-boastful higher knowledge.
- ☆ There are number of wordings which can be addressed. To mention a few 'I have got something for you'
- ☆ This should make a feeling among the participants that they have a person with much greater knowledge and experience to his credit.
- ☆ "I've been through this too" This makes a 'we' feeling among the participants and the trainer. The participants feel that the trainer can understand one's problems and this workload so this helps to bring out their own experience in this field.

- ✰ "I admire you" This greeting puts the participants on a higher stand. This is the way one can express one's admiration over the participant's qualities and deed. It may be on very simple actions of theirs but such an admiration heartens the people to a much higher extent.

4. Getting the Best from the First 30 Minutes of Trainers

- ✰ The first 30 minutes of any classroom period is the most crucial period which a trainer should not trample upon. It is called the 'grave period' according to Napien and Gershenfeld (1983) during which any over hostility or antagonism will be submerged under a veneer of politeness, watchfulness and reserve. It is during this time that the participants perceive what role they expect to play during the training program, what they intent to accomplish during the course.
- ✰ Begin the class at the time intended without creating impatience among the participants, once competence should be made known to the members. One should make himself compatible with the group and create trust. The trainee should be clear on what activities are there for the participants and how and when they can get connected to their home town.

5. Reviewing the Agenda

In the beginning moments of the program, one should be made clear of what is going to be done *i.e.* What is expected of the programme and what is expected of the participants? The training objectives should be given in writing and these should be explained clearly. The list of what is to be accomplished should also be presented. They should be informed of how the program will be done with indications of the stay place, food arrangements, telephone messages, *etc.* A content outline and a description of the activities designed should also be given.

6. Inviting Feedback

- ✰ After reviewing the agenda, one must not fail to get feedback on the agenda. This gives the participants an opportunity to give their views or to tell what they expect more of the programs. The simplest approach is to ask directly "Does this m match what you hope to gain from this program?" "Is there anything you would like to add to it?".
- ✰ The feedback helps the trainer to change his program if feasible to the requirements of the participants and remains compatible with them. Otherwise, the programme will be a waste with content not interested to the participants.
- ✰ After all these steps, one can readily and confidently enter into the actual training programme.

Designing Extension Training Programmes

While a training plan provides a structure for training, the design of a training programme provides its content. A training plan provides broad parameters within which training is required to take place in accordance with the assessed training needs of extension personnel within the framework of extension training policy. The design of the training programme operationalizes the training plan and provides actual training. A well-designed training programme will go a long way in ensuring the success of training intervention. As a corollary, an ill-designed training programme is deemed to be a failure. The following are the steps in designing a training programme:

A. Objectives of Training Programmes

The first step in the design of a training programme is a clear statement of the objectives of the training programme. These objectives have to be based on the Training Needs Assessment (TNA) of extension personnel and stated, in order of priority, from general to specific objectives. These objectives have to be stated in terms of knowledge, Skills, Attitudes and Attributes, which the trainee will gain at the end of the training programme.

A clear enunciation of objectives of training programme will enable the trainees to have a clear idea as to what should they expect from training.

At a simple level of treatment, the objectives of a training programme fall under two categories, namely, General behavioural objectives, and Specific Behavioural objectives. Both types of objectives are required to be stated in the objectives of a training programme.

At a more sophisticated level of treatment, following Bloom's taxonomy of educational objectives, the training objectives can be classified into three classes of Cognitive Objectives, Psycho-Motor Objectives, and Affective Objectives, and their sub-classes.

If the design of a training programme is conceptualized to consist of (a) objectives, (b) learning experience, and (c) evaluation, then the objectives provide the base upon which subsequent edifice of learning experience and evaluation could be built and hence their importance in the design of a training program.

B. Skill - Mix

The second step in the design of the training programme is the determination of an appropriate skill mix for different levels of extension personnel.

Katz postulates three types of skills for a manager, namely (i) Technical Skills, (ii) Human Skills, and (iii) Conceptual Skills as proposed by Misra (1990).

For example, field-level personnel like Village Level Extension workers and Agricultural Extension Officers require technical skills in ample measure, human skills in fairly good measure and conceptual skills in moderate measure.

C. Curriculum Development

The next step in the design of a training programme is the development of an appropriate training curriculum. The curriculum is required to be developed for two reasons. A specific curriculum is required to be developed for a specific training course which is organised in response to assessed training needs of extension personnel which emerges out of the changing needs of farmers. Since no readymade curriculum exists for the purpose, a specific curriculum is required to be developed for a specific training course that has a specific target group. Its challenge lies in the fact that it is interdisciplinary. Curriculum development is required to anticipate future needs.

Components of a Standard Curriculum

- ☆ Course objectives
- ☆ Achievement Targets
- ☆ Course Structure
- ☆ Assessment
- ☆ Course Contents

Developing Standard Curriculum

Objective

While developing the curriculum, we must first determine the training objectives - general and specific based on the assessed training needs of extension personnel. The specific objectives should ultimately be broken into specific information units.

a. Subject Matter - Decision Structure Matrix

For determining the content of the training curriculum, it is essential that the target group is located on a subject matter-decision structure matrix.

b. The Concentric Circles Approach to Curriculum

Thirdly, for making the training objectives, we are required to decide the subjects and their number which are required to be taught. This requires the prioritization of subjects. The 'Concentric Circles Approach to Curriculum' divides the subjects into three categories of what must be known, should be known, and could be known. It is essential to concentrate the efforts on the inner circle of 'must know' not to drift to middle or outer circles at the cost of the inner circle as has often been observed in practice.

c. The Unformation Unit

An information unit is a definition or a description of a single idea. Every aspect of the subject matter is expressed in terms of an identifiable technical unit which may be analysed, improved, and linked to others in logical or psychological sequence. This breaking down of all the subject matter for each activity into information units makes it easy for it to be updated and objectives combined in order to obtain new training responses.

d. Training Methodology

If the curriculum is the 'heart' of a training programme, training methods could be described as the 'arteries' and 'veins' of the training system through which training messages reach the trainees and trainees and trainers receive concurrent feedback on the training programme from the trainees. The choice of appropriate training method is required to be guided by the level and background of trainees as well as by the training curriculum and the time available for training. While the appropriate choice of the training methods will certainly enhance the effectiveness of a training programme, an inappropriate choice of training methods is equally likely to mar or reduce the effectiveness of a training programme.

A training method should never be forgotten. It is only a means, for the attainment of training objectives. It should never be treated as an end and an end in itself. In training practice, such a tendency is often noticed and is required to be curbed as such a tendency displaces the training objectives reducing the effectiveness of a training programme.

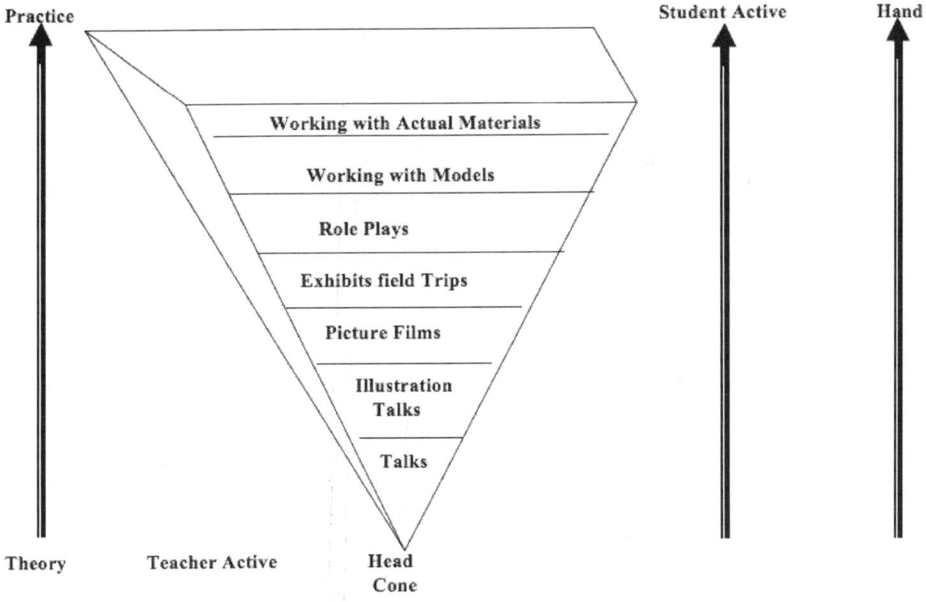

Figure 4: The Extension Teaching Methodology Cone.
(*Source*: Misra, D.C. 1990. New Direction in Extension Training).

e. Training Media

☆ Non-Projected media - Including books, hand-outs, graphics, flipcharts, photographic prints and other printed materials, real objectives, specimens, models and simulation devices, chalkboards, and magnetic and flannel boards.

- Projected media - Including opaque projector (epidiascope), projection, slides and filmstrips, microfilm, movie films, broadcast television, closed circuit television video tape.
- Sound media - such as broadcast radio, reel and cassette tapes, records and public address system.
- The choices of appropriate training media will not only depend upon the nature of audience and the training objectives but also upon the training methods. For group discussion, chalkboard and overhead projector may be adequate but may also be supported by hand-outs. Training media should be used only when required and not used because they are there.

f. ***Training Manuals and Handbooks***

The training manuals and handbooks, in their coverage and content, lie in between the comprehensive textbooks and scattered leaflets and bulletins. Additionally, they focus on a carefully selected subject matter of specialisation. The Design of Training Programmes (DTP) falls within the purview of the training institutions. The training institutions are required to pay special attention to various sub-systems of DTP. It is only when different sub-systems of DTP have been attended to professionally by a training institution, it could be claimed that a training programme has been designed. Training institutions have to realize this responsibility. Reputations are built the hard way and training institutions are no exception to this general precept.

KRISHI VIGYAN KENDRA

Introduction

Indian agriculture has witnessed phenomenal growth during the last four decades with a manifold increase in the production of major commodities like food grains, vegetables, fruits, milk, eggs, and fish. As a result, the per capita availability of important food items has increased despite of increasing population. The ratio of agricultural land to agricultural population has shrunk to 0.3 ha per person in India as compared to 11.0 ha per person in developed countries. The resources are getting marginalized and there is tremendous pressure on natural resources with the divergence of agricultural land and water towards industrial, urban, and non-agriculture sectors. Food security is one of the major concerns. It is estimated that India's population during the last three and half decades has increased from 551 million to 1122 million, which implied that the country had to feed double the population. Food security is attached to national sovereignty and thus, the agriculture sector has to gain strength and vision to uphold the national concern. Knowledge and technology empowerment of farmers is the key to achieving food security. On the recommendation of the Education Commission (1964-66) and discussion of Planning Commission and Inter-Ministerial Committee as well as recommendation by the committee headed by Dr. Mohan Singh Mehta appointed by ICAR in 1973 the idea of establishment of Farm Science Centre (Krishi Vigyan Kendra) was developed. The first KVK, on a pilot basis, was established in **1974** at **Pondicherry** under the administrative control of the Tamil Nadu Agricultural University, Coimbatore. Further in 1976-77, the Planning Commission approved the proposal of the ICAR to establish 18 KVKs during the Fifth Five Year Plan. In 1979, 12 more KVKs were approved and established by Agricultural Produce Cess Fund (AP Cess). Additional 14 KVKs were again approved in 1981 and established during 1982-83 from AP Cess Fund. A High-Level Evaluation Committee on KVK constituted by ICAR in 1984, after thorough review strongly recommended the establishment of more KVKs in the country. Keeping this in view the Planning Commission approved the scheme of ICAR to establish 44 new KVKs during the Sixth Plan. The end of the Sixth Plan saw the functioning of 89 KVKs and during Seventh Plan, 20 new KVKs were established. The success of KVKs at many locations created a great demand for the establishment of more KVKs in the remaining districts of the country so the Planning Commission approved 74 new KVKs to be established during the period 1992-93 than again in the Eighth Plan (1992-97), 78 new KVKs were approved and established making a total number of functional KVKs by the end of the Eighth Plan to 261. The numbers increased to 290 during the Ninth Plan with the establishment of 29 KVKS, an increase of 551 KVKs at the end of the Tenth Plan which includes 371 under State Agricultural Universities (SAU) and Central Agricultural University (CAU), 40 under ICAR Institute, 88 under NGOs, 33 under State Governments, 3 under PSUs and the remaining 16 under other educational institutions.

A Krishi Vigyan Kendra (KVK) is an agricultural extension center in India. The name means "farm science center". Usually associated with a local agricultural university, these centers serve as the ultimate link between the Indian Council of Agricultural Research and farmers and aim to apply agricultural research in a practical, localized setting. All KVKs fall under the jurisdiction of one of the 11 Agricultural Technology Application Research Institutes (ATARIs) throughout India. As of October 2021, there are approximately 722 KVKs throughout India. Krishi Vigyan Kendra (KVK) is the only institution at the district level in India for technological backstopping in agriculture and allied sectors. While some of the KVKs have been effectively contributing to the technology development and promotion process, many are plagued with several problems. All KVKs are envisaged to reduce the time lag between the generation of technology at the research institution and its application to the location-specific farmer fields for increasing production, productivity, and net farm income on a sustained basis.

History

The first KVK was established in 1974 in Pondicherry. Since then, KVKs have been established in all states, and the number continues to grow. The Indian agricultural landscape faces many challenges, including a high percentage of smallholder farmers, a lack of supply chain infrastructure, and extreme weather conditions. A key strategy in addressing these issues, in addition to policy support and a functioning market, is using technology to better understand and adapt to complex challenges. However, higher-level research about modern agricultural trends, such as Climate Change and GMO, takes place in universities. The practical implications of this research, or their relevance to a certain local context, are not readily apparent. For example, academic research on new crop practices or seed types often takes place in centralized testing locations, due to the ease of monitoring and evaluation. The same goes for ground-level innovations which are effective in one local context but may not be extensible to others. Especially in areas with such geographic complexity as India, agricultural extension departments such as KVKs serve to gather, test, and disseminate knowledge between centralized institutions and a geographically-dispersed rural population. In this mandate, the effectiveness of KVKs is difficult to measure, due to large number of farmers served by a single KVK and largely off-line communication between the KVK and farmers. For this reason, research over the past 20 years has focused on the capacity of KVKs to make use of ICT for the purpose of better managing their communications with farmers. A plethora of applications has been developed, sharing advisories such as weather information and market pricing, supplementing the KVK's communication with its beneficiaries. However, many of these initiatives are short-lasted, or have limited impact, since the teams at each KVK often do not have the capacity to maintain software applications or because farmers do not find the information useful. While KVKs are expected to undertake their own projects, they are also expected to serve as a resource center for extending government initiatives to local areas. The current national government's program "Doubling Farmers' Income by 2022" calls for increases in agricultural productivity, development initiatives such as Pradhan Mantri Krishi Sinchai Yojana and Pradhan Mantri Fasal Bima Yojana as well as more focus on technological innovation. The government expects KVKs to aid in the dissemination of information and practices regarding these new government initiatives. In addition to KVKs, there are many local institutions that also interface directly with farmers, such as the Agricultural produce market committee and the Agricultural Engineering Department. As of October 2018, there is an online dashboard that provides updates on the activity of various KVKs.

Mandate

The mandate of KVKs is to provide a key facilitating role in the refinement of technologies to specific conditions, by acting as a two-way link between research and farmers. Application of technology/products through assessment, refinement, and demonstration for adoption, thus, is the main mandate of the KVKs. To achieve this mandate effectively, each KVK is expected to perform the following activities:

- ✰ On-farm testing to identify the location specificity of agricultural technologies under various farming systems

- Frontline demonstrations to establish its production potentials on the farmers' fields.
- Training of farmers and extension personnel to update their knowledge and skills in modern agricultural technologies.
- Work as resource and knowledge centre of agricultural technologies for supporting initiatives of public, private, and voluntary sector for improving the agricultural economy of the district.
- Produce and make available technological products like seed, planting material, bio-agents, young ones of livestock *etc* to the farmers.
- Organize extension activities to create awareness about improved agricultural technologies to facilitate fast diffusion and adoption of technologies in agriculture and allied sectors.

The High Powered Committee on Management of KVKs (ICAR, 2014) in its report has suggested a new vision, mission, and mandate for the KVKs defining its activities as follows:

- On-Farm Testing (OFT) to assess the location specificity of agricultural technologies under various farming systems.
- Out scaling of farm innovations through Frontline Demonstration (FLD) to showcase the specific benefits/worth of technologies on farmers' fields.
- Capacity development of farmers and extension personnel to update their knowledge and skills in modern agricultural technologies and enterprises.
- Work as Knowledge and Resource Centre for improving overall agricultural economy in the operational area. · Conduct frontline extension programmes and provide farm advisories using ICTs and other media on varied subjects of interest to farmers.
- Data documentation, characterization, and strategic planning of farming practices.

Principles

1. Accelerating agricultural productions in the operational area of the Kendra should be the prime goal.
2. Work-experience as the main method of imparting training. Experiential learning *i.e.*, "teaching by doing" and "learning by doing" should be the principal methods of imparting skill training.
3. Training efforts should not be made to make economically good people better but the poor ones good so as to raise the living conditions of the poorest of the poor.

Staffing Pattern

Each KVK has been provided with a team of multi-disciplinary subject matter specialists for taking up the activities of a KVK. The present staffing pattern includes one Sr. Scientist/Head, 6 Subject Matter Specialists in the core areas *viz.* agronomy, plant breeding, horticulture, livestock production, home science, extension education, agricultural engineering, fisheries, soil science, agro-forestry and plant protection depending on the need of the area, 3 Programme Assistants to work as Farm Manager, Computer Assistant and Technician. In addition, the KVKs have been provided with two administrative and 4 supporting staff including drivers.

Research-Extension and Development Linkages

Each KVK has a provision of Scientific Advisory Committee (SAC), the function of this Committee is to provide advice to the KVK in formulation of annual technical programme and also to review the performance mandated activities. The Committee consists of the representatives of National Agriculture Research System, various line departments functioning in the district, media, credit and input organizations/agencies and representative of the farmers and farm women. The idea of constituting this Committee is to see that the Annual Plan of Action is in harmony with the needs of the farmers, resources and location specificity of the technologies, existing government programmes, and support. The committee meets once or twice a year to provide guidance and direction to the KVK to meet the emerging challenge of different crop seasons. Besides, it also serves as a mechanism for research-extension and development linkages. The KVKs also participate in Zonal Agricultural Research and Extension Advisory Committee (ZAREAC), attend workshops and specialized training programmes in the concerned State Agricultural University and ICAR Research Institute. The participation of staff in these forums helps the KVK scientists to get continuous up-to-date technical backstopping and also enables them to help the farmers in the timely adoption of the latest agricultural technologies.

Monitoring Mechanism

The performance of KVK is monitored and reviewed at various levels. At the field level, the Program Coordinator monitors the activities on a day-to-day basis whereas the Head of the host Institution/Organization monitors the technical and financial management of KVK. At the University level, the concerned Director of Extension Education of the State Agricultural University focuses on the functioning of KVKs. The Director of Extension Education has been given the responsibility to provide technologically backstopping and overseeing, irrespective of the host institutions. At the ICAR headquarters, the Division of Agricultural Extension monitors and reviews the functioning of the KVKs. Every year, a National KVK conference is organized where all the KVKs participate to share their experiences, and new approaches followed in implementing the activities of the KVKs. Critical examination of quarterly progress reports, annual reports, visit reports of the Zonal Project Directors and annual meetings/conferences/visits constitute other monitoring and review mechanism. QRTs and Independent Evaluations help improve the quality of functioning of KVKs.

Programmes and Activities

KVK is an agro-based capacity-building institution for the farmers to provide need-based teaching on various aspects of agriculture and allied sectors. KVKs impart the latest technical know-how and do-how to different clientele by formulating various programmes with the principles of learning by doing, seeing is believing, earn while you learn to achieve the desirable changes pertaining to their knowledge, skills, and attitude with a view to helping them live better by improving their farm and allied enterprises. KVK plans to execute and evaluate its programmes with the people (Kokate, 2009). Knowledge "Management" is given emphasis for improved efficiency of KVKs. Knowledge management is all about converting data into information and knowledge and then applying wisely. Knowledge and technology play a key role. In this context, how KVKs act as centres of knowledge and wisdom. The KVKs facilitate translation of data into knowledge leading to effective technology application. KVK can work as effective knowledge and resource center by adopting **KRCTO** model.

- **Knowledge:** The acquisition of competence of KVK on existing farming systems and production systems, agricultural technology, markets (demand and prices), and policy are key to its performance.
- **Resources**: Infrastructure building to provide necessary technological backstopping and capacity building are crucial for knowledge and technology dissemination.
- **Competency:** Evaluating technologies, processing and value addition, weather-based agro advisories, market intelligence, and agro-logistics.
- **Technology products**: Critical technology products and problem-solving consultancy are to be provided by the KVK.
- **Organizing farmer's groups:** Building farmer's network around KVK to exchange information, facilitating learning from experience and support in decision making.

KVKs need to be strengthened by increased technological backstopping by ICAR and SAUs for playing the role of knowledge and resource centres effectively and efficiently. There is also a need for extensive research by ICAR research institutes and SAUs to suggest innovative approaches and methodologies to KVKs for a critical assessment of location-specific technologies, frontline demonstrations, effective capacity building of stakeholders, vocational training, and entrepreneurship development for sharing successful experiences.

Responsibilities

- **On-Farm Testing:** Each KVK operates a small farm to test new technologies, such as seed varieties or innovative farming methods, developed by ICAR institutes. This allows new technologies to be tested at the local level before being transferred to farmers.

- **Front-line Demonstration:** Due to the KVK's farm and its proximity to nearby villages, it organizes programs to show the efficacy of new technologies on farmer fields.
- **Capacity Building:** In addition to demonstrating new technologies, the KVK also hosts capacity-building exercises and workshops to discuss modern farming techniques with groups of farmers.
- **Multi-sector Support:** Offer support to various private and public initiatives through its local network and expertise. It is very common for government research institutes to leverage the network of KVKs when performing surveys with a wide range of farmers.
- **Advisory Services:** Due to the growing use of ICT, KVKs have implemented technologies to provide farmers information, such as weather advisories or market pricing, through radio and mobile phones.

In each of these activities, the KVK focuses on crops and methods specific to the local climate and industry. Some factors which may impact this decision are soil type, crops grown, water availability, seasonal temperatures, and allied sectors such as dairy and aquaculture. In addition to addressing local factors, KVKs are also mandated to increase the adoption of practices that align with remunerative agriculture, climate-smart, and dietary diversification. Some KVKs also host social activities to facilitate rapport between the institutions and the local community.

Krishi Vigyan Kendra Features

1. Creation of valuable resources in terms of technical manpower and assets
2. Confirmation of technologies to suit local specificity
3. Showcasing the frontier technologies
4. Capacity-building among stakeholders
5. Front runner in technological application, information, and inputs
6. Participatory approaches in planning, implementing, executing, and evaluation

KVK – Aims and Objectives

1. Krishi Vigyan Kendra is critical in demonstrating location-specific agricultural technologies through on-farm testing
2. Through technology assessment, refinement, and demonstrations, it aims to assess location-specific technology modules in agriculture and related enterprises. It also aspires to put agricultural research into practice in a small-scale, localized setting
3. Krishi Vigyan Kendras have been serving as agricultural technology knowledge and resource centers, assisting public, private, and non-profit initiatives aimed at improving the district's agricultural economy

Krishi Vigyan Kendra - KVK Portal

The government has launched a Krishi Vigyan Kendra online portal in the year 2018. The KVK portal assists in the national monitoring of all KVK centers and the provision of timely advisories to farmers.

KVK Portal - Objectives

- Observe the activities and resource utilization of various Krishi Vigyan Kendras
- Creating a database of all of the KVK's various programs
- Using web-portal and mobile technology, assisting farmers in resolving their issues
- Providing farmers with timely weather and market information via the portal. In addition, providing information about the KVK's various facilities and activities
- Farmers and Agricultural Officers, for example, can register and gain access to relevant information about different Krishi Vigyan Kendras

Aim of Krishi Vigyan Kendra Portal

The portal's purpose is to use web and mobile technology to quickly and effectively transfer agricultural scientists' technologies to farmers, as well as to track the activities of Krishi Vigyan Kendras (KVKs).

The XII Plan Working Group on Agricultural Extension (Planning Commission, 2012) made following recommendations to make KVKs more effective:

- The KVK Farms should be developed as centres of excellence and a role model for farmers. It needs to be ensured that every extension staff, including supervisory and administrative level officials, possesses superior competency, skills and knowledge.
- Extension support is weak or non-existent in the case of animal husbandry and fisheries. As separate extension machinery for animal husbandry and fisheries are not going to be feasible in many states, this has to be integrated with ATMA. In districts where livestock and fisheries play a major role, staffing structure within ATMA and KVKs should be modified to include more staff with specialization in these sectors.
- The changing roles of and expectations from KVKs necessitate the regular capacity building of its professionals. Hence, exclusive capacity build programmes shall be designed and conducted (like induction training, refresher courses, management, and executive development programmes) for effective implementation of the mandated activities and image building/branding of the KVK system. NAARM, Agricultural Extension Division of ICAR, and Zonal Project Directorates will jointly take up the responsibility for this.

The Way Forward

NSSO, based on a countrywide survey of nearly 35,000 households revealed that farmers continue to remain far removed from new technologies and guide state-run research institutes including KVKs (NSSO, 2014). Over 59 per cent of the farm households received no assistance from either government or private extension services. Of the 40.6 per cent households who received extension assistance, only 11 per cent of the services came from government machinery like extension agents, KVKs and agricultural universities nearly farmers depended on other progressive farmers (20 per cent), media including radio, TV, newspaper (19.6 per cent) and private commercial agents (7.4 per cent).

Some of the suggestions which will be useful towards making KVKs more vibrant and relevant are:

- **Entrepreneurship Development:** KVKs must develop farmer entrepreneurs to help in technology transfer through farmer-farmer extension. KVK training should promote entrepreneurship among rural youth, helping them in gaining self-employment.
- **Promotion of diversified farming systems including Agro-tourism:** KVK farms and KVK adopted villages can be developed as agro-tourism sites, to demonstrate diversifying farm income portfolio and which should be a cost-effective option by farmers.
- **Resource Generation:** KVKs can compete and tap funds available from various government schemes/NABARD/Agricultural Skill Council of India for skill and entrepreneurship development in rural areas as well as generate a part of their resources from the sale of planting materials and other produce from their farms, funding relationship with NGOs, and national and international organizations.
- **ATMA-KVK link:** Lot more needs to be done to achieve the convergence between ATMA and KVK (as envisaged under the joint circular) operational. There might be some success stories, wherein, the proposed changes like quarterly joint meetings, earmarking of funds to KVKs, visit of ATMA staff to the cluster villages of KVKs, and ATMA sponsorship for Technology Weeks (being organized by KVKs) are achieved successfully, which may be shared for wider replication in other districts.

EXTENSION TEACHING AND LEARNING PROCESS

Extension is an educational process in which teaching has a pivotal role. An extension worker can teach people about new varieties of rice, the right method of milking a cow, the implications of export of sugar to a country, climate change and environmental issues, applying the right fertilizers in appropriate quantities, the importance of family planning, *etc.* The list can go on, depending on the field of development. All this makes us understand that teaching helps in transferring knowledge and skills. Teaching can also make differences in the attitude of a learner.

Extension teaching methods are used to improve extension teaching, *i.e.*, to increase the concreteness, clarity, and effectiveness of the ideas and skills to be transferred.

Extension-teaching methods are the tools and techniques used to create situations in which communication can take place between the extension workers and the beneficiaries of development. These methods of extend new knowledge and skills to the beneficiaries of the development by drawing their attention towards them, arousing their interest, and helping them to have a successful new learning experience and practice.

Meaning and Definition

A method is a way of doing something, an orderly arrangement of a set of procedures. Thus it involves a sequence of progressive steps in an orderly and logical regularity in order to accomplish some task or purpose. An extension teaching method may, then, be defined as a sequence of progressive steps, undertaken to create situations that are conducive to effective learning. According to Leagans (1961), extension teaching methods are the devices used to create situations in which communication can take place between an instructor and that learner.

As Ensminger (1957) said, before an extension worker can become efficient in the use of methods, he must know what methods are available, when to use a given method, and become effective in using each. Extension teaching methods may be defined as devices used to create situations in which new information can pass freely between the extension worker and the farming communities. The extension-teaching methods can also be defined as the tools and techniques used to create situations in which communication can take place between the rural people and the extension workers. They are the methods of extending new knowledge and skills to the rural people by drawing their attention towards them, arousing their interest and helping them to have a successful experience of the new practice It is the function of the extension worker to use the extension methods which provide opportunities for rural people to learn and which stimulate mental and physical activities among the people. For extension workers to be successful they must be proficient in technical knowledge and educational process and must also have the right attitude towards rural people.

However, normally no extension worker has ability to use all methods with equal skill. Further, there is no one method that is best for all situations alike and hence calls for different method (s). It is also obvious that no one method can reach all the audience. Behavioural changes required on the part of the learners may also require several exposures with the same, different or a combination of methods. Research bears ample evidence to suggest that a combination of methods or media-mix is required for effective technology transfer.

An Effective Learning Situation Requires the Following Essential Elements

1. An instructor (an extension worker, *e.g.* an extension officer or a village level worker)

2. Learners (the farmers, the processors, the farm women and the youth).
3. Subject-matter (the recommended improved practices, such as the seeds of high-yielding varieties, fertiliser, balanced diet, *etc.*)
4. Teaching material, such as a flannel-board, a black-board, charts, models, samples, slides, film strips, *etc.*
5. Physical facilities, such as sitting accommodation, good visibility, *etc.*

Functions

The following are the functions of extension teaching methods:

- ✰ To provide communication so that the learner may see, hear and do the things to be learnt.
- ✰ To provide stimulation that causes the desired mental and or physical action on the part of the learner.
- ✰ To take the learner through one or more steps of teaching-learning process, *viz.* attention, interest, desire, conviction, action and satisfaction.

Classification of Extension Teaching Methods

According to Use

One way of classifying the extension methods is according to their use and nature of contact *vis-a-vis* interpersonal, group and mass communication. Based upon the nature of contact, they are classified into following categories:

a) Individual-contact Methods

Extension methods under this category provide opportunities for face-to-face or person-to-person contact between the rural people and the extension professionals. These methods are very effective in teaching new skills and creating goodwill between farmers and the extension professionals.

b) Group-contact Methods

Under this category, the rural people or farmers are contacted in a group which usually consists of 20 to 25 persons. These groups are usually formed around a common interest. These methods also involve a face-to-face contact with the people and provide an opportunity for the exchange of ideas, for discussions on problems and technical recommendations. In this way, the future course of action is finalized.

c) Mass or Community-Contact Methods

An extension professional has to approach a large number of people for disseminating information and helping them to use it. This can be done through mass-contact methods conveniently. These methods are more useful for making people aware of the new technologies, quickly.

Classification of Extension Teaching Methods According to Use

Sl.No.	*Individual Contact*	*Group Contact*	*Mass Contact*
1	Farm and home visits	Result demonstration	Farm publications
2	Farmer's call	Method demonstration	Mass meeting
3	Personal letter	Group meeting	Campaign
4	Telephone call	Small group training	Exhibition
5	Adaptive trial	Field day	Newspaper
6		Study tour	Radio
7			Television
8			Posters

Another Classification of Extension Teaching Methods which is very Common in Extension Publications is According to their Form

Sl.No.	*Written*	*Spoken*	*Visual*	*Spoken and Visual*
1	Bulletins	Meetings	Result demonstrations	Method demonstration
2	Leaflets	Farm and home visit	Exhibits	Result demonstrations
3	Personal letters	Office calls	Posters	Television
4	Circular letters	Radio and recordings	Charts	Movies
5	Farm journals	Telephone calls	Slides	Puppets
6			Film strips	Campaigns
7			Flash cards	
8			Flannel graphs	
9			Bulletin boards	

A brief description of some of the extension methods which are commonly used by extension professionals is given below:

Farm and Home Visit

Farm and home visit constitutes the direct or face-to-face contact by an extension professional with the farmer or the members of his family. During these visits, information is exchanged or discussed. The visits may be to get acquainted with the problems of the farmers. Such visits provide an opportunity for a two-way communication.

Method Demonstration

It is used to show the technique of doing things or carrying out new practices *e.g.* clean milk production, paneer making, ghee making, *etc.* This method is usually used for groups of people.

Result Demonstration

Result demonstration is meant for proving the advantages of recommended practices and to demonstrate their applicability to the local conditions. It is conducted by a farmer under the direct supervision of an extension professional. It is designed to teach others, in addition to the person who conducts the demonstration. It helps the farmers to learn by seeing and doing. This method can be used to show the superiority of practices, such as dahi culture, value addition, hygienic handling of animals, *etc.*

Group Discussion

All the farmers cannot be contacted by extension professionals individually because of their large number. It is convenient and feasible to contact them in groups. This method is commonly known as group discussion. It is used to encourage and stimulate the people to learn more about the problems that concern the community through discussion. It is a good method of involving the local people in developing local leadership and in deciding on a plan of action in a democratic way.

Exhibition

An exhibition is a systematic display of information, actual specimens, models, posters, photographs, and charts, *etc.* in a logical sequence. It is organised for arousing the interest of the clientele in the things displayed. It is one of the best media for reaching a large number of people, especially illiterate and semi-literate people. Exhibitions are used for a wide range of topics, such as planning a model village, showing high-yielding breeds of cattle and buffaloes, new agricultural implements and the best products of village industries.

Campaign

Campaign is used to focus the attention of the people on a particular problem, *e.g.* milk adulteration, vaccination and prevention of animal diseases, tick control *etc.* Through this method, maximum number of farmers can be reached in the shortest possible time. It builds up community confidence and involves the people emotionally in a programme.

Field Tour

Conducted tours for farmers are used to convince them and to provide them with an opportunity of seeing the results of new practices and products, skills, *etc.* and to give them an idea regarding the suitability and application of these things in their own area. Such tours may also be arranged to enable the rural people to visit places and institutions (connected with the problems of rural life), such as research institutions, training institutions, agricultural universities, model villages, areas of advanced developments, leading private farms, exhibitions, and agricultural and cattle fairs/dairy mela.

Print Media

Newspapers, magazines, bulletins, leaflets, folders, pamphlets and wall news-sheets are another set of mass media for communicating information to a large number of literate people. They are used for communicating general and specific information on a programme of technology or a practice. Small folders, leaflets and pamphlets are used to give specific recommendations about a practice, such as clean milk production, vaccination schedule, detection of milk adulterants, *etc.*

Radio

It is one of the most powerful media of communication. It is a mass medium of communication and can reach a large number of people at any given time involving the least expense. Extension professionals use the radio for communicating information on new methods and techniques, giving timely information about the control of animal diseases such as foot and mouth disease (FMD), animal pests, weather, market news, *etc.* For this purpose, talks, group discussions, folk-songs, dialogues and dramas are usually broadcast. There are radio programmes broadcast by All India Radio (AIR), FM (frequency modulated) radio, community radio, ham radio, *etc.*

Television

It combines both audio and visual impact and is very suitable for the dissemination of agriculture and dairy information. It is more useful in teaching to do a specific job. A beginning has been made in India for using this medium for development programmes and it is expected that its use will become more extensive in the coming years. At present, along with the Government-owned channel (Door Darshan), several other private channels are telecasting various kinds of entertainment and developmental programmes to reach the viewers.

Leaflet

A leaflet is a single sheet of paper used to present information on only one developmental idea in a concise manner, using simple language.

Folder

A folder is a single piece of paper folded once or twice, and, when opened, the material is presented in sequence.

Pamphlet

A pamphlet is an unbound single sheet of paper that is printed on both sides, printed in colours with action photographs, giving full information about a topic in greater length than in folders or leaflets.

Bulletin

A bulletin is a publication of around 20 pages, with the primary objective of giving complete information which the intended readers can apply to their own situation.

Booklet

When the extension material exceeds 20 pages and is less than 50 pages, it is called booklet.

Extension Learning Process

Extension is an educational process for bringing desirable changes among the different stakeholders, which involves both learning and teaching and needs some tools or methods commonly known as extension-teaching methods. It is, therefore, necessary here to understand what is meant by learning, teaching and extension methods.

Any change of behaviour which takes place as a result of experience may be called learning.

'Learning' is the process by which an individual, through his own activity, attains a change in his behaviour (Leagans, 1961). It is an active process on the part of the learner. The essential role of an extension worker is to create effective 'learning situations'. An effective learning situation requires the following essential elements (Leagans, 1961):

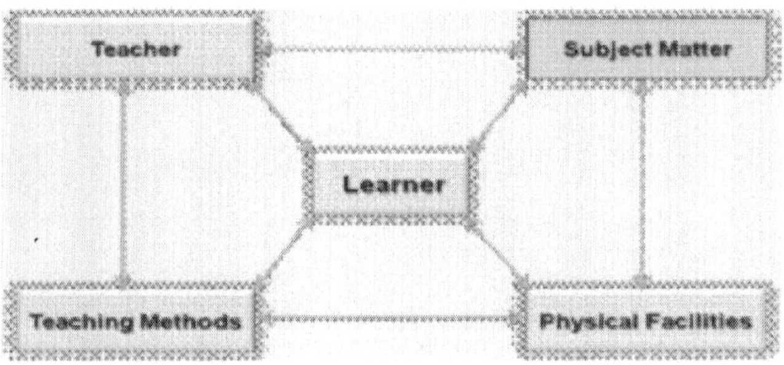

Elements of Learning Situation.

Teacher (Instructor)

A successful teacher is the one who takes into account the following important considerations:

- Understands the nature of the learner and is in a position to evaluate his motives and ability.
- Reacts appropriately to the feelings, emotions and attitudes of the learners.
- Is conscious of the fact that the learner responds to the whole gamut of learning situation.
- Is professionally alert.

- ☆ Is well-composed even in the adverse conditions.
- ☆ Has the ability to maintain better human relations.
- ☆ Always makes efforts for his professional growth.
- ☆ Selects those experiences which suit to the abilities and needs of the learners.
- ☆ Should be close to masses and belong to the same groups for making dialogues easier.

Learners

They should have:

- ☆ Clear objectives regarding the use of the subject matter.
- ☆ Interest in the subject matter.
- ☆ Need for the subject matter.
- ☆ Ability to understand and opportunity to apply the subject matter.
- ☆ Physical and mental preparation.

Subject Matter (Contents)

The subject matter should fulfil the following conditions:

- ☆ Should be in the line with the needs and interests of the learners.
- ☆ Should be valid, authentic, factual and applicable.
- ☆ Should be well-organized.
- ☆ Should be at a level where it could be understood by the learners.
- ☆ Should be timely and appropriate.

Teaching Methods

Subject matter cannot effectively be transferred without the help of suitable teaching methods and aids. Proper selection and skilful handling of teaching methods and aids facilitate creation of desirable learning situation. Teaching methods and aids should be:

- ☆ Readily available.
- ☆ In working conditions.
- ☆ Suitable to the subject-matter.
- ☆ Diversified, flexible and suited to the needs of the clients.

The following points need to be kept in mind while using extension methods for appropriate involvement of different sense organs:

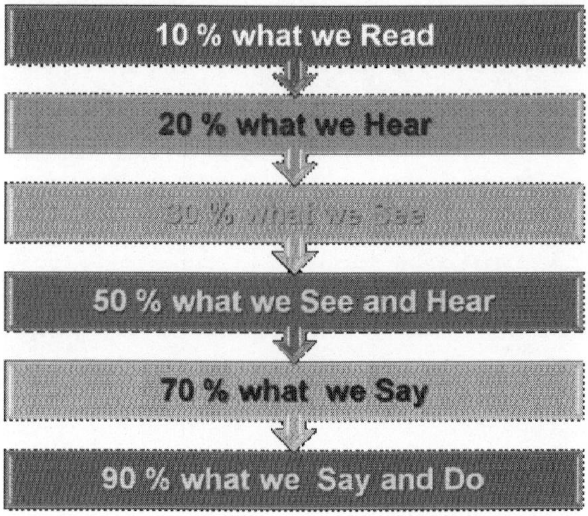

People Remember the Things.

Physical Facilities

Light, ventilation, chair and table arrangements, outdoor facilities, gardening and good land escaping should be a satisfying and comfortable, if possible. The good teacher will make all these arrangements prior to the meetings. The teacher should skilfully manipulate the elements of the learning situations and provide satisfactory learning experiences for the stakeholders. The main aim of the teacher is to bring about a change in the behaviour of the learners with the help of a judicious combination and use of different elements.

TEACHING PROCESS IN EXTENSION

'Teaching' is the process of arranging situations in which the things to be learnt are brought to the notice of the learners, their interest is developed and desire aroused, *i.e.* they are stimulated to act. For example, if we want to teach the dairy farmers the use and advantages of milk processing (Paneer), we do this by conducting demonstrations at the household levels, showing them how to boil the milk and at what temperature citric acid is to be added, *etc.* Then, comparison of the selling cost of paneer and raw milk is to be done to make them understand its benefits. After seeing the beneficial effect of a paneer, the farmer is convinced and motivated for action to start the processing of milk.

Teaching in Extension

The development and progress that we aim for is based on what we know, what we think, what we actually did, and what we can do with our physical, technological and human resources. To progress, we must not stop at the present conditions of living and must take the necessary actions to improve them on a sustainable basis. In this process, we must identify and perpetuate only the useful and successful developmental initiatives from the past, exercise concern only for the promising developmental ideas in the present, and focus on all of these practices on achieving a better and sustainable development for the future. Thereby, we gain the ability to substitute the 'good and new' developmental ideas with the 'old and outdated' through learning. Thus, learning is most effective when done under organized teaching. Teaching is defined as an interactive process primarily involving classroom talk, which takes place between teacher and pupil, and occurs during certain definable activities (Amidon, 1963).

Steps of Teaching

Teaching, in the context of extension and development, can be thought of as providing purposeful direction through learning process. Teaching *per se* is not giving knowledge or skills to people or communities, instead teaching is the process of providing opportunities to people or communities to produce relatively permanent change through their engagement in learning experiences that are provided by teachers.

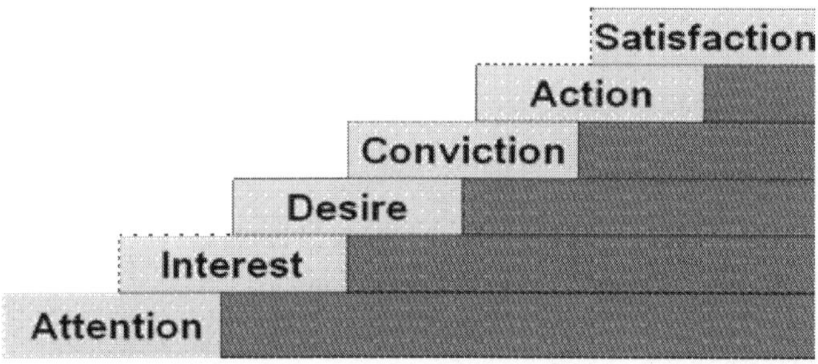

Steps in Extension Teaching (Wilson and Galup, 1955).

Attention

Attention is the starting point to arouse the interest of the learners. Research suggests that the attention of people is attracted by various senses in the following proportions (Reddy, 1998):

Attention of People through Various Sense Organs

Sense	Per cent	Sense	Per cent
Seeing	87.00	Hearing	7.00
Smell	3.50	Touch	1.50
Taste	1.00		

The famous extension saying is, *seeing believes*. Thus, seeing and hearing are the major senses involved to attract attention and enhanced learning.

Interest

Once attention is captured, instructors and/or extension professionals can bring the audience's attention to developmental needs and arouse their interest in further consideration of ideas. Extension professionals should make them understand how development contributes to the overall well-being of the community.

Desire

It concerns about the continuation of the audience's interest in the developmental ideas or better practices, until that interest becomes a desire, or, a motivating force.

Conviction

In this step, people know what action is necessary and just how to take that action. Extension professionals also ensure that people visualize the action in terms of their own situations and acquire confidence in their own ability to participate in the people-centred developmental initiatives.

Action

Unless this conviction is converted into action, the efforts of extension for development will go unrewarded. It is the job of extension professionals to make it easy for the people to act. For example, if the adoption of a new high yielding cattle breed is the action needed by farmers, that breed should be made available within the reach of farming communities, along with other recommended package of practices. If the action does not quickly follow desire and conviction, the new idea may fade away. Therefore, this phase should never be ignored.

Satisfaction

Satisfaction is the end product of extension teaching process. Follow up action(s) by extension professionals helps people to learn and evaluate the development progress. The saying, *satisfied customer is the best advertisement*, also applies to extension and development work. Satisfaction helps the people to continue developmental work with increased attention, interest, desire, conviction and action.

The six steps in teaching discussed as above, often blend with each other. As an extension professional, you need to arrange the learning situations in all the six teaching steps with the help of suitable extension teaching methods and audio-visual aids. Various teaching methods and audio-visual aids are not equally suited for every step in teaching. Every method and aid under certain circumstances makes a contribution to each step. It depends on the extension professional how he handles the situation.

Selection of Extension Teaching Methods

Extension education has its roots in researches related to agricultural sciences, fishery sciences dairy and animal husbandry sciences *etc.*, but the method of sharing this information with stakeholders depends on a social sciences process. The major role of extension professionals in this process is to educate the stakeholders. Hence, it is his responsibility to provide better learning situations so that learners can learn effectively. No single technique alone is suited to all situations. Therefore, a combination of suitable teaching methods may be employed to do the needful, some of which are being discussed below:

1. Methods for Getting Attention

To provide learning experiences for people, extension professionals must get their clients attention. This is the fundamental stage for successfully making the ideas enter the minds of people, and it requires planned and frequent use of a variety of teaching methods, *viz.* pictures, demonstrations, news, success stories, survey results, slogans, posters, radio talks, cartoons, displays, exhibits, *etc.*

2. Methods for Developing Interest

The amount and direction of developmental accomplishments of extension teaching are largely determined by the interests of the people. Learning without interest does not take place to any appreciable degree. People who are interested in solving developmental problems acquire more information about them. Interest usually represents the objectives or goals of an individual. Development means comparatively little to people until they are able to connect it to their interests.

3. Methods for Developing Confidence

Development and maintenance of confidence should be parallel to all other changes in people's attitudes and behaviour. Confidence is the key to achieve results from extension education, because action taken by people is voluntary. Confidence of the client corresponds to the goodwill of the extension professionals. Once broken, confidence is difficult to restore, either in oneself, or in the community as a whole. It is, therefore, necessary for extension professionals to safeguard the factor of confidence in all the steps of teaching.

4. Methods for Creating Desire

Desire has a definite focus, unlike attitudes and interests which are general, which is specific to the objects to which it is attached. Desire comes only when a developmental idea, or intervention, suggested by extension professionals is considered favourable by people. It may be noted here that desire is an outgrowth of interest and confidence. In creating desire, appeal to the feelings and emotions of people should be employed to support the interest they already have. A desire is a want, and to make that want to be felt and to satisfy the want, the right teaching methods must be used to stimulate feelings and emotions, *viz.* exhibitions, demonstrations, pictures, working models. For example: solar water heating devices, circulars suggesting benefits, 'before' and 'after' pictures that demonstrate development, and actual development results presented in all possible ways.

5. Methods to Ensure Action

Action follows desire. If action does not follow soon after the desire has been created, the desire soon fades away and people continue as before. Consequently, a variety of extension teaching methods must be planned and used to ensure action that result in development. The methods useful for promoting action are: use reminders; ensure the participation of leaders in development work; cooperative action; success stories and radio talks about and by people taking action.

6. Methods for Maintaining Satisfaction

Satisfaction depends upon the confidence, pride and success as a result of adoption of developmental initiatives. After a developmental practice has been adopted by people, it is important to remain in touch with them till they experience real satisfaction and continue with the practice. Continuation of the practice and recommendations to others is an indication of satisfaction. Methods that are useful in maintaining satisfaction are: personal contact, whenever possible; timely hints; showing the value of results; giving more information; and publicity in media.

Factors for Selection of Extension Teaching Methods

No single "rule-of-thumb" can be given for the selection and use of the various extension methods for ensuring success in all situations. On one hand, the individual contact methods furnish the most direct opportunities for influencing people effectively; on the other hand all other methods of group and mass procedures are dilutions or compromises created by the pressure of necessity. We must reach more people, teach them more often, and keep down the cost per contact. In order to get most effective results the extension professionals should: select the appropriate methods; have a suitable combination of the selected methods and; use them in proper sequence, so as to have repetition in a variety of ways. For doing this, a number of factors should be considered, such as:

1. Audience

Based upon the principle of *sine qua non*, it is imperative that the instructors must be aware of the type of audience they are going to interact with. In addition, size of the audience coupled with level of their educational qualification also contributes significantly as a factor influencing the choice of extension teaching methods. For instance, group discussion cannot be used effectively when the number of participants exceeds thirty; method demonstration can be used for a relatively small audience, while lecture meetings can be used for large audiences.

2. Teaching Objective

Before starting the selection process for identifying suitable methods, the objectives of the whole process must be very clear in minds of the instructor as well as learners. In this regard, some specific queries need to be addressed: do you want to think about a change in thinking or knowledge or in attitude or feeling or in action or skill? If you want merely to inform or influence a large number of people slightly, you should use mass media. If you want to influence a relatively small number of people to make maximum improvements, resort to individual contact methods. If you want to change attitude or arrive at a consensus of opinion, arrange group discussion or work through village leaders. If you want to teach a skill, use the method demonstration.

3. Subject Matter

Where the new practice is simple or familiar (*i.e.*, similar to those already being followed) the news article, radio or circular letter will be effective, whereas complex or unfamiliar practices will require face-to-face contacts, written materials and audio-visual aids.

4. Stage of Development of Extension Organization

In the initial stages of extension, result demonstrations will be necessary to gain confidence of farmers. But if extension work is already well established and the farmers have confidence in extension services, result demonstrations may not be necessary and local illustrations of adoption by village leaders will suffice,

5. Availability of Communication Media

The availability of certain communication media (*i.e.* news papers, telephones, radio *etc.*) will also have a direct bearing on the extent to which these methods can he used.

6. Relative Cost of the Method

The relative cost of the method (*i.e.*, the amount spent on extension teaching in relation to the extent of practices changed) is also an important consideration in their selection and use.

7. Extension Professionals' Familiarity with Extension Methods

An extension professionals familiarity with, and skill in the use of, several extension methods will also influence his choice and use of the methods.

Considerations in the Selection of Extension Teaching Methods

The following guidelines are helpful in the selection of suitable extension teaching methods:

1. Educational Level of the Audience

- ☆ For illiterates - Personal visits.
- ☆ For educated - Written materials.

2. Size of the Audience

- ☆ For less than 30 - Lecture, Group discussion.
- ☆ For more than 30 - Mass methods.

3. **Teaching Objective**
   - ☆ To bring awareness - Mass methods.
   - ☆ To change attitude - Group discussion.
   - ☆ To impart skill - Demonstration.

4. **Subject Matter**
   - ☆ To prove value of a recommended practice - Result demonstration.
   - ☆ To teach a new skill, or an old one in an improved way - Method demonstration.
   - ☆ To disseminate simple technology - News article.
   - ☆ To teach a complex technology - Face-to-face contact with audio visual aids.

5. **Extension Organization's Credibility**
   - ☆ New organization, yet to gain confidence of people - Result demonstration.
   - ☆ Well established organization with proven success- Circular letter.

6. **Size of Extension Staff**
   - ☆ Few staff members - Group and mass contact methods.
   - ☆ Large number of staff - Individual contact methods.

7. **Availability of Media**
   - ☆ For creating awareness and reinforcement of ideas - Television, radio, News paper.

8. **Time of Dissemination**
   - ☆ Emergency for individual - Phone call.
   - ☆ Emergency for a group of people, or a large number of people - Radio, television, public address system.

Relative cost, familiarity of the extension professionals with teaching methods, the needs of the people, time period the extension programme has been going on in the area, availability of physical facilities, and weather conditions are some of the important factors to be considered while selecting extension teaching methods. Here, one thing is worth mentioning that apart from all afore-mentioned teaching methods for specific categories, there are certain methods which can be universally applied irrespective of any other distinctive feature/factor, *viz.* radio, television, films, *etc.*

ICT APPLICATIONS IN TOT

Information and communication technology in agriculture (ICT in agriculture), also known as e-agriculture, focuses on the enhancement of agricultural and rural development through improved information and communication processes. More specifically, e-agriculture involves the conceptualization, design, development, evaluation and application of innovative ways to use information and communication technologies (ICTs) in the rural domain, with a primary focus on agriculture. ICT includes devices, networks, mobiles, services and applications; these range from innovative Internet-era technologies and sensors to other pre-existing aids such as fixed telephones, televisions, radios and satellites. Provisions of standards, norms, methodologies, and tools as well as development of individual and institutional capacities, and policy support are all key components of e-agriculture.

Many ICT in agriculture or e-agriculture interventions have been developed and tested around the world to help agriculturists improve their livelihoods through increased agricultural productivity and income, or by reducing risks. Some useful resources for learning about e-agriculture in practice are the World Bank's e-sourcebook ICT in agriculture – connecting smallholder farmers to knowledge, networks and institutions (2011), ICT uses for inclusive value chains (2013), ICT uses for inclusive value chains (2013) and Success stories on information and communication technologies for agriculture and rural development have documented many cases of use of ICT in agriculture. Information and communication technologies (ICTs) comprise those networks, mobiles, devices, services, and applications that aid the processing, management, and exchange of data, information, or knowledge with a target audience. They include a broad range of converging technologies, including traditional telecommunications, television and video, radio, CD-ROMs, cell phones and smart devices, and several modern technologies such as computers and the internet, sensors, Geographic Information Systems, satellites, and the like. Essentially, the purpose of ICT is to transfer information from one point to another.

ICTs have largely revolutionised the way people, governments, and businesses, both large and small, function in the modern world. Close to 60 per cent of the global population has access to the internet, and mobile internet is now the most widely-used channel for internet access worldwide. The tremendous adoption of ICTs has made it possible to facilitate better communication and ensure the delivery of services and information to people who previously lacked access.

Empowering Smallholder in Agriculture

In developing countries, ICT in agriculture provides farmers with vital information pertaining to sowing, crop protection, and improving soil fertility that enables them to improve agricultural productivity. Weather-related advisories and alerts help them prepare for sporadic events such as floods, drought, or even pest and disease outbreaks, thus preventing significant crop loss. ICTs also provide them with a reliable channel to seek the best market price in the local markets and other daily updates for their produce to ensure they receive fair returns. The increasing penetration of budget-friendly mobile phones and the internet is an added advantage for farmers living in remote areas in several emerging nations. Access to inexpensive mobile devices has now made it easier for them to acquire additional information and services that enable informed decision-making.

Now, the use of ICT has ensured that with a touch of a few buttons, agriculturists can connect with the global network of farmers, agronomists, businesses, and other service providers to stay up-to-date on the latest crop cultivation practices. For policy makers, the advantage of adopting ICT in agriculture is that information sharing enables them to gain a better understanding of the situation at the ground level, which will contribute to the designing and implementation of agrarian and rural development policies that benefit the farmers. The ICT's extensive reach to even the remotest location can also help them address issues concerning gender bias, women empowerment, and other socio-economic concerns.

The Role of Agricultural Extension

According to IFPRI, agricultural extension (also known as agricultural advisory services) plays a crucial role in promoting agricultural productivity, increasing food security, improving rural livelihoods, and promoting agriculture as an engine of pro-poor economic growth.

Yet, the specific roles of agricultural extension services were enumerated here defined extension as aiming to,

- ✰ Improve the wellbeing of individuals and communities
- ✰ Change production systems so that they improve rural livelihoods and sustain the resource base
- ✰ Improve agriculture and the social, economic and political status of rural communities
- ✰ Improve the wellbeing of farm families
- ✰ Improve productivity and livelihoods for farmers
- ✰ Increase and improve farmers' incomes and productivity on a sustainable basis
- ✰ Enhance farmers' production
- ✰ Attain higher levels of efficiency in the farm enterprise
- ✰ Attain food security and improve rural livelihoods.

For a comprehensive study of the role and functions of agricultural extension, see an elaborate review here.

However, extension alone cannot lift millions out of poverty unless there is the right mix of policies, technologies, and market opportunities –laments Fancis (ibid).

ICTs in Extension Work

There has been a major shift in modernizing extension and advisory services and key studies were done by Bell (2011) and Vignare (2013). Mark Bell and Judith Payne (2011) provide an interesting mapping of ICT options to extension functions, see the table given on next page.

Commonly Applied Technologies in Extension Services

- ***Radio and Television*** - Radio and Television have played, and still continues to play, an important role in agricultural extension and in getting farmers in touch with key services. The Farm Radio international has played a key role in this regard and has documented many such use cases. Another example is the African Farm Radio Research Innitiative (AFRRI).
- ***Videos*** - Videos also continue to be useful in extension services, for example Access Agriculture offers an internet-based platform for agricultural research and development (R and D) staff and other stakeholders including farmer organisations. They have branded this service AgTube. Another example is Digital Green in India which trained rural communities to produce videos by farmers, of farmers, and for farmers to exchange best agricultural practices to boost farm productivity and improve nutrition
- ***Mobile phones*** - There are many examples of the use of mobile phones in agriculture, the e-agriculture platform constantly cover these examples. Mobile phones either use voice, text or photos within the farmer/ extensionists interaction. There is a need to formulate the right messages for and with farmers, addressing illiteracy and empowering farmers to use mobile phones can lead to increased adoption of new technologies and improved practices.

ICTs and Strengthening Linkage between Extension, Research and Farmers

ICTs have a possibility of strengthening the linkage between extension, research and farmers. This weak linkage between the tripartite parties can be strengthened by technology. The eAfghan Ag was developed with this objective in mind - eAfghan Ag "provides credible relevant information to those helping farmers in Afghanistan".

What is a Media Mix?

A media mix is the combination of communication channels your business can use to meet its marketing objectives. Typically, these include newspapers, radio, television, billboards, websites, email, direct mail, the Internet and social media, such as Face book or Twitter. Combining these channels in a media mix enables you to communicate in the most effective way with different types of customers and prospects at different stages of the purchase decision, according to Entrepreneur.

Right Message to the Right Audience

An effective media mix delivers the right marketing message to your customers and prospects at the lowest cost and with minimal waste. If you want to reach a consumer audience across the country, you might use a media mix that includes national newspapers, radio or television. If you wanted to reach a specific group

Information Communication Technology and Tools

Extension Function	*Radio*	*TV and Video*	*Cell Phones*	*Feature and Smart Devices*	*Computer and Internet*
Identifying farmers' problems and opportunities – What do they need and want?					
Diagnose problems	Some potential if dealing with general problems, or if capacity for interaction and expertise available	Visuals are very helpfulas "seeing is believing."Even better if combinedwith ways to receivefeedback.	Some potential if farmers can call or text in and sufficient expertise is available.	Additional potential to a simple cell phone as it enables web or App access to special diagnostic tools.	Good comprehensive diagnostic tools are available
Collect information	Some potential if capacity for interaction		Can use for data collection.	Good for data collection with GPS.	Some potential if internet available.
Promoting behavior change – What is practical and relevant to meet the needs?					
Raise aware of general opportunities or needs; convince farmers to try something new	Very good especially with persuasive programming	Visuals are usually very helpful as "seeing is believing"	Is an option if users are registered to receive such messages (SMS)	Is an option if users are registered to receive such messages (SMS, email)	Is an option if users are registered to receive such messages (email)
Provide specific information needed for change. What is involved? What are the benefits/ Demonstrate or train	Some potential – but limited information delivered. Can be enhanced with call in.	Good option as "seeing is believing	Potential if farmers can call or text in and sufficient expertise is available	Additional potential to a simple cell phone as it enables web access and plays videos.	Good option for intermediaries to seek information and videos.
Facilitate access to credit and inputs	Can be used to inform of available services, but one-way communication	Can be used to inform of available services, but one-way communication	Mobile banking and negotiate directly with the suppliers	Mobile banking and negotiate directly with the suppliers	Mobile banking and negotiate directly with the suppliers

Extension Function	*Radio*	*TV and Video*	*Cell Phones*	*Feature and Smart Devices*	*Computer and Internet*
Link farmers to markets	Good for providing general price reports		Access to price information (call in, subscription)	Can bring potential buyers and producers together; access price information	Can bring potential buyers and producers together; price info.
Collect feedback – How can each step be improved?					
Collect and respond to farmer feedback	Good if producers can call or text and sufficient expertise is available	Good if producers can call or text and sufficient expertise is available	Some potential if farmers can call or text in and sufficient expertise is available	Good option for intermediaries to seek information (if optimized for smart devices)	Good option for intermediaries to seek information
Assist with business planning	Some potential	Some potential		Simple farm management "Apps"; record keeping	farm management tools; record keeping

Source: Mark Bell and Judith Payne (2011).

of business decision-makers, such as technical directors, your mix might include specialist business magazines or exhibitions aimed at those directors. To reach a small number of key executives who influence a major purchasing decision, you might include personalized direct mail or an executive briefing session in your mix.

Aligning Media Mix with Buying Stages

Entrepreneur notes that the emphasis in the media mix changes at different stages in the buying cycle. When prospects are looking for information, they may read publications covering their interests, search websites, visit trade shows or check product review sites. So, it's important that you have information in the places they are likely to visit. The emphasis in your media mix would be on raising awareness through advertisements, press releases, product pages on your website, participation in trade shows or comments on social media.

Integrated Media Work Harder

The components of a media mix are more effective when they are integrated. The benefit of an integrated campaign is that the media mix is more effective when the components work together and communicate consistent messages each time, according to MMC Learning. In practical terms, that means using the same creative themes and marketing messages across all elements of your media mix. Prospects viewing an advertisement, website page, direct mail piece or product guide from an integrated campaign would receive consistent messages, with each element of the mix reinforcing the others.

Fine-Tuning the Mix

Analytical tools are available to assess your media mix and improve the results you achieve. These tools identify the strengths and weaknesses of your marketing programs and your media mix. By demonstrating how changes in the mix can affect results, the tools help you to reallocate your budgets and create a better mix to improve marketing performance.

Importance of Media Mix Strategies

- ☆ It is a powerful media to help advertise the goods and services of a particular business entity so that its sales revenue and profit can grow at a rapid rate. The importance of the media mix can be stated from the fact that if it is used in the right manner, the companies are able to deliver exactly the same kind of message that they want to.
- ☆ It prospects lower cost and minimum waste. it holds importance as it lets the company reach to even those parts of the locations which are quite separated from the rest of the world.
- ☆ No to reach for a particular ours which includes all kinds of decisions makers, the technical analyst then it is possible for the media mix to consist of business magazines and even a few of the exhibition which is entirely aimed at those directors.

- ☆ Also, a well-planned media mix helps in strengthening the media channel so that more and more people can get influenced. The media mix helps in coordinating creativity with the right kind of timing and an apt frequency.
- ☆ This creates a powerful impact on the demographic and encourages them to purchase the goods or the services that are being advertised. The media mix is also important because it does not remain fixed with time.
- ☆ It understands the need for continues changes that people are going through and hence is adjustable according to the demand of the time. Also for a media mix to be effective is essential that continuous research is done and progressive changes are eking made to keep in touch with the latest trends and create an impact on the masses.

Media Mix Marketing Strategy for the 21st Century

Marketing has undergone many changes over the past few decades. As the Internet has grown, the ability for marketers to increasingly capture and analyze patterns of customer behavior has brought new insights to the way people consume products and services both in-store and online. The availability of these new data points has changed the way businesses must both plan and execute their marketing strategies. The ability to dynamically deliver optimized cross-channel content to users based on a unified profile allows for new strategies to be used.

An Outdated Approach

A typical marketing cycle consists of two main phases: a planning phase and an execution phase. During the planning phase, the marketing team decides both the financial and creative components for the following period, which may be any length of time, such as year-long or quarter-long. The financial component consists of determining what resources the department will need to execute the marketing plan. Meanwhile, the creative component includes identifying the content, audience segments, channels, and timeframe for the plan. Finally, the execution phase involves implementing the marketing plan as close to schedule and budget as possible. This includes adjusting content, plans, and budgets as situations arise or markets change during the time period.

Media-Mix Planning

Media-mix planning helps to determine the right budget mix to maximize the effectiveness of a marketing program. There are two main approaches to media-mix marketing: top-down and bottom-up.

The top-down approach involves looking at high-level spend and historical revenue data to predict how your plan will perform moving forward. This type of approach is considered easier to execute than the bottoms-up approach, as it looks at higher-level, concrete data as a basis for its predictions.

The bottom-up approach is considered to be more per-formant, as it provides both high-level and more-detailed insights into strategy. However, it is more difficult to execute this approach based on its inputs. The bottom-up approach looks at more granular cookie-, user-, and creative-level data points for its predictions. These data points are difficult to capture and correlate, especially in offline scenarios. It is also difficult to connect economic data with the bottoms-up approach, making it challenging to accurately assess the potential ROI for your program.

Understanding the Advantage of an Ad-Stack

The ad-stack consists of a mixture of creative, designed to be executed across multiple channels and to capture important, real-time analytics on the behaviors of the people interacting with it. It allows a multitude of content to be managed in a single space from online e-commerce to mobile sites and apps, across your social outlets, and even through specialty devices such as in-store kiosks. The ad-stack allows these data points to be captured in real time, which enables faster response times when changing content within a marketing program. A faster reaction to changing markets equals greater conversions on campaigns that, historically, had the potential to waste large amounts of company resources and result in negative brand impact. If real-time markets need real-time analytics, then the real-time analyzation capabilities of the ad-stack allow media-mix planning to meet that need.

Factors Influencing in Selection of Extension Teaching Methods

A. Selection of Methods

1. *The Audience*

(a) **Individual and collective differences***:* People vary greatly in their knowledge, attitudes, skills, their position in the "diffusion process", and in the "adoption categories", their educational training, age, income level, social status, religious beliefs *etc.* Some are progressively seeking change, others are slow to change. Some are "eye-minded" while others are "ear-minded".

These individual and collective differences influence the teaching approach. For instance, people with little or no education and low incomes may respond to personal visits and result demonstrations. The better educated and the more progressive elements of the population usually respond well to methods like group meetings and discussions, exhibits and written materials. A man in "awareness stage" cannot straight away jump to "adoption stage" but can be gradually brought to the adoption stage by using suitable methods. For "late adopters" (conservatives), direct approach may not yield so good results as approaching through the "early adopters" and "informal leaders".

(b) **Size of Audience***:* It is also a factor influencing the choice of extension methods. For instance, group discussion cannot be used effectively when the number of participants exceeds thirty; method demonstration can be used for a relatively small audience, while lecture meetings can be used for large audiences.

2. ***The Teaching Objective (or nature of change aimed at)***

Do you want to bring about a change?

a. In thinking or knowledge?
b. In attitude or feeling?
c. In action or skill?
d. If you want merely to inform or influence a large number of people slightly, you should use mass media.
e. If you want to influence a relatively small number of people to make maximum improvements resort to individual contact methods.
f. If you want to change attitudes or arrive at a consensus of opinion, arrange group discussion or work through village leaders.
g. If you want to teach a skill, use the method demonstration.

3. ***The Subject Matter***

Where the new practice is simple or familiar (*i.e.*, similar to those already being followed) the news article, radio or circular letter will be effective, whereas complex or unfamiliar practices will require face-to-face contacts, written materials and audio-visual aids.

4. ***The Stage of Development of Extension Organisation***

In the initial stages of extension, result demonstrations will be necessary to gain confidence of farmers. But if extension work is already well established and the farmers have confidence in extension services, result demonstrations may not be necessary and local illustrations of adoption by village leaders will suffice.

5. ***Size of Extension Staff***

In relation to the size of extension clientele: The larger the number of extension workers, the greater is the scope for direct or personal contact method.

6. ***The Availability of Certain Communication Media***

Such as newspapers, telephones, radio, internet *etc.*, will also have a direct bearing on the extent to which these methods can be used.

7. ***The Relative Cost of the Method***

i.e., the amount expended on extension teaching in relation to the extent of practices changed) is also an important consideration in their selection and use.

8. ***An Extension Worker's Familiarity***

With and skill in the use of the several extension methods will also influence his choice and use of the methods.

B. Combination of Methods

Extension field studies conducted in U. S. A. over a long period of years show that people are influenced by extension education to make changes in behaviour in proportion to the number of different teaching methods with which they come in contact. The number of methods of exposure to extension information increases from 1 to 9, the number of farm families changing behaviour increases from 35 per cent to 98 per cent. Therefore, if widespread response is desired, people must be exposed to teaching effort in several different ways (*i.e.* repetition but in a variety of ways).

C. Using the Methods in Proper Sequence

To answer our teaching needs, our extension plans of work must include methods that,

- ✰ Enable our farmers to see, hear and do the thing to be learnt
- ✰ Enable us to reach large numbers of people and
- ✰ Create confidence – building situations.

Our completed plans should provide not only for doing each of these three things but must be so organized that the completed plan, as a unit, does all three of these things. For instance, a personal contact is made through an office call or farm visit. A leader is visited. A demonstration is established. A meeting is held to discuss the demonstration. The meeting is advertised by circular letters. A news story is written on the results of the demonstration as seen at the meeting. These happenings and results are broadcast over the radio.

AUDIO-VISUAL AIDS

Audio Visual Aids are also called instructional material. Audio literally means "hearing" and "visual" means that which is found by seeing. So all such aids, which endeavor to make the knowledge clear to us through our sense are called "Audio Visual Aids" or Instructional Material. All these learning material make the learning situations as real as possible and give us first-hand knowledge through the organs of hearing and seeing. Therefore, any device which can be used to make the learning experience more concrete and effective, more realistic and dynamic can be considered audio visual material.

We learn through our sense organs. Senses are the ways of knowledge. All the sense organs help us in understanding the environment. Most of the knowledge, which we acquire from the school, comes through our ears and eyes.

Meaning

Audio-visual aids are instructional devices which are used to communicate messages more effectively through sound and visuals.

Audio-visual aids help in stimulating the sensory organs like ears and eyes and facilitate quick comprehension of the message by the audience. These may be used for literate as well as for illiterate people.

What is Audio?

Audio means what we hear. The five senses audio, visual, touch, smell and taste play an important role in communicating message. Hearing plays an important role in receiving and sending a message effectively. The most basic form of communication is oral and face to face contact. Hearing plays an important role in oral-face to face communication. In recent days due to the invention of modern gadgets like radio, tape recorder, public address system telephones and mobile phones the type of communication is more of an indirect type as the individuals do not face each other. People in such situations communicate without coming into close proximity.

What are Audio Aids?

Audio Aids are the instructional devices through which the message can only be heard.

Or

An audio aid is an instructional device in which the message can be heard but not seen.

What is Visual?

A visual is what can be seen.

Visual helps one to communicate more effectively. Out of the five physical senses through which we learn, the eye is the **most helpful** in learning. Words are not enough for communicating an idea. The same word may even mean different things to different people. We speak different languages and so, many times communication becomes difficult.

What are Visual Aids?

Visual Aids are the instructional devices which help to visualize the message.

Or

A visual aid is an instructional or communicating device in which the message can be seen but not heard.

What is Audio-visual?

Audio-visual means the things which we hear as well see.

What are Audio-visual Aids?

Audio-visual aids or devices or technological media or learning devices are added devices that help the teacher to clarify, establish, co-relate and co-ordinate accurate concepts, interpretations and appreciations and enable the learner to make learning more concrete, effective, interesting, inspirational, meaningful and vivid.

Or in other words

Audio-visual aids are used to improve teaching, *i.e.* to increase the concreteness, clarity and effectiveness of the ideas and skills being transferred. They enable the audience to LOOK, LISTEN and LEARN (*by doing*); to learn faster, to learn more, to learn thoroughly and to remember longer.

According to an old Chinese proverb the importance of audio visual aids is indicated by the saying that *"if I hear I forget, if I see I remember, if I do I know."*

The audio-visual aids help in completing the triangular process of learning, **motivation, clarification-stimulation.**

The aim of teaching with technological media is clearing the channel between the learner and the things that are worth learning. The basic assumption underlying Audio-Visual Aids is that learning and clear understanding-stems from sense of experience. The teacher must 'show' as well as 'tell'.

Audio-Visual Aids provide significant gains in informational learning, retention, recall, thinking, reasoning, activity, interest, imagination, better assimilation, personal growth and development. The aids are the stimuli for learning **'why', 'how', 'when' and 'where'.** The 'hard to understand principles are usually made clear by the intelligent use of skillfully designed instructional aids. It must be remembered that Audio-visual aids can only **supplement** the teacher but can never **supplant** the teacher.

Definition

According to Audio-Visual Aids: The instructional devices through which the message can be heard and seen simultaneously are known as Audio-Visual Aids

According to Burton: Audio-Visual aids are those aids which help in completing the triangular process of learning *i.e.*, motivation, classification and stimulation

According to Edgar Dale: Audio-Visual are those devices by the use of which communication of ideas between persons and groups in various teaching and training situations is helped. These are also termed as multi-sensory materials

Good's Dictionary of Education: Audio-Visual aids are anything by means of which learning process may be encouraged or carried on through the sense of hearing or sense of sight

According to Kinder; S. James: Audio-Visual aids are any device which can be used to make the learning experience more concrete, more realistic and more dynamic

According to Mcknown and Roberts: Audio-Visual aids are supplementary devices by which the teacher; through the utilization of more than one sensory channels is able to clarify, establish and correlate concepts, interpretations and appreciations.

Audio-visual aids are also known as Teaching aids, Educational Communication Technology, Audio-Visual Media, Learning Resources and Instructional or Educational Media.

Objectives of Teaching Aids

1. To enhance teachers skills which help to make teaching-learning process effective
2. Make learners active in the classroom
3. Communicate them according to their capabilities
4. Develop lesson plan and build interest
5. To make students good observer
6. Develop easy and understandable learning material
7. Follow child cornered learning process
8. Involve intimation in objectives
9. To create interest in different groups
10. To make teaching process more effective

Types

It can be classified simply on the bases of sensory experience, because human beings derive their experiences mainly through direct sensory contact. Keeping this in view, it can be classified in to three main groups:

Classification of Audio-Visual Aids

1. **Audio aids:** The instructional device through which message can be only heard are known as audio aids

   Examples: Audio cassette recorder, radio, telephone, gramophone, language laboratory and linguaphone
2. **Visual aids:** The instructional device through which helps to visualise the message is known as visual aids
   a. **Projected visual aids**: examples - slides, over-head projector, power point slides
   b. **Non projected visual aids**: examples-Poster, charts, graphs, models, specimens, chalkboards, picture and photographs
   c. **Display type**: Visuals are those which are spread before the audience for viewing, who get the message by looking at them. Examples; Poster, models, exhibits and specimens

d. **Presentation type**: Visuals are those presented or projected before the audience for viewing but at the same time, one explains or presents the message of the visuals, so that the messages gets a meaningful understanding of them. Example; Slides, Over-head projectors, charts and power point slides

3. **Audio visual aids:** The instructional devices through which the message can be heard and seen simultaneously are known as audio aids

Projected Audio visual aids: examples; LCD projector, Film projector, TV, Computer, VCD player, Virtual Classroom, Multimedia Video and cinema *etc.*

Non project Audio visual aids: examples; Drama, Puppet show and street play

Advantages

1. Its helps to make learning process more effective and conceptual.
2. Its helps to grab the attention of students
3. It builds interest and motivation teaching students learning process
4. It enhance the energy level of teaching and students
5. It is even better for over burden classrooms
6. It provides students a realistic approach and experience

Disadvantages

1. Technical Problems
2. Students Distractions
3. Expensive
4. Time consuming
5. Need Space
6. Convenience

Characteristics

1. Relevancy
2. Useful and purposeful teaching
3. Accuracy
4. Interest
5. Minimize verbalism
6. Comprehensibility
7. Motivation
8. Realism

Who can take Benefit from Visual Aids

Though every children can take benefit while using visual aids for learning. But it is more helpful for those students:

- Having Language Disorder
- Autism Spectrum Disorder
- Down Syndrome
- Those who have Learning Disabilities
- Student who have English as a secondary Language
- Those having Oppositional Defiant Disorder
- Personality Development Delay
- Have the problem of Hearing Impairment
- Have the symptoms of Attention Deficit Hyperactivity Disorder ADHD

There are some features, which can be used to ascertain the extent to which audio visual aids are of great value in a learning procedure. And their values depend upon the extent to which they assist in achieving the objectives of instructions and maybe indicated briefly under the following heads:

1. Relevancy

The aid must be relevant to the topic which needs to be improved. Very important standard value is the extent to which any aid is directly related to the understanding of the subject knowledge. A visual aid could be accurate to the best details comprehensible and interesting and yet to be of little value because of its irrelevancy. Let us see a specimen; a teacher is trying to make clear the pattern of sodium atom with the help of chart. Suppose if the teacher uses accurate and beautiful chart on which there are so many other atomic structures drawn. Automatically it will lose much of the teaching value.

2. Accuracy

It is very important to make the audiovisual aids are exact and accurate but it is difficult to make them as accurate as desired and expected. The chart is always missing the third-dimensional attribute and must be viewed from a fixed standard place. The models overcome both these difficulties but are seldom as accurate in the details as images. The accuracy of understanding is often considering as important rather than the accuracy of the specimen. For example; if someone is teaching the principal parts of flowers by drawing an apple blossom. It is of some little importance whether the diagram resembles the apple blossom in every detail but the important thing will be to clarify the proper relation of such parts as petals, stamens, *etc.* for this reason a good diagram might be of even greater value than an accurate photograph. Hence it is the accuracy of the concept produced rather than the accuracy of details (exactness) in visual aids, moreover which is important. And on the other hand, a teacher uses a large model of a leaf plant in his classroom and then observes that the students when given the actual plant do not recognize it as being the same thing.

3. Interest

Audio-Visual utilization makes the learning procedure very easy and its absence makes it disgusting its presence makes the work a privilege and its absence makes it a punishment or lack of source of knowledge. The main job of the teacher is to create interest in the students for learning any subject taught. Almost all the academics should be taught with the help of audio-visual aids so many concepts in secondary level classes that can be made interesting with the help of chats, images, and films. For example atomic structure, patterns, modules, and digestive systems, mathematical figures, historical events geographical maps and graphs, *etc.*

4. Comprehensibility

In selecting and using visual aids we must try to choose those things as aids which relate the new experience with past experience and which are within the comprehension of the students who are to use it. For instance; lee's take a teacher finds that his class is experiencing substantial problem in locating and understanding the connection between the parts of a circulatory system. And the instructor then draws purely a diagrammatic draft on the black board very clearly. This clarified the difficulties and issues because it is simple and readily comprehended by the students.

5. Motivation

How to motivate students for learning language? Audio visual aids motivate and act as a catalyst for learners as they make the learning process in the classroom interesting especially in activity-oriented subjects for effective learning.

6. Realism

A visual aid must necessarily possess the quality and niche of developing real concepts of different areas of life. Instead, we may try to develop in our students the concept very similar to that of the movement of electrons in the atom. We the teachers, in this case, try our best to describe accurately with the help of diagrams and charts but the students are shown the actual film of the movements of electrons; an entirely very different, but effective concept will emerge.

Advantages of Audio-Visual Aids

1. **To Challenge the Attention of the Pupils:** The teacher who uses devices can usually see that the attention of the whole classroom is on the lesson and that should not be distracted. This is generally true when teachers are dealing with the lower classes. Devices should never be used by the teacher as mere attractions or entertainment. Exposure to visual or audio-visual material and nothing more is not an educative procedure.

2. **To Stimulate the Imagination and Develop the Mental Imagery:** Devices stimulate and increase the imagination of the pupils. Intellectual imagery can be used as a vehicle of thought and as a means of clarifying ideas and concepts. As imagination plays a vital role in any innovation and for any learning and almost all the innovations are in the form of imaginations and they evolved to be the theories and principles.
3. **To Facilitate the Understanding of the Pupils:** The most widely acknowledged use of aids, whether visual or audio-visual, its use in aiding understanding. As we learn everything and anything after understanding otherwise there can be any learning without understanding. Language learning can be acquired up by using models, filmstrips, movies, and pictorial material to supplement textbooks and printed materials.
4. **To Provide Incentive for Action:** The use of devices in education, such as pictures and objects will arouse emotion and incites the individual to action or learning. The teacher must select the right kind of device to excite the students to a worthwhile intellectual activity during the class. Asking the pupils to collect pictures representing water, air, land transportation which stimulates them to action or immediate learning.
5. **To Develop the Ability to Listen:** The ability to listen can be developed best through the use of audio-visual materials. It is also the responsibility of the schools colleges and other educational institutions, to provide training for our students to be good listeners first. And listening is the best and first way of learning in the procedure. And as we knew the best saying, „one who is a good listener can be a good speaker or orator too". And as we know Training in the art of listening is one of the aims of audio-visual education.

Importance of Audio-Visual Aids

Dale has listed the following importance of using audio-visual aids:

1. Reduce the time lag
2. Make learning permanent
3. Add interest and involvement
4. Stimulate self activity
5. Provide direct interest
6. Develop continuity of thought.
7. Develop meaningful vocabulary
8. Enlarge the range of possible experience
9. Teach efficiently
10. Add highly useful variety
11. Improve the effectiveness of other materials.
12. Multiply messages.

Further Hoban, Finn and Dale (1950) on the basis of a review of research studies, draw the following conclusions about the contribution of audio visual aids to teaching.

- ☆ They supply a basis for conceptual thinking and hence reduce meaningless word responses of learners
- ☆ They have a high degree of interest for learners
- ☆ They make learning permanent
- ☆ They offer a reality of experience which stimulates self-activity on the part of learners
- ☆ They develop a continuity of thought, which is especially true of motion pictures, television *etc.*
- ☆ They contribute to growth of meaning and hence to vocabulary development
- ☆ They provide experience not easily obtained through other materials and contribute to the efficiency, depth and variety of learning

Selection of Audio Visual Aids

1. Teaching objective
2. Nature of subject matter
3. Nature of audience
4. Size of audience
5 Availability of equipment's, materials and funds
6. Skill and experience of extension agent in preparation and use of audios visual aids

Combination of (Media mix) Teaching Methods

Use of Audio-Visual Aids

There are numerous audio-visual aids that you could use in your lessons. This unit will focus will on selected audio-visual aids that are normally available in your school. Some of these aids could be made by you. However, the focus of this unit will not be on how to make these aids but on their use.

Need to Use Audio-Visual Aids in Lessons

a. To maintain a high level of interest in the lesson
b. To get students to use the language, especially at the beginning stages
c. To promote greater student participation
d. They can be used at all levels of learning

Guidelines for Using AV Aid

1. Planning AV Aids

Audiovisual aids, including overhead transparencies, flip charts, slides, chalkboards (and occasionally film and video)—can supply added impact and clarity to your, presentation. You can use AV aids to visually reinforce the points made verbally in your presentation, to summarize the points you will make, as well as those you have already made, and to visually clarify important concepts and analogies. Audiovisual aids not only help your target group follow your presentation, but they provide support to the presenter by helping to keep, you on track.

When you plan for audiovisual aids, follow these simple guidelines:

1. DO use them to summarize or show the sequence of content.
2. DO use them to visually interpret statistics by preparing charts and graphs that illustrate what you will say.
3. DO use them to illustrate and reinforce your support statements.
4. DO use them to add visual clarity to your concepts and ideas.
5. DO use them to focus the attention of the target group on key points.
6. DON'T project copies of printed or written text. Instead, summarize the information and show only the key points on the visual aids.
7. DON'T put yourself in the role of aiding your visuals: A presentation is primarily an oral form of communication. If your only function is to read the information on your overheads or slides, the target group will become easily bored.
8. DON'T use copies of your transparencies as handouts. They reinforce what you are saying— they don't say it for you. If you want your target group to remember what you meant, you'll need to provide written text in addition to any key point summaries or charts that you need for your transparencies.
9. DON'T use charts, graphs, or tables that contain more information than you want to provide.

2. Choosing Appropriate Media

You may wish to combine different types of media in a single presentation. The media used most often are overhead projection transparencies, flip charts, slides, blackboards, and handouts. You may occasionally use videotapes and/or films. In the future, use the following guidelines to help you choose the media best suited to your presentation.

In selecting aids, take into account the following

- ✰ Practicability
- ✰ Attractiveness and interest; vividness

- Suitability
- Complexity
- Clarity
- Portability
- Serviceability
- Availability
- Location
- Preparation and presentation
- Time factor

Types of Media and their Uses

a) Overhead Transparencies

- Used for displaying charts and graphs.
- High visibility for large groups.
- Showing summaries, sequences, relationships, and process steps.

b) Flip Charts

- Used for small groups (under 20).
- Ad lib charting and graphing.
- Showing summaries and sequences.
- Listing, recording, outlining.
- Developing charts.
- Instructions.
- Reinforcing group contributions.

c) Blackboards/Whiteboards

- Spontaneous presentations.
- Listing/recording.
- When no other media is available.

d) Slides

- Displaying high quality artwork.
- Displaying a series of photos.
- Demonstrating- a "hands-on" process.
- Creating a mood, or inspiring.

e) Videotapes and Films

- ☆ Demonstrating a process.
- ☆ Creating a mood, or inspiring.
- ☆ Presenting
- ☆ Testimonials.
- ☆ Supplementing presentation ideas.

f) Written Handouts

- ☆ Providing background information.
- ☆ "Sending home" your ideas.
- ☆ Reminder of follow-up activities.
- ☆ Providing a record of presentation.
- ☆ Clarifying abstract or complex topics.

Over Head Transparencies

Overhead transparencies are inexpensive, easily prepared, and highly effective visual aids for target groups of five or more people. With smaller target groups, choose overhead transparencies as your primary A/V medium if the target group prefers a somewhat formal mode of communication. With small groups, overhead projection may tend to distance you from the target group and limit personal interaction. If you are presenting to a relatively informal group of less than five people, you may wish to use transparencies only to display charts and graphs.

a) Visibility

Lettering should be neat and at least a quarter inch high. You should be able to read the original at a distance of seven feet. Computers make very effective, highly visible transparencies. If a computer is not available, use a typewriter equipped with a special "Orator" typeface, or a special lettering system designed for overhead transparencies, or "rub-off' lettering (available at any art supply store) to provide readable lettering. If you must hand-print, be sure that your printing is neat and even. Use all capital letters, and write with a black, felt-tipped pen.

b) Clarity

Everything displayed on an overhead transparency should be instantly recognized by the target group. Label all elements of your charts and graphs. Use color or shading Ito emphasize and clarify key elements. Be sure to title every transparency.

c) Simplicity

Try to limit each original to one point or comparison. Use a maximum of six or seven words per line. No more than six lines should be included in a single transparency. Use a horizontal format whenever possible so that the projected image will fit the shape of most standard screens.

Flip Charts and Boards

Flip charts are often the ideal visual aid for small target groups. You can prepare your chart pages in advance for maximum clarity and impact. The flip chart can also be used to "ad lib," or give the appearance of spontaneity. Prepared flip charts can be used to show pre-drawn charts, graphs, and diagrams, or you can draw them "on the spot."

Chalkboards may be used in the same ways as flip charts. But the space they provide is more limited and the medium is less permanent. If possible, draw important charts on chalkboards before the presentation and mask them with paper. Be sure to consider what information you will be able to erase during the presentation and what you will save for reference at a later point.

The guidelines for preparing effective flip charts and boards include the following:

a. Always title each page (or diagram).
b. Print clearly. For maximum visibility, letters should be at least one inch high for each ten feet of distance from viewers.
c. Use pictures, diagrams, and symbols whenever possible.
d. Skip a sheet between prepared flip pages to avoid "bleed through."
e. Leave room for "ad lib" flip pages (plan when you'll ad lib).
f. If you plan to ad lib, make sure you have at least two markers with you—one will always run out of ink.

Slides, Video and Films

Slides, videotapes, and films are included in the same category because they all require professional preparation. These media can have a very dramatic effect if properly selected. However, unless you are using media that is available through the company library or training center (or your public library), all three of these media types are expensive. Slides may occasionally be in your budget, but very few people have the budget to produce quality videotapes or, even more expensive, films to supplement a presentation. If you can afford to use slides, video, or film in your presentation, be sure the media you select reinforce and enhance what you have to say, and that the media do not contradict your presentation points. The guidelines for using slides, tapes, and films include the following:

1. Check the room layout and locate light switches and dimmers. Films and slides must be shown in a darkened room.
2. Make sure the media are up to date, unless you intend to contrast old and new.

3. Make sure that the images (and sound) are clear—free of static and other interference.
4. Know how to use the equipment—or enlist the aid of a projection assistant for your rehearsal and delivery.
5. Make sure that the media are compelling and interesting to the target group.
6. If you are using slides, check to make sure that they are all in order and facing in the right direction (not upside down).
7. Check your equipment. Make sure it can project the media you want to use, in the manner that you want to use it. Verify that it is in good working condition and that all cables, wires, plugs, *etc.*, are in place and ready to go.
8. Load and advance the tape or film to the starting point before the meeting begins. Check sound levels and image clarity.
9. Plan what you will do if the equipment breaks down.

Written Handouts

When providing written text, remember that you won't be there to explain or remind the target group of what you mean. Written communications should be as brief as possible, but they must also be meaningful to the reader, whether they are read a week before your presentation or a year after. They should always be:

a) Clear

Written in language that the target group can easily understand and follow.

b) Relevant

Do not simply provide data. Explain why the information should be read and what it means. Explanations should not repeat your entire presentation, but should capsulate it.

c) Specific

Do not give your target group a mass of reading material that they do not need. Make sure that all documentation is absolutely necessary.

COMMUNICATION

According to Leagans (1961), Communication is the process by which two or more people exchange ideas, facts, feelings or impressions in ways that each gains a common understanding of the meaning, intent and use of messages. The term 'communication' stems from the Latin word 'communis'-meaning' 'common'. Communication, then, is a conscious attempt to share information, ideas, attitudes and the like with others. In essence it is the act of getting a **sender** and a **receiver** tuned together for a particular message, or a series of message. "Communication means the movement of knowledge to people in such ways that they act on that knowledge to achieve some useful result. This result may range all the way from a small improvement in doing some productive task, to the generation of a sense of national unity and strength in a country. Communication in this sense includes the whole learning process. It encompasses the teacher- the message or material to be taught- the means or media used to carry the message -the treatment given by those media- the learning achieved by the audience or student- and the actions by which the learning is put into practice" (Winfield (1957).

Good communication does not consist merely of giving orders, but of creating understanding. It does not consist merely of imparting knowledge, but of helping people gain a clear view of the meaning of knowledge. Most of the progress in the future will stem from better technology and greater skill in communicating it to others. Too many people saying the wrong things at the wrong time, in the wrong way, to the wrong people, slows down progress. What is needed is more people saying the right things at the right time, in the right way, to the right people. This is the formula for good communication.

Importance of Communication

1. Communication establishes a favourable climate in which development can take place.
2. Communication has multiplier effect
3. Communication varies the aspirations of the people
4. Communication is essential for all human activities
5. Communication is essential for good leadership

Elements of Communication

1. The Communicator

This is the person who starts the process of communication in operation. He is the source or originator of messages. He is the sender of messages. He is the first to give expression to message intended to reach an audience in a manner that results in correct interpretation and desirable response. When a communicator does not hold the confidence of his audience, communication as conceived will not take place.

The following are the good qualities of a good communicator

a. The Communicator Knows

- ☆ the specifically defined objectives.
- ☆ the needs, interests, abilities *etc.* of the audience.
- ☆ the content, validity, usefulness and importance of the message
- ☆ the channels that will reach the audience and their usefulness
- ☆ the tactics of organising and treating the message
- ☆ his/her professional abilities and limitations.

b. *The Communicator is Interested in*

- the intended audience and their welfare
- the specific message and its effects in helping the people
- the entirety of communication process
- the proper use and limitation of communication channels.

c. *The Communicator Prepares*

- a specific teaching plan for communication
- materials and equipments needed for communication
- a plan for evaluation of results.

d. *The Communicator has skill in*

- selecting messages
- treating messages
- expressing messages in verbal and written forms
- the selection and use of channels
- understanding the audience
- collecting evidence of results.

In contrast to the above, the following are the **qualities of a poor communicator**

- Fail to have ideas to present that are really useful to the audience.
- Fail to give the complete story and show its relationship to people's problems
- Forget that time and energy are needed to absorb the material presented.
- Feel they are always clearly understood.
- Refuse to adjust to 'closed' minds.
- Talk while others are not listening.
- Get far too ahead of audience understanding.
- Fail to recognize others' view - point and develop presentation accordingly
- Fail to recognize that communication is a two-way process.
- Let's their own biases over-influence the presentation.
- Fail to see that everyone understands questions brought up for discussion
- Fail to provide a permissive atmosphere.
- Disregard the values, customs, prejudices and habits of people with whom they attempt to communicate.
- Fail to start where people are, with respect to knowledge, skill, interest and need.

2. Message or Content

A message is the information communicator wishes his audience to receive, understand, accept and act upon. Messages, for example, may consist of statements of scientific facts about agriculture, sanitation or nutrition; description of action being taken by individuals, groups or committees; reasons why certain kinds of action should be taken; or steps necessary in taking given kinds of action.

Features of a Good Message

A good message must be:

- ☆ in line with the objectives to be attained.
- ☆ clearly understandable by the audience
- ☆ in line with the mental, social, economic and physical capabilities of the audience
- ☆ Significant-economically, socially or aesthetically to the needs, interests and values of the audience.
- ☆ specific in terms of audience and locale
- ☆ accurate so as to be scientifically sound, factual and current in nature.
- ☆ appropriate to the channel selected.
- ☆ appealing and attractive to the audience signifying the utility values and immediacy of use.
- ☆ adequate in such a way to have effective proportionate combination of principle and practice manageable so as to be handled by the communicator within the resources availability.

In contrast, poor communicators often commit the following which mar the effectiveness of message sending

- ☆ Fail to clearly separate the key message from the supporting content or subject-matter.
- ☆ Fail to prepare and organise their message properly.
- ☆ Use inaccurate of 'fuzzy' symbols-words, visuals, or real objects-to represent the message.
- ☆ Fail to select messages that are sharply in line with the felt needs of the audience.
- ☆ Fail to present the message objectively-present the material, often biased, to support only one side of the proposition.
- ☆ Fail to view the message from the standpoint of the audience.
- ☆ Fail to time the message properly within a presentation or within a total programme.

Selecting and 'packaging' messages so they have a good chance of being understood, accepted and acted upon when received is a crucial step in the communication process. It is one of the six keys to success in efforts to influence people to change their ways of thinking and of doing that lead to social and economic improvement.

3. Channels of Communication

The sender and the receiver of messages must be connected or 'tuned' with each other. For this purpose, channels of communication are necessary. They are the physical bridges between the sender and the receiver of messages-the avenues between a communicator and an audience on which messages travel to and fro. They are the transmission lines used for carrying messages to their destination. Thus, the channels serve as essential tools of the communicator.

A channel may be anything used by a sender of messages to connect him with intended receivers. The crucial point is that he must get in contact with his audience. The message must get through.

However, channels are no good without careful direction or use in the right way, at the right time, to do the right job, for the right purpose with the right audience, all in relation to the right message. So, proper selection and use of channels constitutes a third determinant of successful communication. Without proper use of channels, messages, no matter how important, will not get through to the intended audience. Many obstructions can enter channels. These are often referred to as **'noise'** - that is, some obstruction that prevents the message from being heard by or carried over clearly to the audience. 'Noise' emerges from a wide range of sources and causes.

Sources and Causes for Noise

- ☆ Failure of a channel to reach the intended audience. Usually, no one channel will reach an entire audience. Some examples: Meetings-all people cannot or may not attend. Radio-all people do not have access to a receiving set or may not be tuned in if they did. Written material-many people cannot read, and others may not.
- ☆ Failure on the part of a communicator to handle channels skillfully. If a meeting, tour, radio programme or any other channel is not used according to good procedure and technique, its potential for carrying a message is dissipated.
- ☆ Failure to select channels appropriates to the objective of a communicator. All channels are not equally useful in attaining a specific objective.
- ☆ Failure to use channels in accordance with the abilities of the audience. Written materials, for example, cannot serve as useful channels for communicating information to people who are unable to read or to understand the level of complexity or abstraction of the message.

- ☆ Failure to avoid physical distraction. When using the channel of meetings, for example, distractions including people moving in and out, loud noises in or out of the group, heat, lighting, crowded condition and many other forms of distraction often obstruct successful message sending.
- ☆ Failure of an audience to listen or look carefully. The only messages that get through to an audience are those which are heard, seen or experienced. An unfortunate tendency of people is not to give undivided attention to the communicator. This is a powerful obstruction that prevents messages from reaching desired destination.
- ☆ Use of too many channels in a series. An important principle of communication is that the more channels used in a series the less chance a communicator has for getting his message through to the intended audience. In this context, the following principles are to be borne in mind: (1) The more steps by which the communicator is removed from his intended receiver, the greater are his chances of losing the proper message. (2) When lines of communication get too long for assured communication they can be improved in two primary ways: (a) by using additional channels in parallel, and (b) by eliminating some of the channels in the series.

4. Treatment of Messages

Treatment has to do with the way a message is handled to get the information across to an audience. It relates to the technique, or details of procedure, or manner of performance, essential to expertness in presenting messages. Hence, treatment deals with the design of methods for presenting messages. Designing the methods for treating messages does not relate to formulation of the message or to the selection of channels, but to the technique employed for presentation within the situation provided by a message and a channel.

The purpose of treatment is to make the message clear, understandable and realistic to the audience. Designing treatment usually requires original thinking; deep insight into the principles of human behaviour and skill in creating and using refined techniques of message presentation.

5. The Audience

An audience is the intended receiver of message. It is the consumer of messages. It is the intended respondent in message sending, and is assumed to be in a position to gain economically, socially or in other ways by responding to the message in particular ways. In good communication the communicator already identifies the audience aims.

The importance of clearly identifying an audience cannot be over-stressed. Likewise, the more a communicator knows about his audience and can pin-point its characteristics the more likely he is to make an impact.

The following are some of the issues to clarify the nature of audience:

- Communication channels established by the social organisation.
- The system of values held by the audience-what they think is important.
- Forces influencing group conformity-custom, tradition *etc.*
- Individual personality factors-susceptibility to change *etc.*
- Native and acquired abilities.
- Educational, economic and social levels.
- Pressure of occupational responsibility-how busy or concerned they are.

It is useful to a communicator to understand these and other traits of an audience in making his plan for communication.

6. Audience Response

This is the terminating element in communication applied to rural development programmes. Response by an audience to messages received is in the form of some kind of action of some degree, mentally or physically. Action, therefore, should be viewed as a product, not as a process; it should be dealt with as an end, not as a means. Consequently, the five elements hither to analysed *viz.,* communicator, message, channel, treatment and audience are intended to be viewed as an organised scheme (means) for attaining the desired action (end) on the part of an intended audience.

The number of possible kinds and degrees of response to messages received are almost infinite. The following gives an idea of possible variety in response that may result when a useful message is received by the intended audience:

1. **Understanding Vs knowledge**. Knowledge of facts alone does not constitute understanding. It is only the first step. Understanding is attained only when one is able to attach meaning to facts, see the relationship of facts to each other and to the whole of a proposition and the relationship of the total body of facts to the problem under consideration.
2. **Acceptance Vs rejection**. A free, alert and thinking human mind requires that understanding precede acceptance of facts and propositions. In turn, it insists on mental acceptance before resorting to action.
3. **Remembering vs. forgetting**. When opportunity for action is not immediately available or action is delayed, the factor of forgetting what was learned influences the kind and extent of action taken at any point of time in the future.
4. **Mental Vs physical action.** Changes in the mind of man, must always precede changes in the actions of his hands. In short, man's mind controls his overt behaviour. Consequently, a message suggesting physical action could receive all the mental action required, except the final decision to act. This is sometimes referred to as 'lip service'.

5. **Right vs. wrong.** The intent of a communication is to promote desirable action by an audience as determined by the communicator and expressed in his objectives. Consequently, resulting action in line with the intended objectives is assumed to be 'right' action.

Communication Models

In social science research, a model is a tentative description of what a social process, say the communication process or a system might be like. It is a tool of explanation and analyses, very often in a diagrammatic form, to show how the various elements of a situation being studied relate to each other. Models are not statements of reality. Only after much further research and testing would the model is considered viable. It could then be developed into a theory. The term model can also refer to a particular process or object, which is used as a point of reference, when an attempt to explain the unknown is being made. It comprises involving an analogy to throw up the similarities between the phenomena to be explained and one, which is well known, *i.e.* the model.

The advantage of models in communication research is that it allows the researcher to account for different variables in different communication situations. Models only represent systems or processes. Since they are not real, they are just symbolic ways of looking at systems to help us to think about them more lucidly. Again since models do not show every part of a system, they are usually incomplete in that sense. Even those that are shown are represented only in enough detail to help us look at the processes or features in which we are interested.

1. Aristotle's Model of Communication (Devito, 1978)

The earliest model of communication was the symmetrical and simple model developed by the great Greek philosopher Aristotle some 2000 years before. Aristotle in his model includes the five essential elements of communication, *i.e.*, the speaker, the speech or message, the audience, the occasion, and the effect. In his rhetoric, Aristotle advises the speaker on constructing a speech for different audiences on different occasions for different effects. This model is most applicable to public speaking.

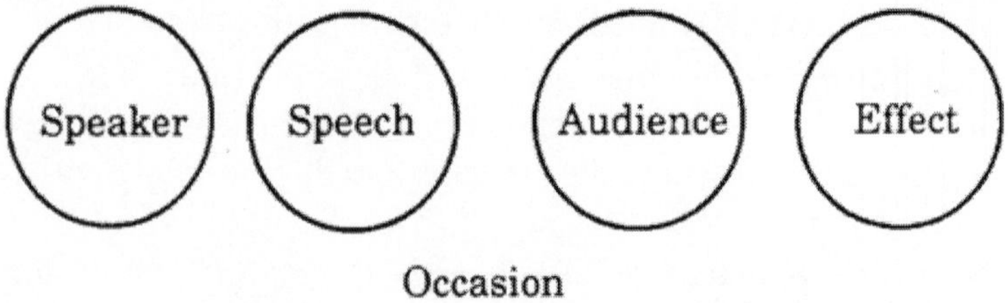

Aristotle's Model of Communication.

2. Lasswell's Model (1948)

Lasswell has given us another simple model. His model belongs specifically to the area of mass communication. He argued that to understand the process of mass communication one needs to study each of the stages in his mode. "Who says what, in which channel, to whom, and, with what effect."

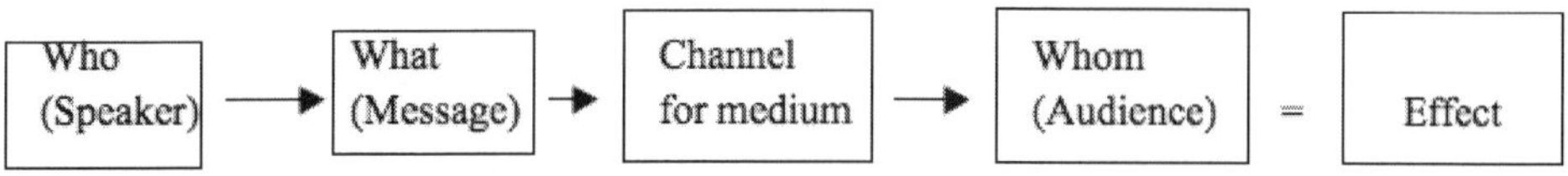

Lasswell's Model.
(*Source*: Public Opinion and Propaganda by Harold Lasswell, 1948).

This is the verbal version of Shannon and Weaver's original model. It is linear. It sees communication as the transmission of message. It raises the issue of the effect rather than meaning. Effect implies an observable and measurable change in the receiver that is caused by identifiable elements in the process. A change in one of these elements will change the effect. We can change the encoder. We can change the message. We can change the channel, and each one of these changes would produce the appropriate change in the effect. Most mass communication research has implicitly followed this model.

3. Lasswell's Model - Comments

Until the 1960s Lasswell's four questions (of who says what, by what channel, to whom and with what effect) dominated studies of the mass media in France. Not only his exemplary expression defines the different research areas for communication investigations, but also seemed to prescribe the appropriate concepts and methodological orientation to be followed. Thus, Lasswell's paradigm served the entire scientific community of communication scholars.

This Lasswell model was represented by Michael Buhler.

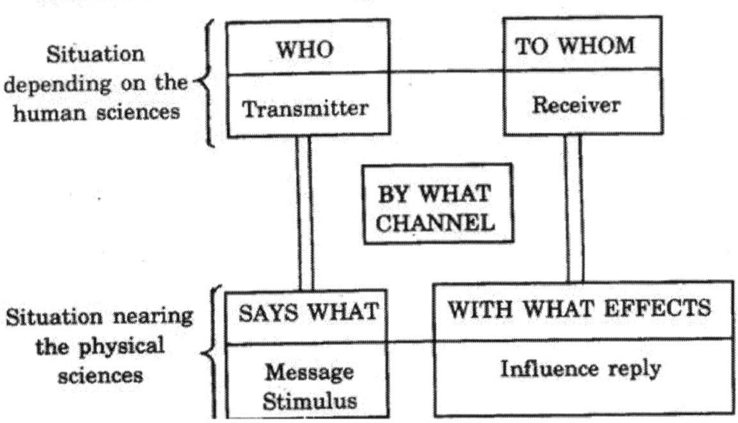

Lasswell's Communication Model.

4. Katz and Lazarfeld's Model (1955)

The model of 1955 was based on earlier research in which they found that information presented on the mass media did not have the reach and impact upon the receivers as previous views of communication seemed to suggest it would. Specifically their research indicated that political radio and print messages seemed to have a negligible effect on individuals voting decisions. Their research also indicated that some people were consistently more influential than others, leading them to conclude that ideas often seemed to flow from radio to print to opinion leaders and from them to the less active sections of the population - in a two-step flow.

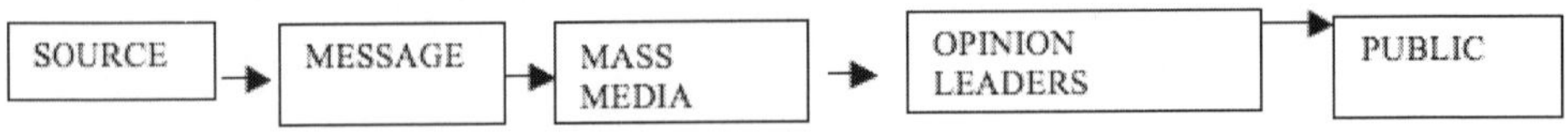

Katz and Lazarfeld's Model.
(*Source*: Personal influence by Eliha Katz and Paul Lazarfeld, 1955).

5. David Berlo's Model (1960)

In the last twenty-five years, a number of models of communication have been developed by communicologists, each one expanding the earlier presentation. David Berlo's model is one among them, which has been profusely quoted and frequently mentioned in discussions. In his book, "The Process of Communication" written in 1960, he presented this model which has a close similarity to the earlier Aristotelian model, including the traditional elements of source, message, channel and receiver. Berlo's model attempts to explain the various components in the communication process. For each of these basic components, controlling factors were listed.

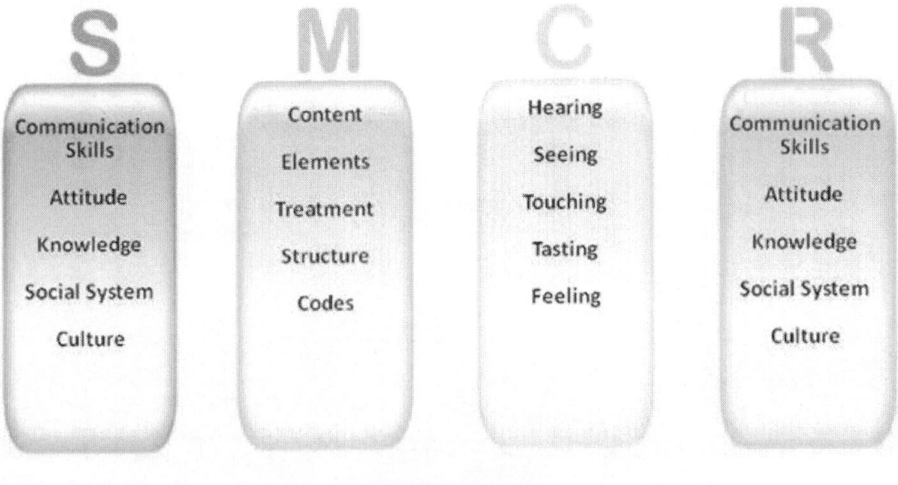

David Berlo's Model.
(*Source*: "The process of communication - An Introduction to Theory and Practice" by David Berlo, 1960).

His model reinforced a shift away from views of communication that emphasized the transmission of information to perspectives that focused on the interpretation of information. Berlo writes, "People can have similar meanings only to the extent that they have had similar experiences". Berlo also felt that human communication always had a purpose "our basic purpose in communication is to become an affecting agent, to affect others, our physical environment, and ourselves. We communicate to influence to affect with intent".

Feed Back in Communication

Feedback as a concept was developed in the 1940s and 1950's, when scientists began looking at the world in terms of systems model. They were interested in the nature of the systems, the applicability of system model to both the physical and social world. Feedback is thus defined as information regarding actual performance or the result of the activities of a system. Not all information is feedback; only information which is used to control the future functioning of the system is considered feedback.

Importance of Feedback

1. Feedback is a basic component of self-regulating system
2. Feedback is information about the output of a system which controls the system input or transformation processes
3. Feedback is any information about the system functioning which has the potential of being used to change the operation of the system.
4. Viewing organizations as open systems, feedback is a necessary component, enabling the correction of errors, the adaptation to environmental change and learning.
5. Since in social systems such as work organizations, feedback does not automatically create change in the system operation, the process of obtaining, interpreting and using feedback information is important.
6. Since organizations often ignore feedback or do not make an effort to use feedback effectively, organizational development activities serve an important function of facilitating feedback processes, thus helping organizations to correct errors, adapt, learn and grow.

Forms of Communication

The oral and written communication may take a number of shapes or forms depending upon the situation. The following list presents some possible and commonly used ones in communication in different directions.

Forms of Communication

	Oral		*Written*
	1. Downward Communication		
1.	Personal instructions	1.	Orders and instructions
2.	Lectures, conferences and meetings	2.	Letters, memos, circulars
3.	Interviews, counseling	3.	Bulletin and notice boards
4.	Social and cultural affairs	4.	Posters
5.	Grapevine; rumors	5.	Hand books and manuals
		6.	Annual reports
		7.	House organs
		8.	Union publications
	2. Upward Communication		
1.	Face-to-face conversation	1.	Reports
2.	Interviews	2.	Personal letters
3.	Meetings and conferences	3.	Grievances
4.	Social and cultural affairs	4.	Suggestions system
5.	Grapevine, rumors	5.	Complaints system
6.	Union channels	6.	Attitude and moral surveys
7.	Telephone and other devices	7.	Union publications
	3. Horizontal Communication		
1.	Lectures, conferences, meetings	1.	Letters, memos, reports
2.	Telephone and other devices	2.	House organ
3.	Social and cultural affairs	3.	Handbooks and manuals
4.	Grapevine, rumors	4.	Annual reports
5.	Grapevine, rumors	5.	Union publications

Of these, it is not necessary that all of these would be used in a particular organisation, but it will be determined by the organisational needs, its size and resources, management attitude towards these forms, *etc.*

Direction of Communication

Communication requires two parties, the sender and the receiver, their juxtaposition vis-à-vis each other determining in the organisation context, the particular dimension of the communication flow. Thus, the communication flow within the organisation may be inter-scalar when two parties are at the different managerial levels or inter-scalar when they are at the same hierarchical level - this is also known as horizontal communication. The inter-scalar communication may again be classified as downward, when the message flows from the higher to lower level, or upward when it is the other way round.

i) Downward Communication

Downward Communication within the organization flows from a superior either in the same line of command or in the different one It stands out as a great force for controlling influencing and indicating activities of organization members. Communication in this category includes (1) orders and instructions about job, (2) directions about understanding of job and its relationships with other jobs, (3) organisational policies and procedures, (4) feedback of subordinates' performance, (5) reprimands, criticisms, *etc.*, and (6) questions inviting upward communication.

ii) Upward Communication

Upward communication flows from a subordinate position to a superior position. It includes information about (i) subordinates' work performance (ii) problems relating to work, (iii) performance appraisal of their subordinates, (iv)feedback of understanding of orders, instructions, *etc.* (v) clarifications of orders, *etc.* (vi) opinions, attitude, feeling, *etc.* (vii) procedures, methods, practices followed in doing the work, (viii) criticisms, (ix) new ideas and suggestions, and (x) personal and family problems, Upward communication is more susceptible to various obstructions and bottlenecks discussed later on, because of its special nature. Managers, often times, fail to realize that upward communication cannot be taken for granted, as is the case with downward communication. This is so because, unlike downward communication, upward communication is devoid of any support of managerial hierarchy.

iii) Horizontal Communication

It is the flow of information between persons of the same hierarchical level. Formal organisation provides for horizontal communication by means of right of persons at any level to consult or work with others at the same level. Communication among peers, in addition to providing task coordination, also furnishes emotional and social support to the individual. Horizontal communication is impeded in the organization that overstresses functional departmentalization. The creation of functional departments of units creates problems of coordination and communication between members of such units.

Barriers to Communication

In between the communicator and the receiver, certain barriers considerably affect the quality of information transmission. Some of the major barriers are discussed below.

1) Expression
2) Interpretation
3) Response

These are the crucial points in communication. If the expression is not clear, the interpretation will be inaccurate and the response improper, thus one's effort to communicate will not succeed. In other words, if the source does not have adequate or clear information if the message is not encoded fully, accurately, effectively in transmittable signs; if these are not transmitted fast enough and accurately enough, despite interference and competition, to the desired receiver; if the message is not decoded in a pattern that corresponds to the encoding; and finally if the destination is unable to handle the decoded message so as to produce the desired response, then, obviously, the system is working at less than top efficiency.

Barriers to Communication in an Organization

- Lack of communication policy
- Past attitude and lapses on the part of the management to keep promises. This gives rise to credibility gap.
- Legal limitations
- Unfavourable climate
- Poor listening
- Poor strategy of communication
- Allowing the 'grape vine' to go wild
- Over confidence in one's ability to know what others are thinking
- Underestimating the intelligence and overestimating the fund of knowledge of the listeners
- Lack of clarity
- Too many items in the same message
- Too many links in the circulation chain resulting in loss of details and in distortion.
- Lack of motivation
- Failure to seek an instant response and lack of follow up
- Neglect in the use or misuse of available media.

AGRICULTURAL JOURNALISM

Definition and Meaning

Journalism is that part of social activity, which is concerned, with the dissemination of news and views about the society. Modern journalism feeds five departments of mass communication.

1. Newspaper and periodicals,
2. Radio,
3. Television,
4. Films and
5. Advertising.

Journalism is the systematic and reliable dissemination of public information, public opinion and public entertainment by modern mass media of communication.

In modern societies, journalism has become the media of mass education providing supplementary education to students at all stages and to the general masses - educated and uneducated.

All activities concerned with the communication of mass media is not journalism but the part of that activity involving writing, preparation and production of the communication messages is journalism. Thus, essentially journalists are writers, authors, reporters, correspondents, editors, sub-editors, interviewers, storywriters, scriptwriters, scenario editors and allied specialists.

Principles of Modern Journalism

The main function of journalism is to bring to light to the people what its members feel and think journalist are therefore supposed to possess a good deal of intelligence, knowledge and experience as well as natural and trained powers of observation and reasoning. A journalist has the same obligations as a teacher or interpreter as an educationist has. A journalist who misuses his power for any selfish or unworthy purpose is to a high trust. While the freedom of the press needs to be guarded as a vital right of humankind, it has to suffer certain explicit restrictions to subserve social needs. Freedom of the press does not imply promotion of any private interest of the journalist contrary to the general welfare. That is the way the so-called news communications from sundry sources are not published by sensible journalists without substantiating their value as news both in form and substance.

In addition, partisanship in editorial comments, which is knowingly based on a lie, does violence to the best spirit of journalism.

Rules to be followed by a Journalist

1. Study the field and have an estimate of what type of articles is published in newspapers, magazines, books *etc.*
2. Look ideas
3. Develop the idea
4. Keep an unbiased mind
5. Write clearly and meaningfully
6. Remember the people to whom the articles are written
7. Write with a purpose
8. Write on timely subjects
9. Keep in touch with editors
10. Watch out for important events/occasions
11. Establish credibility, politeness and do not assume familiarity

12. Understand the communication process and the elements of communication.

Agricultural Journalism

Agricultural journalism is a specialized branch of journalism which deals with the techniques of receiving, writing, editing and reporting from information through the media like newspapers, periodicals, radio, TV, advertising *etc.* and the management processes connected with such production. It is the timely reporting and editing with words and photography of agricultural news and information for newspaper, magazine, radio and television.

Importance

The mass media are capable of reaching vast widespread audiences, thanks to fast moving newspapers/radio broadcasts, TV telecast and the celluloid films. A single broadcasting network today can reach millions of people at the same time. The world stands of the threshold of new communication systems which enable large number of citizens to regularly and effectively interact with each other. To make full use of the interactive information systems made possible by the computer technology, citizens can remain so well informed that they will be able to perform their duty adequately and efficiently and accelerate the process of development in different social fields.

Modern communication systems reflect the philosophy and achievements of society in all spheres by fast flow and pave the way for the homogeneity of culture-not only within its geographical unit but also beyond. Journalism is the publication of news and views on various aspects of human activities in newspapers and periodicals. Ina broader sense, the functions of journalism are to convey national policies to the public and to keep the governments at local state and central levels, informed of public needs. It also brings to the notice of the government public reaction to government policies and decisions. In addition, it keeps the public and the government informed of events and happenings at home and abroad.

On one hand, it performs the utilitarian purpose of information and on the other; it provides entertainment to its readers by publishing short stories, poems, sports, cinema, features, *etc.* A clever journalism by his penmanship creates interest in day-to-day political happenings. He presents the drop events and incidents in an artistic way, which provides the reader information.

In modern times, the horizon of journalism has widened and it has transcended the limits of mere reporting of political as well as entertainment and economic news. As it is a vehicle of mass-communications, it is performing the function of social intercourse between the people having identical interests. Hence, we find literary, political, economic and scientific magazines as also household or industrial magazines, which cater to the interests of their respective readers.

Scope

The farmers are information hungry and present public extension system is not able to meet the demand of the farmers for information. The farmer and extension worker ratio is widening. On the other side, communication tools development is enormous. Private extension is also coming into picture. Today, journalism in India has got lot of scope with media barons opening new channels or newspapers or publishing houses on a regular basis. The competition is so rife that each channel or newspaper tires to produce something exclusive, which in turn has given the audience a great deal of variety.

Sources of News

1. Result demonstrations
2. Research Stations
3. Research publications – Annual reports, highlights
4. Kisan melas
5. Farmers fields
6. Agricultural Universities/State Departments
7. Other extension activities like field days/training programmes, *etc.*
8. Plan estimates related to agriculture and allied activities
9. Agriculture finance institutions
10. Agriculture input agencies
11. Agriculture Market committees
12. Electricity and irrigation sectors
13. Farmers committees and associations
14. NGOs *etc.*

Types of News

- Before and after event stories
- Experience and success stories
- New development – such as pest outbreaks, scientific discoveries, weather and crop conditions, progress made on plans
- Predictions – such as long range crop and livestock report economic outlook information, long range weather forecasts
- Subject matter - when tied to an event, situation, development or problem

Purposes of Journalism

1. They may inform, disseminate news and miscellaneous non-news items.
2. They may influence, giving the public either a social or commercial message.

3. They may be entertaining, presenting features, fiction, humour, comics and similar methods.

Functions of Journalism

1. News Function

The primary function of press is to inform. Examining the glut of public occurrences, ideas and situation, newspapermen must determine which will interest the public. Apart from factual presentation of news, for the complex situation the interpretation and explanation are also required.

2. The Opinion Function

Modern man frequently finds himself in the midst of confusion which product to purchase? What decision to take? Whom to vote? He requires a medium of communication, which will guide him to understand the positive and negative points of the situations because of logical arguments. Thus, the modern press has to be both a daily teacher and a daily tribute. Therefore, the editorial is the only means of building public opinion.

3. The Entertainment Function

Entertaining the public is the function and a business too. Since it is too big a job for the local staff, newspaper relies upon syndicated materials. Entertainment is where you find it. It pops up in human-interest stories and news features. Public interest in various features, comics in particular sometimes determine the choice of a newspaper.

Principles of Farm Journalism

Our purpose in writing is to communicate information. Therefore, our first consideration is our reader audience. If you were writing for a scientific paper, you would use a vocabulary and style different from what you would use when writing for the general public. How clearly you communicate information to average readers depends on how well you select, sift and sort your facts.

1. Select Facts

a. Suitable subject matter. Does it meet a need? Is it timely? Is it of current interest? Does it

b. Readers: Who are the people you want to reach? What are their problem, interests and educational levels, Do they have the environment and capacity to make use of the information?

c. Purpose of Publication: What do you want it to teach and accomplish? Do you want to stimulate interest in a programme or do you want to influence the people to do something?

2. Sift Facts

a. Sift essential facts necessary to give information clearly.
b. Screen out difficult concepts, which are beyond reader's experience or understanding (*e.g.*) pH value, calorific value.
c. Give layman an application of subject rather than a detailed explanation.
d. Express highlights.
e. Do not try to impress the lay reader with all you know.
f. Do not document everything.

3. Sort Facts

a. Arrange facts in logical order
b. Set out important points in 1-2-3 order(step y step)
c. Guide reader with attractive subheads and suitable illustrations and pictures.

4. ABC's of Journalism

Accuracy, Brevity and Clarity which are fundamentals of good writing.

5. Adopt the Following Tips for Readability

1. Short sentences, clear in meaning, simple in construction with few prepositional phrases and dependent clauses.
2. Simple words which are familiar and concrete words.
3. Personal or human - interest words.

Photo-Journalism

It is a process of creating a mental picture through the combined use of visual and verbal medium that could come about on the printed page.

Elements of Photo-journalism

1. Possesses a visual impact
2. Defines life in the picture form
3. It represents reality.

DIFFUSION AND ADOPTION OF INNOVATION

Diffusion is the process by which an innovation is communicated through certain channels over time among the members of a social system. It is a special type of communication, in that the messages are concerned with new ideas.

Elements in the Diffusion of Innovations

The four main elements in diffusion of innovations are:

1. Innovation
2. Communication channels
3. Time
4. Social System.

The descriptions for these elements are presented below:

1. The Innovation

An innovation is an idea, practice or object that is perceived as new by an individual or other unit of adoption. The perceived newness of the idea for the individual determines his or her reaction to it. The "newness" aspect of an innovation may be expressed in terms of knowledge, persuasion or a decision to adopt.

In this context, to know about the perceived attributes of innovation would be appropriate which are described in the succeeding paras:

a. Relative Advantage

It is the degree to which an innovation is perceived as better than the idea it supersedes. The degree of relative advantage may be measured in economic terms, but social-prestige factors, convenience and satisfaction are also often the important components.

b. Compatibility

It is the degree to which an innovation is perceived as being consistent with the existing values, past experiences and needs of potential adopters.

c. Complexity

It is the degree to which an innovation is perceived as difficult to understand and use. In general, new ideas that are simpler to understand will be adopted more rapidly than innovations that require the adopter to develop new skills and understandings.

d. Trialability

It is the degree to which an innovation may be experimented with on a limited basis. An innovation that is trialable represents less uncertainty to the individual who is considering it for adoption, as it is possible to learn by doing.

e. Observability

It is the degree to which the results of an innovation are visible to others. The easier it is for individuals to see the results of an innovation, the more likely they are to adopt.

2. Communication Channels

A communication channel is the means by which messages get from one individual to another. The following classification of channels would help the communicator to use them appropriately:

i) **Interpersonal channels -** It refers to those which are used for face to face communication between two or more individuals.

ii) **Mass media channels - These** enable the messages to reach a larger, diverse audience simultaneously in a relatively shorter time. *e.g.*: Radio and T.V.

iii) **Localite channels -** They originate within the social system of the receiver. eg: neighbours, relatives, opinion leaders *etc.*

iv) **Cosmopolite channels -** They originate outside a particular social system. eg: Extension worker, sales personnel *etc.*

3. Time

It is an important element in the diffusion process. Time is an obvious aspect of any communication process. Time does not exist independently of events, but it is an aspect of every activity. The time dimension is involved in diffusion (i) in the innovation - decision process, (ii) in the innovativeness of an individual or other unit of adoption, and (iii) innovation's rate of adoption in a system.

4. Social System

It is defined as a set of interrelated units that are engaged in joint problem solving to accomplish a common goal. The members or units of a social system may be individuals, informal groups, organisations and/or subsystems. The social system constitutes a boundary within which an innovation diffuses.

Innovation Decision Process

As an alternative to the "Stages in the adoption process" *viz.,* Awareness, Interest, Evaluation, Trial and Adoption, due to the advancements in diffusion research, currently" Innovation - Decision process" is proposed which enlightens the sequential stages in the adoption - decisions made by individuals or other units of adoption. The "Innovation - Decision Process" is the process through which an individual (or other decision - making unit) passes from first knowledge of an innovation, to forming an attitude towards the innovation to a decision to adopt or reject, to implementation of the new idea, and to confirmation of this decision.

This process consists of a series of actions and choices over time through which an individual or an organisation evaluates a new idea and decides whether or not to incorporate the new idea into ongoing practice. The conceptualization of the model of the innovation decision process consists of the following five stages.

Stages of Adoption Process

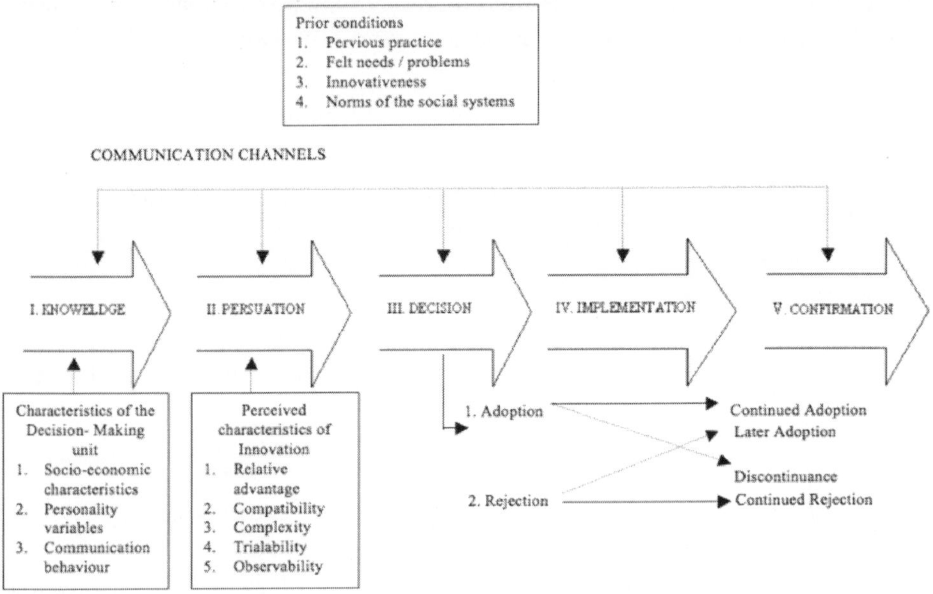

Paradigm on Stages in the Innovation-Decision Process.

1. Knowledge Stage

Knowledge occurs when an individual (or the decision - making unit) is exposed to the innovation's existence and gains some understanding of how it functions. The following three types of knowledge possessed by an individual influence the decisions:

a. **Awareness - knowledge** motivates an individual to seek "how-to" knowledge and principles knowledge. This type of information - seeking is concentrated as the knowledge stage of the innovation - decision process, but it may also occur at the persuasion and decision stages.

b. **How-to knowledge** consists of information necessary to use an innovation properly. When an adequate level of how-to knowledge is not obtained prior to the trial and adoption of an innovation, rejection or discontinuance is likely to result. Change agents could perhaps play their distinctive role to concentrate on "how-to knowledge" at the trial and decision stage in the process.

c. **Principles knowledge** consists of information dealing with the functioning principles underlying how the innovation works. It is usually possible to adopt an innovation without principles knowledge, but the danger of misusing the new idea is greater, and discontinuance may result. The long-range competence of individuals to judge future innovations is facilitated by principles knowledge.

2. Persuasion Stage

Persuasion occurs when an individual (or other decision - making unit) forms a favourable or unfavourable attitude toward the innovation. While the mental activity as the knowledge stage was mainly cognitive (or knowing), the main type of thinking at the persuasion function is affective (or feeling). At this stage, a general perception of the innovation is developed. The individual becomes more psychologically involved with the innovation and hence he or she seeks information about the new idea.

3. Decision Stage

Decision occurs when an individual (or other decision - making unit) engages in activities that lead to a choice to adopt or reject the innovation.

Adoption is a decision to make full use of an innovation as the best course of action available. Rejection is a decision not to adopt an innovation.

The small - scale trial is often part of the decision to adopt, and is important as a means to decrease the perceived uncertainty of the innovation for the adopter.

4. Implementation Stage

Implementation occurs when an individual (or other decision - making unit) puts an innovation into use. Until the implementation stage, the innovation-decision process has been a strictly mental exercise. But implementation involves overt behaviour change as the new idea is actually part into practice.

Problems of implementation are likely to be more serious when the adopter is an organisation rather than an individual. Reason is that in an organisational setting, a number of individuals are usually involved in the innovation - decision process, and the implementers are often a different set of people from the decision makers.

5. Confirmation Stage

Confirmation occurs when an individual (or other decision - making unit) seeks reinforcement of an innovation - decision already made, but he or she may reverse this previous decision if exposed to conflicting messages about the innovation. The confirmation stage continues after the decision to adopt or reject for an indefinite period in time. At this stage, the change agents have the additional responsibility of supporting messages to individuals who have previously adopted.

As a sequential effect, there is a possibility for "discontinuance". Discontinuance is a decision to reject an innovation after having previously adopted it. There are two types of discontinuances:

a. Replacement discontinuance - is a decision to reject an idea in order to adopt a better idea that supersedes it.
b. Disenchantment discontinuance - is a decision to reject an idea as a result of dissatisfaction with its performance.

Model of Adoption Process

1. **Need**: This is a stage when an individual wishes to change his existing practices, express dissatisfaction and develops a compromise.
2. **Awareness:** The individual just comes to know about an innovation without knowing the details of it.
3. **Interest:** He makes an attempt to know more about the innovation. Asks extension agents/friends and seeks information and sees the innovation.
4. **Deliberation:** This is a stage of deliberation and mental evaluation. The individual mentally examines the possibility of application of the innovation under own condition. He seeks advice of opinion leaders, observes the performance at different places and discusses with family members. The individual then takes a decision to try out or reject the idea.
5. **Trial**: An individual uses an innovation in part or sometimes in full. The individual applies the practice on a limited scale to observe the performance under own conditions.
6. **Evaluation**: The individual evaluates the performance of the innovation. The individual observes the performance of an innovation on various dimensions. Collects data on the performance of an innovation on others' situations. Compares the performance of the new with the old one and figures out changes which will be necessary if innovation is to be adopted. Calculates input- output, risks, uncertainties *etc.*
7. **Adoption**: It is a decision to use the practices on continued basis.

Adopter Categories

There are different categories of farmers. According to Rogers (1971), the farmers based on their innovativeness can be classified as

1. Innovators (Venturesome)
2. Early adopters (Respectable)
3. Early majority (Deliberate)
4. Late majority (Skeptical)
5. Laggards (Traditional)

Characteristics of Farmers

All individuals in a social system do not adopt an innovation at the same time. Rather, they adopt in an ordered time sequence, and they may be classified into adopter categories on the basis of when they first begin using a new idea. In technology transfer programme, it is of great practical utility for the extension workers to identify the individuals who are likely to adopt innovations early and who may lag behind.

The adoption of an innovation over time follows a normal, bell-shaped curve when plotted over time on frequency basis. If the cumulative number of adopters is plotted, it results in an S-shaped curve. The S-shaped curve rises slowly at first when there are few adopters in a time period, accelerate to a maximum when about half of the individuals in the system have adopted and then increases at a gradually slower rate as the few remaining individuals finally adopt (Following Figure). The S-shaped curve is like that of a 'learning curve' as propounded by the psychologists. Each adoption in the social system is in a sense equivalent to a learning trial by an individual.

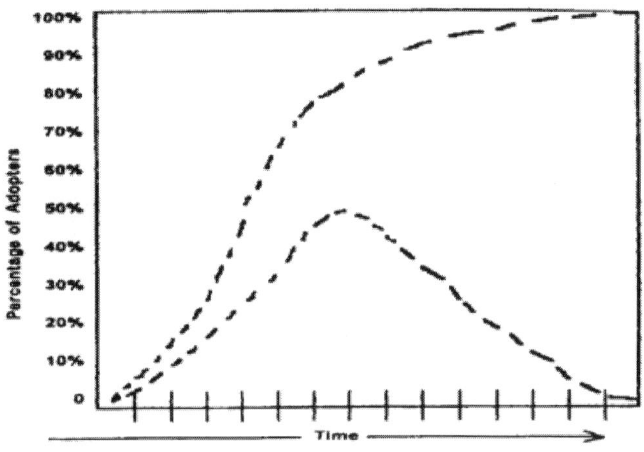

The Bell Shaped Frequency Curve and the S-shaped Cumulative Curve for Adopter Categories.

Both of these curves are for the same data, the adoption of an innovation over time by the members of a social system. But the bell-shaped curve shows these data in terms of the number of individuals adopting each year, whereas the S-shaped curve shows these data on cumulative basis.

The distribution of adopters over time closely approaches normality, and may be explained by the statistical concept of normal curve. The distribution of the adopters may be partitioned into five adopter categories by using the mean (x) and standard deviation. The area lying to the left of the mean time of adoption minus two standard deviations includes 2.5 per cent of the individuals who are the first to adopt an innovation and are known as innovators. The next 13.5 per cent between the mean minus one standard deviation and the mean minus two standard deviations to adopt the new idea are called as early adopters. The next 34 per cent of the adopters between the mean date of adoption and minus one standard deviation are known as early majority. Between the mean and one standard deviation to the right of the mean are located the next 34 per cent to adopt the new idea, the late majority. The last 16 per cent to the right of mean plus one standard deviation are the last to adopt the innovation the laggards.

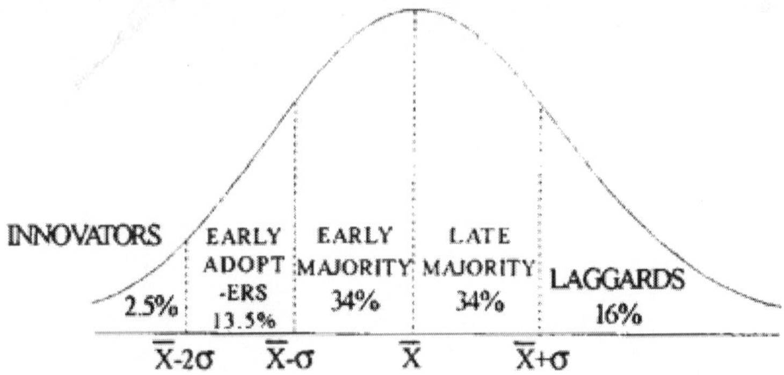

Adopter Categorization on the Basis of Innovativeness.

The innovativeness dimension, as measured by the time at which an individual adopts an innovation, is continuous. However, this variable may be partitioned into five adopter categories by laying of standard deviations from the average time of adoption.

The detailed information on the characteristics of adopter categories is presented in the succeeding pages:

Innovators: Venturesome

Observers have noted that venturesomeness is almost an obsession with innovators. They are eager to try new ideas. This interest leads them out of a local circle of peers and into more cosmopolite social relationships. Communication patterns and friendships among a clique of innovators are common, even though the geographical distance between the innovators may be great. Being an innovator has several prerequisites. These include control of substantial financial resources to absorb to understand and apply complex technical knowledge. The salient value of the innovator is venturesomeness. He desires the hazardous, the rash, the daring, and the risky. The innovator also must be willing to accept an occasional setback when one of the new ideas he adopts proves unsuccessful.

These are the first people to adopt a new idea, much ahead of other people. They are very few in numbers, probably not more than one or two in a community.

Characteristics

1. Have larger farms.
2. High net worth and risk capital.
3. Willing to take risks.
4. Usually not past middle age
5. Generally well educated

6. Have respect and prestige in progressive communities but not in conservative type of communities.
7. Mentally alert and actively seeking new ideas.
8. Their sphere of influence and activity often goes beyond the community boundaries.
9. They have many formal and informal contacts outside the immediate locality.
10. They often by-pass the local extension worker in getting information from the originating sources, and may learn about new things even before he does. They sometimes manage to get samples of seeds or chemicals even before they are released for public use.
11. They subscribe to many farm magazines and specialised publications.
12. Other farmers may watch the innovators and know what they are doing but the innovators are not generally named by other farmers as "neighbours and friends" to whom they go for information.

Early Adopter: Respectable

Early adopters are a more integrated part of the local social system than are innovators. Whereas innovators are cosmopolites, early adopters are localities. This adopter's category, more than any other, has the greatest degree of opinion leadership in most social systems. Potential adopters look to early adopters for advice and information about the innovation. The early adopter is considered by many as "the man to check with" before using a new idea. This adopter category is generally sought by change agents to be a local missionary for speeding the diffusion process. Because early adopters are not too far ahead of the average individual in innovativeness, they serve as a role model for many other members of a social system. Members of a social system respect the early adopter. The early adopter is respected by his peers. He is the embodiment of successful and discrete use of new ideas. And the early adopter knows that he must continue to earn this esteem of his colleagues if his position in the social structure is to be maintained.

Characteristics

1. Younger than those who have a slower adoption rate, but not necessarily younger than the innovators
2. They are not the persons who test the untried ideas but they are quickest to use tried ideas in their own situations.
3. Have large farms.
4. Higher education than those who adopt more slowly.
5. High income.
6. They participate more in the format activities of the community.
7. They also participate more in government programmes.

8. This group usually furnishes a disproportionate amount of the formal leadership (elected positions) in the community.
9. They read papers and farm journals and receive more bulletins than people who adopt later.
10. They may be regarded as community adoption leaders.

Early Majority: Deliberate (Local Adoption Leaders)

The early majority adopt new ideas just before the average member of a social system. The early majority interacts frequently with their peers, but leadership position; are rarely held by them. The early majority's unique position; between the very early and relatively late to adopt make; them an important link in the diffusion process. The early majority may deliberate for some time before completely adopting a new idea. Their innovation-decision is relatively longer than that of the innovator and the early adopter. "Be not the last to lay the old aside, nor the first by which the new is tried", might be the motto of the early majority. They follow with deliberate willingness in adopting innovations, but seldom lead.

Characteristics

1. Slightly above average in age, education and farming experience.
2. They take a few more farm journals and bulletins than the average.
3. They have medium high social and economic status.
4. Less active in formal groups than early adopters, but more active than those adopting later.
5. In many cases, they are not formal leaders in the association
6. They also attend extension meetings and farm demonstrations.
7. They are most likely to be informal resources than early adopters and innovators, and so cannot afford to make hasty or poor decisions.
8. They associate mainly with people of their own community.
9. They value highly the opinions their neighbours and friends hold about them; for this is their main source of status and prestige.
10. They are mostly mentioned as "neighbours and friends" from whom the majority of farmers seek information.

Late Majority: Skeptical

The late majority adopt new ideas just after the average member of a social system. Adoption may be both an economic necessity and the answer to increasing social pressures. Innovations are approached with a skeptical and cautions air, and the late majority do not adopt until most other in their social system have done so. The weight of system norms must definitely favour the innovation before the late majority is convinced. They can be persuaded of the utility of new ideas, but the pressure of peers is necessary to motivate adoption.

Characteristics

1. Those in this group have less education and are older than the early majority.
2. They form the major part for formal organisational membership, although they participate less in such formal groups.
3. They take fewer leadership roles than the earlier adopters.
4. They take and read fewer papers, magazines and bulletins, than the early majority.
5. They do not participate in as many activities outside the community as do people that adopt earlier.

Laggards: Traditional

Laggards are the last to adopt an innovation. They possess almost no opinion leadership. They are the most localite in their outlook of all adopter categories, many are near isolates. The point of reference for the laggard is the past. Decisions are usually made in terms of what has been done in previous generations. This individual interacts primarily with others who have traditional values. When laggards finally adopt an innovation, it may already have been superseded by another more recent idea which the innovators are already using. Laggards tend to be frankly suspicious of innovations, innovators, and change agents. Their tradition direction slows the innovation decision process to a crawl. Adoption lags far behind knowledge of the idea. Alienation from a too-fast-moving world is apparent in much of the laggard's outlook. While most individuals in a social system are looking to the road of change ahead, the laggards has his attention fixed on the rear-view mirror.

Characteristics

1. Least education.
2. Oldest.
3. Participate least in formal organisations, cooperatives and government programmes.
4. They hardly read farm magazines and bulletins.

Factors Influencing Adoption Process

The Nature of the Practice

The speed with which adoption will take place is partly dependent on the nature of practice itself.

A) Complexity

Generally speaking, the more complex a practice and the more change it requires in the existing operations, the more slowly it will be adopted. The following classification of practices in terms of their complexity roughly represents the decreasing order of speed with which acceptance may be expected to occur:

a. **A simple change**: A change in materials and equipment only, without a change in technique or operation (e. g., new variety of seed).

b. **Improved practice:** Change in existing operation with or with-out a change in materials or equipment (e. g., change in rotation of crops).

c. **Innovation**: Change involving new technique or operation (e. g., contour cropping).

d. **Change in total enterprise:** *e.g.*, from crop to livestock farming.

B) Cost

Those practices which cost little seem to be adopted more rapidly than those which are more expensive.

C) Net Returns

Those practices which yield, the greatest marginal returns per rupee invested, and in the shortest time seem to be adopted most readily. The above two characteristics *viz.*, cost and net returns are also referred to as "relative advantage" or "Profitability".

D) Compatibility

Is the degree to which an innovation is consistent with existing values and past experiences of the adopters. An idea that is not compatible with the cultural norms of a social system will not be adopted so rapidly as an idea that is compatible *e.g.*, the lack of compatibility of beef production with cultural values in India.

E) Divisibility (Trialability)

Is the degree to which an innovation may be tried on a limited basis. New ideas that can be tried on a small scale or on the instalment plan will generally be adopted more rapidly than innovations that are not divisible. *e.g.*, new seeds or fertilizers can be tried on a small scale, but new machinery or a thing like cow dung gas plant cannot be so tried.

F) Communicability (Observability)

Is the degree to which the results of an innovation may be diffused to others. The results of some practices are easily observed (e. g., application of nitrogenous fertilizer to plants), while the results of some innovations are not easily observed (*e.g.*, pre-treatment of seeds, or soil conservation measures).

Chapter 3

Rural Sociology and Educational Psychology

SOCIOLOGY AND RURAL SOCIOLOGY

Sociology is the study of society, human social behaviour, patterns of social relationships, social interaction, and culture that surrounds everyday life. It is a social science that uses various methods of empirical investigation and critical analysis to develop a body of knowledge about social order and social change. Sociology can also be defined as the general science of society. While some sociologists conduct research that may be applied directly to social policy and welfare, others focus primarily on refining the theoretical understanding of social processes.

Traditional focuses of sociology include social stratification, social class, social mobility, religion, secularization, law, sexuality, gender, and deviance. As all spheres of human activity are affected by the interplay between social structure and individual agency, sociology has gradually expanded its focus to other subjects and institutions, such as health and the institution of medicine; economy; military; punishment and systems of control; the Internet; education; social capital; and the role of social activity in the development of scientific knowledge.

The range of social scientific methods has also expanded, as social researchers draw upon a variety of qualitative and quantitative techniques. The linguistic and cultural turns of the mid-20th century, especially, have led to increasingly interpretative, hermeneutic, and philosophic approaches towards the analysis of society. Conversely, the turn of the 21st century has seen the rise of new analytically, mathematically, and computationally rigorous techniques, such as agent-based modelling and social network analysis. Social research has influence throughout various industries and sectors of life, such as among politicians, policy makers, and legislators; educators; planners; administrators; developers; business magnates and managers; social workers; non-governmental organizations; and non-profit organizations, as well as individuals interested in resolving social issues in general. As such, there is often a great deal of crossover between social research, market research, and other statistical fields.

Rural sociology is a field of sociology traditionally associated with the study of social structure and conflict in rural areas although topical areas such as food and agriculture or natural resource access transcend traditional rural spatial boundaries (Sociology Guide 2011). It is an active academic field in much of the world, originating in the United States in the 1910s with close ties to the national Department of Agriculture and land-grant university colleges of agriculture.

The sociology of food and agriculture is one focus of rural sociology, and much of the field is dedicated to the economics of farm production. Other areas of study include rural migration and other demographic patterns, environmental sociology, amenity-led development, public-lands policies, so-called "boomtown" development, social disruption, the sociology of natural resources (including forests, mining, fishing and other areas), rural cultures and identities, rural health-care, and educational policies. Many rural sociologists work in the areas of development studies, community studies, community development, and environmental studies. Much of the research involves developing countries or the Third World.

Meaning of Rural Sociology

India is basically a land of villages. Rural sociology is the sociology of the village or village society. It is a branch of sociology which studies rural society. Rural sociology studies the relations of the people who live in the villages. It is just like a mirror of the rural social life. It provides a detailed study of knowledge about different aspects of rural life, its problems, its culture, its religion, its economic and political life. The Indian society has euphoria for village life. The village society is considered as the backbone of Indian society. Millions of money has been spent on the development of villages. The basic aim of the study of rural sociology is to make the village people self-sufficient and also link them with the wider society at regional and national levels.

- Sociology is the science of society.
- Society is a group of people who live together long enough, sharing common values and general interests.
- Sociology is a body of facts and principles which are based on scientifically organized knowledge.
- Rural Sociology has two terms- Rural +Sociology
- Rural geographically refers to the villages and demographically to the people living in villages.

- ☆ Rural Sociology is the study of life in the rural environment, which systematically studies rural communities to discover their conditions and tendencies to formulate principles of progress.
- ☆ It is also the study of rural population, social organizations and the social process operative in the rural society.

Definition

Many Western and Indian sociologists have given their definitions regarding rural sociology in their own point of view. Among the sociologists, T.L. Smith, A.R. Desai, Dwight-Sanderson, Chapin *etc.* are very important.

"...... the prime objective of rural sociology should be to make a scientific, systematic and comprehensive study of the rural social organisation of its structure, functions and objective tendencies of development and on the basis of such study, to discover the laws of its development." — **A.R. Desai**

"Rural sociology is the sociology of life in the rural environment." —**Sanderson**

"The sociology of rural life is a study of rural population, rural social organization and the rural social processes operative in rural society." —**F. S. Chapin**

"Such sociological facts and principles as are derived from the study of rural social relationships may be referred to as rural sociology." —**T. L. Smith**

The above definitions have given a clear cut Idea about rural sociology. The definitions no doubt, include the theoretical as well as applied aspect of rural sociology. Rural sociology clarifies the laws of rural social life and points out the necessary conditions of its progress.

Nature of Rural Sociology

In India, rural sociology is treated as a part of sociology. Like sociology it is also considered as a social science. It is precisely a specialized field of sociology. Its nature is scientific. Actually to say, by nature rural sociology is a science. Science in simple words refers to a systematic body of knowledge. In case of rural sociology the rural problems are systematically and logically studied. As a result of which the study becomes universally applicable. Rural sociology fulfils the different criteria of science. But the whole controversy centres round some problems which deny the rural sociology the status of a science. Generally the term 'Nature' refers to essential qualities or characteristic features of a phenomenon.

Characteristics of Rural Sociology

1. Rural sociology is of recent origin and a very new discipline. Here sociological principles are applied for studying rural society.
2. Rural sociology is multi-dimensional in character.
3. Rural sociology is scientific and systematic in character.
4. It has acquired an inter-disciplinary status over a period of time.

5. It emphasises on micro studies.
6. It employs comparative method.
7. The development of rural sociology in India has not attained a scientific level. It is to be hoped that there will be a development in this direction in the future.

Different Scope of Rural Sociology

(i) A Study of Rural Community

Rural sociology is concerned with the characteristics, features, nature and human ecology of village community. It is also the study of activities of rural people. Through these activities we know the institutional objects of the community.

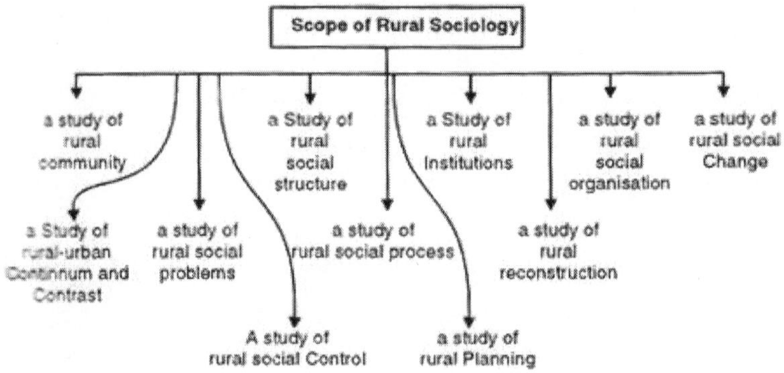

(ii) A Study of Rural Social Structure

Rural sociology studies the various components of rural social structure. For example, village community, caste, class, dominant caste, jajmani system, caste and politics, backward class *etc.*

(iii) A Study of Rural Institutions

Rural sociology studies the structure, characteristics and functions of rural social institutions. It studies the institutions like family, marriage, kinship, religion, caste *etc.* in the rural context.

(iv) A Study of Rural Social Organization

In rural sociology rural social organisations are unavoidably studied. It includes the study of rural family pattern, marriage, rural social stratification, educational system, religion, cultural institutions *etc.*

(v) A Study of Rural Problems

The subject matter of rural sociology includes the problems of rural life such as social, economic, political and cultural problems. Rural social problems are increasing day-by-day. It is essential that these should be solved in a planned way.

(vi) A Study of Religion and Culture

In rural society religion and culture of rural people are given more emphasis. Religion plays an important role in rural society. Rural society is basically religion oriented society. It acts as an informal means of social control. Culture includes old customs, traditions, folkways, norms, values, *etc.*

(vii) A Study of Rural Social Process

Social process indicates the fundamental ways through which these people can interact with other groups. Co-operation as an associative social process can be observed more than any other processes. Competition and conflict can also be seen in rural society. These processes take place in peculiar way in rural society and studied separately from a particular angle.

(viii) A Study of Rural Social Control

Rural sociology is also a study of rural social control. Social control is the control over the society. It starts from the family level. Rural sociology significantly studies the infernal means of social control like customs, traditions, folkways, mores, norms, religion *etc.* Social control may be conducted in various ways. It varies from society to society.

(ix) A Study of Social Change

Social change is the change in the social relationships; the changing aspect of any system of the society. Due to the impact of modern education, means of transport and communication, modernisation, urbanisation, industrialisation, the rural society is now undergoing tremendous changes.

(x) A Study of Rural Reconstruction

Rural reconstruction is an important aspect of life and without which development is impossible. Rural sociology studies various plans and programmes relating to community development programmes. It also studies different laws of government regarding the reconstruction of village community. Rural society needs rural planning and reconstruction in a systematic and planned manner.

(xi) A Study of Rural – Urban Continuum and Contrast

Rural sociology studies rural-urban contrast and continuum on the basis of social, economic, religious and cultural point of view. Both village and city hypothetically are two contrasting modes of community life. It has to be studied in a scientific manner. By studying the comparison between these two, it can meet different plans and programmes for the development of rural community.

(xii) A Study of Planning

Rural sociology is a study of social planning relating to rural society. Community development projects, Panchayati Raj, Co-operative movements are coming under this study. Their success and failure can only be measured by the study of rural sociology.

Importance and Significance of Rural Sociology

After Independence it is very much necessary for our country to have a systematic study of the rural social organization, its structure, function and evolution.

Prof. A. R. Desai, while highlighting the importance of rural sociology, observes, "It is, however, urgently necessary to make a scientific and systematic study of rural society; of its economic foundations, social and cultural superstructure, of its institutions and functions, of the problems arising from the rapid process of disintegration which is undergoing and which even threatens its breakdown."

The practical value of the study of rural sociology is widely recognized today. Rural sociology which aims at providing systematic and scientific approach to rural problems and life is assuming great significance because in India rural society after independence has acquired new significance. India is a classic land of agriculture; the economic foundation of village community centres on agriculture. It is with the help of rural sociology that it can be possible to find out what are the causes of break-up of the self-sufficiency of the village community. It is necessary to understand and to act according to the laws prevailing in the village in order to achieve progress of the villages. This necessitates a special study of the rural society. In our country, the emphasis of rural sociology is to analyze the implications of traditional rural society in the modern context.

Importance of Rural Sociology

Rural sociology lays great stress on systematization and scientific analysis. Many studies have been made but many of them now are in monograph or article form. Some of the studies seriously lack proper analysis and solution.

1. Vast majority of population lives in the villages which has its own problems. Even today, two-third of the world population lives in rural areas. Rural sociology aims at studying all those problems and life of rural people.
2. Rural society is the fundamental basis of human civilization and culture. People who are living in urban areas they are mainly the migrants from the rural areas. So rural area or village is the well-spring of our culture and civilization.
3. The basic importance of rural sociology has to find out the laws of development and those principles only can be discovered by studying rural communities in detail.
4. The importance of rural society cannot be ignored in studying rural problems in India. India is a country of villages, therefore, rural sociology is comparatively more important here than in any other country of the world.

5. Now India is facing with problems of social, economic and political reconstruction. Unless, one understands what rural India is, one cannot understand the problems of India as a whole and its ancient tradition, customs, culture and ways of life.

The importance of rural sociology is increasing day by day. The reasons are:

1. Social life In India is mostly the rural life.
2. About 80 per cent of Indian population lives in villages.
3. Rural sociology provides us knowledge about the study of rural problems and their solutions.
4. Rural society is the store house of ancient cultural heritage. Its study is having importance.
5. Agriculture is the livelihood of the country.
6. Unique nature of transformation of Indian society.
7. Greater interest of democratic decentralization. Scientific study of village community is a pre-requisite for it.
8. Village is the basic unit of study.
9. Growing influence of modernization, industrialization and urbanisation.
10. Rural sociology is a means of mass media exposure with regards to rural reconstruction.
11. In recent years, the rural sociologists have attached supreme importance to the study of agrarian relations, land reforms, peasant movements, rural stratification system, rural leadership *etc.*

Origin and Development of Rural Sociology in India

Rural sociology is relatively novel branch of sociology. It is originated in the United States of America in the form of systematic science in the year 1820. It has taken more than half a century to become established as a distinct academic need or professional study.

In the year 1907, rural sociology seems to be started its growth. At that time, American President Roosevelt emphasized on the study of rural sociology and rural social problem.

In 1911, a Rural Sociology Department was established in American Sociological Society. In 1919, a Rural Sociology Department was established under the auspices of the Bureau of Agricultural Economics under the leadership of Dr. C.J. Golpin. The name was later changed to "Rural Population and Rural Life." A quarterly named "Rural Sociology" was published in 1936.

In India, only after the independence, literature on rural society has been on the increase. After independence, during the first five-year plan, strategies were made for the upliftment of the condition of rural people.

In the year 1955, prominent village studies by famous sociologists were brought out in India. It was suggested that there was a need of rural reconstruction and welfare. Therefore community development programmes came into the scene. Planning Commission has introduced several projects, land reforms, co-operative movement, five-year plans, Panchayat Raj *etc.*

These plans and programmes have contributed a lot towards the development of rural sociology in India. Different sociologists have made their original contributions in the field of rural sociology in India. Among them are Dr. D. N. Majumdar, Prof. N. K. Bose, Prof. M.N. Srinivas, S. C. Dube *etc.* are the pioneers.

Before independence, also, many studies were conducted about village life in India. In 1926, Royal Commission on Agriculture was set up. In the meanwhile, Viswa Bharati set up a Rural Reconstruction Board. In 1932 Scottish Church College, Calcutta published accounts of village life. But all these studies before independence remained isolated. Only after 1947, our nation decided to have planned development.

SOCIAL ECOLOGY

Rural sociology is a study of rural happenings, relationships and interaction in the village society. From this point of view its scope is very wide and vast.

1. Designing policies and programmes for rural development is not possible without the knowledge of rural social life
2. The findings of social life in rural areas itself is a guide and a package of suggestion, to bring out reforms for reconstruction of rural society to meet the days challenges
3. The socio psychological aspects of rural life provides adequate information to study the non-cultural material of the society
4. The study of rural social organizations provides wide scope regarding value systems, which is very much essential to introduce any structural reforms.

Principles of Social Ecology

- ☆ **Interdependence and the principle of unity in diversity**. Social ecology seeks to oppose the standardization of beings and thoughts, promoting the importance of diversity and the organic unions between different parts of society.
- ☆ **Decentralization**. A social ecology society would take the form of a confederation of decentralized municipalities linked to each other by commercial and social ties.
- ☆ **Direct democracy**. Structured around the principle of a form of communalism called libertarian municipalism, social ecology advocates the development of municipal assemblies, a modernized version of the

type developed by the Athenians in Antiquity or implemented during the Paris Commune for political decision-making.

- ✰ **A renewal of citizenship**. At the base of the social ecology system are the citizen and the community. All people must relearn to participate in the decision-making process concerning local life, specifically by learning to come to these decisions through a communal process.
- ✰ **A liberating technology**. Social ecology is not opposed to modern technologies but is in favor of developing them solely to be used in service of human beings. Science must regain a moral foundation and develop for the benefit of humans, not to enslave them.
- ✰ **A social vision of work**. Developing machines have, in social ecology, the aim of freeing human beings from a large part of manual work (factory work) that can be done by machines, in order to leave human beings to more creative work and reduce working time.
- ✰ **Dialectical naturalism**. Dialectical naturalism is a dialectical philosophy developed to serve as an ethical foundation for a society based on the principles of social ecology.

Rural Society

Rural society, society in which there is a low ratio of inhabitants to open land and in which the most important economic activities are the production of foodstuffs, fibres, and raw materials. Such areas are difficult to define with greater precision, for, although in non-industrialized nations the transition from city to countryside is usually abrupt, it is gradual in industrialized societies, making it difficult to pinpoint the boundaries of rural places. A second, related problem is that governments do not use the same statistical criteria for rural and urban populations; in Japan, for instance, any cluster of fewer than 30,000 people is considered rural, whereas in Albania a group of more than 400 inhabitants is regarded as an urban population.

Collectively, the characteristics are often designated by the term gemeinschaft, an expression introduced by the German sociologist Ferdinand Tönnies. Tönnies described the contrasting nature of urban life with the term gesellschaft, a state characterized by impersonal bureaucracy, rationalized specialization, and mechanization. Gesellschaft is typically associated with modern industry, where people are employees who perform specific, goal-oriented functions in a rational and efficient, as opposed to a traditional and organic, manner. The two terms are sometimes translated as "community" and "society." Rural inhabitants work with people they know well and are accustomed to relationships of great intimacy and small scale, whereas urban dwellers know each other in narrow, segmented ways that have little to do with family or friendship. According to Tönnies and subsequent sociologists, all societies are characterized by mixtures of gemeinschaft and gesellschaft; the United States, where even agriculture is almost completely mechanized, is closer to the gesellschaft end of the spectrum, whereas rural India, which is still heavily guided by tradition, is an example of gemeinschaft.

Historically, farming societies have had higher birthrates than urban societies; their populations have also tended to be younger, to live in larger families, and to include slightly greater percentages of males. These phenomena were related: it was to a farmer's advantage to have many offspring, especially males, who could work in the fields as children and then would support their parents as they grew older.

Social Groups

In the social sciences, a social group can be defined as two or more people who interact with one another, share similar characteristics, and collectively have a sense of unity. Other theorists disagree however, and are wary of definitions which stress the importance of interdependence or objective similarity. Instead, researchers within the social identity tradition generally define it as "a group is defined in terms of those who identify themselves as members of the group." Regardless, social groups come in a myriad of sizes and varieties. For example, a society can be viewed as a large social group.

Social Cohesion Approach

A social group exhibits some degree of social cohesion and is more than a simple collection or aggregate of individuals, such as people waiting at a bus stop, or people waiting in a line. Characteristics shared by members of a group may include interests, values, representations, ethnic or social background, and kinship ties. Kinship ties being a social bond based on common ancestry, marriage, or adoption. In a similar vein, some researchers consider the defining characteristic of a group as social interaction. According to Dunbar's number, on average, people cannot maintain stable social relationships with more than 150 individuals.

There are four main types of groups: 1) primary groups, 2) social groups, 3) collectives, and 4) categories.

1) Primary Groups

Primary groups are characterized by high amounts of cohesiveness, member identification, face-to-face interaction, and solidarity. Such groups may act as the principal source of socialization for individuals as primary groups may shape an individual's attitudes, values, and social orientation. Primary groups are small, long-term groups.

Three sub-groups of primary groups are:

1. Kin (relatives)
2. Close friends
3. Neighbours.

2) Social Groups

Social groups are also small groups but are of moderate duration. These groups are often formed due to a common goal. In this type of group, it is possible for out group members (*i.e.*, social categories of which one is not a member) to become in b group members (*i.e.*, social categories of which one is a member) with reasonable ease. Social groups, such as study groups or co-workers, interact moderately over a prolonged period of time.

3) Collectives

In contrast, spontaneous collectives, such as bystanders or audiences of various sizes, exist only for a very brief period of time and it is very easy to become an in group member from an out group member and vice versa. Collectives may display similar actions and outlooks.

4) Categories

Categories consist of individuals that are similar to one another in a certain way, and members of this group can be permanent ingroup members or temporary ingroup members. Examples of categories are individuals with the same ethnicity, gender, religion, or nationality. This group is generally the largest type of group.

Social Stratification

It refers to the arrangements of the individuals or groups of people into hierarchically arranged strata in a community. A community may appear as homogenous, ex: urban, rural *etc.*, but there may be internal inequalities, divisions and distinctions. These distinctions become patterned and stabilized with unequal distribution of privileges, power and status positions.

The determinants of status are:

- Wealth- its quantity and quality. Ex: Income, standard of living.
- Ancestry- nationality, ethnic background, length of residence in the area, family reputation.
- Functional utility of the individual- education, occupation, skill.
- Religion- the kind and degree of religion followed.
- Biological characteristics- male, female, age.

Social stratification refers to a society's categorization of its people into groups based on socioeconomic factors like wealth income, race, education, ethnicity, gender, occupation, social status, or derived power (social and political). As such, stratification is the relative social position of persons within a social group, category, geographic region, or social units.

Principles of Social Stratification

Four principles are posited to underlie social stratification. First, social stratification is socially defined as a property of a society rather than individuals in that society. Second, social stratification is reproduced from generation to generation. Third, social stratification is universal (found in every society) but variable (differs across time and place). Fourth, social stratification involves not just quantitative inequality but qualitative beliefs and attitudes about social status.

Complexity

Although stratification is not limited to complex societies, all complex societies exhibit features of stratification. In any complex society, the total stock of valued goods is distributed unequally, wherein the most privileged individuals and families enjoy a disproportionate share of income, power, and other valued social resources. The term "stratification system" is sometimes used to refer to the complex social relationships and social structures that generate these observed inequalities. The key components of such systems are: (a) social-institutional processes that define certain types of goods as valuable and desirable, (b) the rules of allocation that distribute goods and resources across various positions in the division of labor (*e.g.*, physician, farmer, 'housewife'), and (c) the social mobility processes that link individuals to positions and thereby generate unequal control over valued resources.

Social Mobility

Social mobility is the movement of individuals, social groups or categories of people between the layers or within a stratification system. This movement can be intragenerational (within a generation) or intergenerational (between two or more generations). Such mobility is sometimes used to classify different systems of social stratification. Open stratification systems are those that allow for mobility between, typically by placing value on the achieved status characteristics of individuals. Those societies having the highest levels of intra-generational mobility are considered to be the most open and malleable systems of stratification.

Culture

Culture is a term that refers to a large and diverse set of mostly intangible aspects of social life. According to sociologists, culture consists of the values, beliefs, systems of language, communication, and practices that people share in common and that can be used to define them as a collective. Culture also includes the material objects that are common to that group or society. Culture is distinct from social structure and economic aspects of society, but it is connected to them both continuously informing them and being informed by them. Culture is one of the most important concepts within sociology because sociologists recognize that it plays a crucial role in our social lives. It is important for shaping social relationships, maintaining and challenging social order, determining how we make sense of the world and our place in it, and in shaping our everyday actions and experiences in society. It is composed of both non-material and material things.

Culture is important to sociologists because it plays a significant and important role in the production of social order. The social order refers to the stability of society based on the collective agreement to rules and norms that allow us to cooperate, function as a society, and live together (ideally) in peace and harmony. For sociologists, there are both good and bad aspects of social order.

Culture consists of the beliefs, behaviors, objects, and other characteristics common to the members of a particular group or society. Through culture, people and groups define themselves, conform to society's shared values, and contribute to society. Thus, culture includes many societal aspects: language, customs, values, norms, mores, rules, tools, technologies, products, organizations, and institutions. This latter term institution refers to clusters of rules and cultural meanings associated with specific social activities. Common institutions are the family, education, religion, work, and health care.

Popularly speaking, being cultured means being well educated, knowledgeable of the arts, stylish, and well mannered. High culture generally pursued by the upper class – refers to classical music, theater, fine arts, and other sophisticated pursuits. Members of the upper class can pursue high art because they have cultural capital, which means the professional credentials, education, knowledge, and verbal and social skills necessary to attain the "property, power, and prestige" to "get ahead" socially. Low culture or popular culture generally pursued by the working and middle classes refers to sports, movies, television sitcoms and soaps, and rock music. Remember that sociologists define culture differently than they do cultured, high culture, low culture, and popular culture.

Sociologists define society as the people who interact in such a way as to share a common culture. The cultural bond may be ethnic or racial, based on gender, or due to shared beliefs, values, and activities. The term *society* can also have a *geographic* meaning and refer to people who share a common culture in a particular location. For example, people living in arctic climates developed different cultures from those living in desert cultures. In time, a large variety of human cultures arose around the world.

Culture and society are intricately related. A culture consists of the "objects" of a society, whereas a society consists of the people who share a common culture. When the terms *culture* and *society* first acquired their current meanings, most people in the world worked and lived in small groups in the same locale. In today's world of 6 billion people, these terms have lost some of their usefulness because increasing numbers of people interact and share resources globally. Still, people tend to use *culture* and *society* in a more traditional sense: for example, being a part of a "racial culture" within the larger "U.S. society."

Basic Concept of Rural Sociology

Structural Concepts

- Family,
- Group,

- ☆ Community,
- ☆ Society

Functional Concepts

- ☆ Socialization,
- ☆ Social system,
- ☆ Social stratification,
- ☆ Social institutions,
- ☆ Social interaction,
- ☆ Culture,
- ☆ Values and beliefs,
- ☆ Culture

Structural Concepts

Family

Family is a group defined by a sex relationship sufficiently specific and enduring to provide for the procreation and upbringing of children. According anthropologists, a family is "a group characterized by common residence, economic cooperation and reproduction. It includes adults of both sexes, at least two of whom maintain a socially approved sexual relationship, and one or more of their children of their own or adopted by the sexually cohabit adults." May 15 is celebrated as the International Day of the Family. This day highlights the importance of families. It aims at promoting equality, bringing about a fuller sharing of domestic responsibilities and employment opportunities

The Family as a Functional Unit

Every association of people; it be a state, a nation, or a tribe has its own distinctive culture, its modes of living and thought, which are developed as a response to the peculiar circumstances of the environment, natural and ideological. Family is the agency through which the impressionable rising generation is made familiar with such traditions. It teaches the individual what situations to anticipate, how to behave and what behaviour to expect, by giving one the gifts of language and dress which integrate within one's cultural ethos. It facilitates adjustment to people and groups outside the family circle. Family plays an important role in transmission of the cultural traditions from one generation to another. It acts as an educative unit and a socio-cultural agency. The importance of this aspect lies in the fact that children all over the world get their earliest instruction in the family beginning with language.

The Main Functions of Family

Procreation- Reproduction is the main function of any society, either socially or in a biological way, or both. Providing sustenance and care to the dependants, especially children and the aged. It acts as an educative unit and a socio-cultural agency. The importance of this aspect lies in the fact that children all over the world get their earliest instruction in the family beginning with language. Furnishing education to the young, which were passing down the accumulated knowledge, traditions, values and techniques?

Family Plays an Important Role

Family plays an important role in transmission of the cultural traditions from one generation to another.

- Furnishing status to the family members
- It teaches the individual what situations to anticipate, how to behave and what behaviour to expect, by giving one the gifts of language and dress which integrate within one's cultural ethos.
- It facilitates adjustment to people and groups outside the family circle.
- Providing cooperative interaction necessary for production or earning a living, Consumption, recreation, worship and companionship.

Categorization of Families

According to blood relations living together

Nuclear or primary family consisting of husband, wife and their unmarried Children Joint family, which is an aggregation of more than one primary family, on the basis of close blood relationships and common residence.

- A number of generations may be living together.
- According to stage of life
  - Family of origin or family of orientation is the family in which one is born and gets initial orientation to the norms of the family and the society
  - Family of procreation is the family which one sets up after marriage.
- According to numerousness of husbands and wives
  - Polygamous family is the family where a man marries more than one wife
  - Polyandrous family where a woman is married to more than one man, usually several brothers.
- According to rule of lineage
  - Patri-lineal family, in which determination of descent and property inheritance take place along the male line

- Matri-lineal family, in which determination of descent and property inheritance take place along the female line

- According to rule of residence
  - Patri-local family is one where the married couple and their off springs put up with the husbands family or in a new household set up by the husband.
- Matri-local family is one where the husband goes to live with the family of his wife and the family line goes with the mother.

Characteristic Features of Groups

- Promotion of group discipline
- Existence of loyalty among the members
- Adherence to group responsibility
- Group or peer pressure.

Factors in Organizing Groups

External Factors

- A group may be organized when the individuals feel the need, that cannot be satisfied individually and when there is no other such group meeting the need.
- A group may be organized when there is similar group existing elsewhere and not serving their interest.
- A group may be organized with or without outside stimulus.

Internal/Personal Factors

- A group may be organized when the members are obedient to each other and have channels to meet the needs and satisfy the interest.
- A group may be organized when there is cooperation among members to hold them together to act as one unit.

A group may be organized when members are willing to compromise with each other and make decisions with consensus.

Involuntary and Voluntary Groups

- Involuntary groups are those groups in which the individual has no choice or choice is not required for membership.

  Ex: Membership by birth-family, community
- Voluntary groups are those where the individual has membership with deliberate choice.

  Ex: youth club.

An in group is one to which the person feel they belong and with which they identify themselves strongly.

- Not only they themselves, but others also feel them to be an integral part of the group.
- Matters of vital interest to the group are confined within the group.

An out group is one with which other individuals cannot identify themselves.

- The individuals feel themselves as outsiders to the group and the group also reciprocates the same towards the individual.
- Matters of vital interest are not shared with the out group members.

Community

Community refers to groups of mutually dependent people, living in a more or less compact continuous geographical area, having a sense of belonging and sharing common values, norms and some common interests and acting collectively in an organized manner to satisfy their chief needs through a common set of organizations and institutions.

There are two dimensions in community

- Territorial dimension- geographical area
- Social dimension- group of people

  Ex: Village, Town, City

Any area of common life like village, town, district, country or an even wider area may be considered as community.

Society

- Society refers to people in a community with similar culture and civilization
- Structural elements of Society
  - Identity: It refers to the characteristics that differentiate one society from the other. Ex: Residence, caste, religion, progressiveness *etc.* distinguishes society to society.
  - Composition: It refers to making up of the society. Society is composed of male-female, educated-uneducated, rural-urban, traditional-modern *etc.*
  - Intergroup relation: It refers to relation between groups-relationship between villages, castes, classes.
  - Intra-group relation: It refers to the relationship within the group-relationship between husband and wife within the family or between families in a village.

Functional Elements of Society

Objectives or ends: These refer to the changes or things which the members of the group are expected to be accomplished through their interaction and activities.

- Ex: Attainment of higher productivity, increase in the production *etc.*

Norms: These refer to the rules or the guiding standards which prescribe what is socially acceptable or unacceptable.

- Norms govern the application of means in the attainment of goals.
- Determining the size of the family and methods by which it may be achieved, is generally guided by the norms of the society.

Leadership

- It refers to the process through which a person directs guides and influences the thought, feeling and behavior of other members of the society.
- The nature of the leadership generally determines the behavior or level of achievement of the members of the society.

Resources

- It refers to the various types of resources and potentialities a society possesses.
- Resources are important because they set the limit of achievement of the society.
- Ex: Human resources, financial resources, natural resources *etc.*

Social System

- It is defined as a set of interrelated units that are engaged in joint problem solving to accomplish a common goal.
- The members or units of a social system may be individuals, informal groups, organizations and subsystems.

Social Interaction

It refers to dynamic interplay of forces in which contact between persons and groups result in modification of attitudes and behavior of the participants.

The interaction processes have been classified on the basis of their cooperation and oppositional nature. Cooperative interaction is working together toward common objectives or goals. It may be informal as in the case of family and neighborhood or formal legal type like in organizations and registered societies offices. Accommodative interaction is accepting each other's views and developing common understanding with compromises and tolerance. It is a process of getting along despite differences. Assimilative interaction is fusing and blending of divergent cultural group, thereby cultural differences disappear. The concept of accommodative interaction and assimilative interaction is the concept of acculturation which is referred as the changes in culture that result from continued contact between two cultures. Modern transport, communication, developmental programmes facilitate the processes of acculturation and assimilation. Competitive interaction is the in which persons and groups compete with each other to reach a goal- competition for profit, ownership *etc.* It is both negative and positive aspects. Too much competition leads to generation of tension and a feeling of insecurity. On the other competition may lead to motivation for excellence and achievement of goals. Conflict interaction is a struggle in which the competent parties, attempting to reach a goal, try to eliminate, abolish, wipeout or make other part ineffective. Conflict is intermittent, emerge as a result of opposing interests and arise in periods of crises.

Social Institutions

It is an organized system of social relationships which represents certain common values and procedures and meets certain basic needs of the society. They are culturally approved patterns of behavior including prescribed roles and procedures and are grouped to satisfy some basic social needs. They have continued long enough to be considered as permanent institutions.

Basic Social Institutions

- ✰ Family
- ✰ Religion
- ✰ Government
- ✰ Economy
- ✰ Education

Family

- ✰ It performs the functions of procreation, upbringing and socialization.
- ✰ It provides economic and social security to the family members.
- ✰ The members perform different roles in decision making, participation in physical activities relating to earning and home making.

Religion

- ✰ Religion constitutes a set of beliefs regarding the ultimate power in the universe, the ideal and proper pattern of behavior and ceremonial ways of expressing these beliefs.
- ✰ It provides a foundation for mores and sanction of taboos in certain societies.
- ✰ It provides a means of strength and determination to withstand crisis and ups and downs in life.

Government

- As a political institution it administers the regulatory functions of law and order, thus maintain security in the society.
- Moulds the behavior of the society through its form of governing.
- Plans and executes development programmes for the welfare of the people.
- Seeks the participation of people for effective implementation of the programmes.

Economy

- Provides basic physical sustenance of the society by meeting the need for food, shelter and clothing and other necessary supplies and services.
- Promotes and strengthens economic institutions like agriculture, industry, marketing, credit and banking.

Education

- It socializes the individual right from family to formal educational institutions.
- It develops knowledge, skill and attitudes of the individuals and moulds them as competent members of the society.
- It tremendously influences the behaviour of the individuals, by widening scope for receptive ideas.

Culture

- It is a complex whole of knowledge, belief, art, morals, laws, customs, and other capabilities and habits acquired by the people as member of society.
- It is social heritage, inclusive of material non material aspects.
- House, food, clothing, tools, equipment and other material items are material aspects while family, kin group, caste group, education, political organizations, government, economic system, religion *etc.* are non-material aspects.

Characteristics of Culture

- Culture is learned through process of socialization, communication and training.
- Culture is transmitted from generation to generations and it is learned from parents and through to children.
- **Culture is universal and unique -** there is culture everywhere, but every society has its own culture.
- **Culture is static as well as dynamic -** it has permanence, but at the same it changes over the years.

- **Culture is integrative** - different aspects of culture may pull in different directions. Still there is consistency and integration so that the society is held together.
- **Culture builds conformity** - the patterns of behavior in a culture are considered as ideal towards which people are expected to strive.
- **Culture is relative** - there is nothing like good culture or bad culture. It is interpreted according to a person or society's own experience.
- **Culture is diverse** - culture varies from country to country and different areas in the same nation. Ecological diversity is an important source of cultural difference.

Implications of Culture

- **Ethnocentrism**- the tendency to consider their own culture is of high value and superior to all others. This exists worldwide.
- **Cultural change**- Culture changes based on two factors- i) Internal- discovery and invention within the culture. ii) External- diffusion and borrowing from outside the society.
- **Cultural lag**- some parts of the culture of people change at faster rate than the other parts.
- Traditional culture is the foundation of indigenous knowledge and technology systems.

Values and Beliefs

- Values refer to what people considered as valuable and desirable.
- It can be defined as expressions of preferences.
- Values are learned and tend to be linked together amicably to form patterns.
- It varies from culture to culture in accordance with relative worth given to them by each culture.

Constraints in the Change of Values

Familism- subordination of individual goals to family

Fatalism - value for fate than in spite individuals abilities.

Conservation - emphasis on the past to preserve and continue it are many a times constraints in changing the values.

Beliefs are closely related to values. They are the mental convictions one has about the truth of something. They can be referred as to what people believe as true.

Social Control

- Social control is the pattern of influence the society exerts on individuals and groups to maintain order and establish rules in the society. Social control helps in maintenance of desired social values. The means of social control are as detailed below.

- ☆ **Folkways**- They are the informal ways of behaviour based on habits and traditions, mostly found in usages. Violations of folkways are viewed seriously by the society.

  Ex: Dressing, eating habits *etc.*

- ☆ **Mores**- They are socially acceptable ways of behaviour which involves moral standards. They are the group shared understandings of what to do and what not to do in a situation. They exert some amount compulsion and violations leads to social action. They are more powerful than folkways in controlling the behaviour of the individual.
- ☆ **Ex**: Removing chappal while entering into the house.
- ☆ **Taboos**- The word taboo in a strict sense refers to prohibition of types of behaviour because of some magical, supernatural or religious sanction. Generally the term more use for the positive action *i.e.* things ought to be done and taboo in for a negative action *i.e.* things ought not to be done

  Ex: Muslim girls should not go to mosque.

- ☆ **Ritual** – Ritual may be defined as a pattern of behaviour or ceremony which has become the customary way of dealing with certain situations, or is the pattern that has been established by law as in the case of governmental affairs or is a part of the rules of a particular organization. Many of the ritualistic rights in primitive cultures are a part of the religion or magic of that group.

  Ex: Ceremony of naming the baby - Namaskaram

- ☆ **Norms-** They are established behaviour patterns for the members of a social system. The members are expected to follow them. A member who deviates them is considered as deviant of the system.

  Ex: Creating the elder ones

- ☆ **Laws-** They grow from mores. As the society becomes larger and more complex, codification of mores takes place for strict adherence and compliance.

  Ex: Prohibition on dowry, child labour, child marriages *etc.*

Social Roles

- ☆ Each person is expected to behave in accordance with socially prescribed patterns of behavior that relate to the individuals social position and society in which he lives. Two types of roles are
- ☆ Ascribed roles: roles are assigned by culture and society to an individual.

  Ex: Head of the family.

- ☆ Achieved role: these roles are earned or achieved by the people through their individual effort and competition.

  Ex: Elected leaders

SOCIAL CHANGE AND DEVELOPMENT

Social change may refer to the notion of social progress or socio-cultural evolution, the philosophical idea that society moves forward by evolutionary means. It may refer to a paradigmatic change in the socio-economic structure, for instance the transition from feudalism to capitalism, or hypothetical future transition to some form of post-capitalism.

Social Development refers to how people develop social and emotional skills across the lifespan, with particular attention to childhood and adolescence. Healthy social development allows us to form positive relationships with family, friends, teachers, and other people in our lives.

Social Development encompasses a commitment to individual and societal well-being, and the opportunity for citizens to determine their own and their society's needs and to influence decisions that affect these. Social change incorporates public concerns in developing social policy and economic initiatives.

Until relatively recently, social development was conceived in terms of a set of desirable results - higher incomes, longer life expectancy, lower infant mortality, more and better education *etc.* Recently emphasis has shifted from the results to the enabling conditions, strategies and public policies for achieving those results. But still little attention has been placed on the underlying social process of development that determines how society formulates, adopts, initiates, and organies; and few attempts have been made to formulate such a framework. However, there are some recognized theories and principles, which will be examined briefly.

Accordingly, it may also refer to social revolution, such as the Socialist revolution presented in Marxism, or to other social movements, such as Women's suffrage or the Civil rights movement. Social change may be driven through cultural, religious, economic, scientific or technological forces.

Role of the Individual in Social Change

Society has no direct means to give conscious expression to its subconscious collective aspirations and urges. That essential role is played by pioneering conscious individuals - visionary intellectuals, political leaders, entrepreneurs, artists and spiritual seekers who are inspired to express and achieve what the collective subconsciously aspires to and is prepared for. Where the aspiration and action of the leader do not reflect the will of the collective, it is ignored or rejected. Where it gives expression to a deeply felt collective urge, it is endorsed, imitated, supported, and systematically propagated. This is most evident at times of war, social revolution, or communal conflict.

For example, India's early freedom fighters consciously advocated the goal of freedom from British rule long before that goal had become a felt aspiration of the masses. The leaders spent decades urging a reluctant population to conceive of itself as a free nation and to aspire to achieve that dream. When finally the collective endorsed this conception, no foreign nation had the power to impose its will on the Indian people.

All human creative processes release and harness human energy and convert it into results. The process of skill formation involves acquiring mastery over our physical-nervous energies so that we can direct our physical movements in a precisely controlled manner. In the absence of skill, physical movements are clumsy, inefficient, and unproductive, like the stumbling efforts of a child learning to walk. Whilst the energies are the motivating force, it is strategic ability and professional skills that turn energy into effective action. Often the high emotions of social change movements are a facet of this undirected energy. Strategic advocacy helps to direct this energy, focusing it in directions where change can most effectively be triggered.

Importance of Rural Sociology in Extension Education

1. In the context of community development and rural development programs deliberate efforts are made to bring about social change in rural areas. This change is brought about not in a vacuum but in a structure of human relations, which necessitates the study of rural sociology
2. An extension worker is a change agent. He has to bring about changes in the rural culture. Therefore, he has to understand the culture in which he has to bring about changes
3. In the cross-cultural situation, it is likely that the change agent may develop the ethnocentric attitude, which makes him difficult to work successfully. Study of rural sociology helps to overcome this difficulty.
4. Sometimes The change agent, while working in rural communities, does not keep the established pattern of hierarchy, this creates problems in his working. Rural sociology helps him to understand the same and its importance
5. The value system of individuals, families, groups and communities is an important factor to be kept in mind while trying to bring about changes in the farming communities Therefore, it is necessary for a student of rural development to study the subject of rural sociology to bring about planned change in the rural communities effectively and without resistance.

Interrelationship between Rural Sociology and Agricultural Extension

Extension – Meaning

The word 'extension' is derived from the Latin roots, 'ex' – meaning 'out' and 'tensio' meaning 'stretching'. Stretching out is the meaning of extension. Extension is an education and its purpose is to change the attitude and the practice of the people with whom the work is done (Ensminger). Extension Education is a process of teaching rural people how to live better by learning ways that improve their farm, home, and community. Extension Education is a defined as an educational process to provide knowledge to the rural people about the improve practices in a convincing manner and to help them to take decisions within their specific local condition. Agricultural Extension is the application of scientific research and new knowledge to agricultural practices through farmer education.

Interrelationship between Rural Sociology and Extension

Sl.No.	*Rural Sociology*	*Extension*
1	It is a scientific study of the laws of the structure and development of rural society	It is informal(actually non-formal) education for the rural people with a view to develop rural society on desirable lines
2	It studies the attitudes and behavior of rural people	It seeks to modify or change for the better, the attitudes and behavior of village people
3	It studies the needs and interests of rural society	It helps rural people to discover their needs and problems and builds educational programs based on these needs and wants
4	It analyses rural social relationships, or group organizations and leadership in rural areas, the social processes like cooperation, association, competition *etc.*, among village people	It fosters(develops) and utilizes village Organizations and leadership and Favorable social processes, to achieve its objectives of rural development
5	It studies social situations and assembles social facts or rural society	It makes use of such social data as a basis for building up its extension programs for rural areas
6	It investigates the social, cultural, political, and religious problems of rural society	It also studies these problems with reference to their impact on extension work in villages

EDUCATIONAL PSYCHOLOGY

Father of Psychology is Sigmund Freud. Psychology has its origin from two Greek words psyche (soul) and logos means science (a rational course or a study) Thus, literally it means study or science of soul.

Definitions of Psychology

Psychology is the scientific study of the behavior of an individual – Daniel Psychology is the science of mental activity of an organism – Guilford Psychology is a science of human behavior Psychology is the science that studies the responses which living individuals make to their environment – Murthy Psychology is the science of behavior and adjustment.

Definitions of Educational Psychology

Educational Psychology is the branch of psychology that describes and explains the learning experiences of an individual and the progress in his educational development from birth to old age. Crow and Crow stated that educational psychology is the study of the psychological aspects of educational situation. According to Trow educational psychology is that branch of psychology, which deals with teaching and learning. It takes its meaning from education, social process and from psychology, a behavioural science (Skinner). Educational Psychology is the discipline concerned with teaching and learning processes; applies the methods and theories of psychology and has its own as well (Woolfolk, 1995). Educational Psychology is the systematic study of the development of the individual within educational setting.

Aim of Educational Psychology

The goal of educational psychology is to understand prediction and control of human behavior. Therefore, educational psychology has the following aim:

1. To furnish students with the knowledge and understanding this will help them to improve the quality of instruction.
2. To enhance and enrich the lives of the learner.
3. To study group behavior environmental adjustments improvement of the environment.

Scope of Educational Psychology in Agricultural Extension

The application of psychology has a very wider area in the study of human behavior as follows:

1. To identify the nature and characteristics of learner
2. The nature of learning process
3. The principles of teaching and learning process
4. The human growth and development
5. The techniques employed in teaching
6. Identification of personality traits
7. Development and adjustment of psychological traits
8. Scientific measurement and evaluation of psychological traits
9. To value the scientific attitude towards education.
10. The educational significance of individual differences in rate and limit of learning.
11. The inner change that occur during learning.
12. The relation of teaching procedures to leaning outcomes.
13. The most effective techniques for evaluating progress in learning.

14. The relative effect upon an individual of formal learning as compared with incidental or informal learning experiences.
15. The psychological impact upon learner's attitude of sociological conditions. Agricultural extension is education and its main purpose is to change the behaviour of farmers. Therefore, knowledge of educational psychology is useful to extension agent for understanding the factors affecting the teaching and learning process.

Importance of Educational Psychology in Extension Education

The importance of educational psychology in agricultural extension is immense as both disciplines deal with human behaviour in educational environment. Following are the some of the reasons which explain the importance of educational psychology in agricultural extension.

1. It helps the teacher to understand the developmental characteristics of children
2. To understand the nature of classroom learning
3. Help to understand individual differences
4. Help to understand effective teaching methods
5. Problem of children
6. Provide knowledge of mental health
7. Help in curriculum construction
8. Help to measurement of learning out come

Role of Educational Psychology in Extension Education

1. Educational psychology helps the extension agent to know the learner, his interest, attitudes, aptitude, level of aspiration, intelligence, interests, individual behaviour in group, *etc.* which plays a major role in one's learning.
2. Its main concern is on teaching and learning. This helps in formulating training programmes for improving the knowledge and skill of extension agent and farmers. It also helps in selection of teaching methods and aids for organizing effective learning situations and suggests technique of learning as well as teaching.
3. It helps in imparting better education by organising the subject matter of learning experience, preparation of different text books, development of assessment patterns, *etc.* for heterogeneous learners.
4. Educational psychology helps in acquainting learner with the mechanism of heredity and environment.
5. It also deals with the problem-solving which is very important for extension agent to develop problem-solving skills amongst farmers.

6. It helps extension agent to find causes of prejudices, the habit of sticking to old practices of farming and ways of doing things, the doubts and lack of confidence and factors affecting motivation.
7. It also helps them to know the emotions and feelings of farmers, how farmers learn new practices.

Behaviour: Cognitive, Affective, Psychomotor Domain

Behavior (American English) or behaviour (British English; see spelling differences) is the actions and mannerism made by individuals, organisms, systems or artificial entities in conjunction with themselves or their environment, which includes the other systems or organisms around as well as the (inanimate) physical environment. It is the computed response of the system or organism to various stimuli or inputs, whether internal or external, conscious or subconscious, overt or covert, and voluntary or involuntary. Taking a behavior informatics perspective, a behavior consists of actor, operation, interactions, and their properties. This can be represented as a behavior vector.

Although there is some disagreement as to how to precisely define behavior in a biological context, one common interpretation based on a meta-analysis of scientific literature states that "behavior is the internally coordinated responses (actions or inactions) of whole living organisms (individuals or groups) to internal and/or external stimuli".

A broader definition of behavior, applicable to plants and other organisms, is similar to the concept of phenotypic plasticity. It describes behavior as a response to an event or environment change during the course of the lifetime of an individual, differing from other physiological or biochemical changes that occur more rapidly, and excluding changes that are result of development (ontogeny). Behaviors can be either innate or learned from the environment. Behavior can be regarded as any action of an organism that changes its relationship to its environment. Behavior provides outputs from the organism to the environment.

Human Behavior

Human behavior is believed to be influenced by the endocrine system and the nervous system. It is most commonly believed that complexity in the behavior of an organism is correlated to the complexity of its nervous system. Generally, organisms with more complex nervous systems have a greater capacity to learn new responses and thus adjust their behavior.

Animal Behavior

Ethology is the scientific and objective study of animal behavior, usually with a focus on behavior under natural conditions, and viewing behavior as an evolutionarily adaptive trait behaviorism is a term that also describes the scientific and objective study of animal behavior, usually referring to measured responses to stimuli or trained behavioral responses in a laboratory context, without a particular emphasis on evolutionary adaptivity.

Consumers Behavior

Consumer behavior refers to the processes consumers go through, and reactions they have towards products or services. It has to do with consumption, and the processes consumers go through around purchasing and consuming goods and services: Consumers recognise needs or wants, and go through a process to satisfy these needs. Consumer behavior is the process they go through as customers, which includes types of products purchased, amount spent, frequency of purchases and what influences them to make the purchase decision or not.

Domains of Learning

Initially developed between 1956 and 1972, the domains of learning have received considerable contributions from researchers and experts in the field of education. Studies by Benjamin Bloom (on cognitive domain), David Krathwohl (affective domain) and Anita Harrow (Psychomotor domain) have been encompassed into the three domains of learning (Sousa, 2016).

A holistic lesson developed by a teacher requires the inclusion of all the three domains in constructing learning tasks for students. The diversity in such learning tasks help creates a comparatively well – rounded learning experience that meets a number of learning styles and learning modalities.

Cognitive Domain

Learning helps develop an individual's attitude as well as encourage the acquisition of new skills. *The cognitive domain aims to develop the mental skills and the acquisition of knowledge of the individual. The cognitive domain encompasses of six categories which include knowledge; comprehension; application; analysis; synthesis; and evaluation.* Knowledge includes the ability of the learner to recall data or information. This is followed with comprehension which assesses the ability of the learner to understand the meaning of what is known. This is the case where a student is able to explain an existing theory in his or her own words (Anderson *et al.,* 2011). This is followed by application which shows the ability of the student to use the abstract knowledge in a new situation. A typical case is when an Economics student is able to apply the theory of demand and supply to the changing market trend of clothing during a particular season. The analysis category aims to differentiate facts and opinions. *The synthesis category shows the ability to integrate different elements or concepts in order to form a sound pattern or structure to help establish a new meaning.* The category of evaluation shows the ability to come up with judgments about the importance of concepts. A typical scenario is when a manager is able to identify and implement the most cost effective methods of production in the bid to increase profits whilst sustaining a high level of competitive advantage.

Affective Domain

The affective domain includes the feelings, emotions and attitudes of the individual. The categories of affective domain include receiving phenomena; responding to phenomena; valuing; organization; and characterization (Anderson *et al.,* 2011). The sub domain of receiving phenomena creates the awareness of feelings and emotions as well as the ability to utilize selected attention. Valuing involves the ability to see the worth of something and express it. This includes the ability of a learner to share their views and ideas about various issues raised in class. *The ability of the student to prioritize a value over another and create a unique value system is known as organization. This can be assessed with the need to value one's academic work as against their social relationships.* The sub domain of characterization explains the ability to internalize values and let them control the behavior of the individual. In view of this, a student considers the academic work highly important as it plays an important role in deciding the career path chosen rather than what may be available.

Psychomotor Domain

The psychomotor domain includes utilizing motor skills and the ability to coordinate them. The sub domains of psychomotor include perception; set; guided response; mechanism; complex overt response; adaptation; and origination. Perception involves the ability to apply sensory information to motor activity. For instance, a student practices a series of exercises in a text book with the aim of scoring higher marks during exams. Set, as a sub domain, involves the readiness to act upon a series of challenges to overcome them. In relation to guided responses, it includes the ability to imitate a displayed behavior or utilize a trial and error method to resolve a situation (Sousa, 2016). The sub domain of mechanism includes the ability to convert learned responses into habitual actions with proficiency and confidence. *Students are able to solve exams questions after they have confidently been able to answer some past questions.* Complex Overt responses explain the ability to skilfully perform complex patterns of actions. A typical instance has to do with the ability of a student to have an increased typing speed when using a computer.

Learning is an integral part of every individual's life. It is very key to growth and development and hence requires the need for both students and teachers to be committed to the process. It is further necessary to ensure that the delivery of learning combines generally different facets which have been identified to be the domains of learning. With the continually increasing need to ensure that students are taught with varying strategies and techniques, it is important for teachers to adopt a teaching strategy that combines various domains of learning to enable teaching and learning to be considered as effective.

Personality

Meaning

The word Personality originated from the Latin word Personare which used to mean the voice of an actor speaking through a mask. Later it came to be applied to the actors themselves

Definitions of Personality

Personality is the sum total of an individual's behavior in social situations. -Trainer 1957 Personality is most comprehensive and explains different dimensions of personality in terms of traits and the adjustment of individual to his environment. -Hilgard Personality is the dynamic organisation within the individual of those psychophysical systems that determine his unique adjustments to his environment. Personality acts as the quality of an individual's total behavior. Woodworth Personality comprises an individual's experience, his knowledge, skill, temperament, attitude, habits, character, and physical traits. Personality is the sum total of ways in which individual reacts to and interacts with others. It is most often described in terms of measurable traits that a person exhibits (Robbins, 2001). Personality can be defined as a dynamic and organised set of characteristics possessed by a person that uniquely influences his or her cognitions, motivations, which they respond to various situations.

Types of Personality

There are three types of personality – C. J. Jung

a. Extrovert
b. Introvert and
c. Ambivert

A. Extrovert

1. An extrovert is socially adaptable and interested in people.
2. He likes to make friends and very soon creates a circle of friends around him.
3. He prefers working in company with other people, is talkative and fond of talking.
4. He is self assertive and generally takes things lightly.
5. He never feels embarrassed.
6. He has a keen sense of observation and is attentive. Reformers and social workers are generally extroverts.

B. Introvert

1. An introvert limits his acquaintance to a few.
2. This person is very conservative and suspicious of the motives of others.
3. He is not social and prefers to remain in the background on certain occasions.
4. He avoids embarrassment and public speaking.

5. He is very reserved, self centre, introspective, absent minded, remains worried and is always day dreaming.
6. He is generally slow and hesitant to take the initiative.
7. Philosophers, poets, and scientists are generally introverts.

C. Ambivert Personality

1. Ambivert types are placed in between extrovert and introverts.
2. Their behaviour is balanced. Their psychic energy is partially directed inwards and partly outwards.
3. They are interested in their own thoughts and emotions and also in other persons and their action.
4. Most of us belong to ambivert type

Role of Personality in Extension Education

1. The knowledge of human personality enables the extension agent to judge and follow the method of guiding by selecting suitable teaching methods.
2. This helps the extension agent to properly plan the educational programme to accomplish the objectives for a desirable change in the farmers.
3. By studying the personality of farmers, extension agent can understand the values and value systems of its clients and can precede his work accordingly.
4. By studying the personality of a particular farmer, the extension agent can get a clear idea about his various traits such as sociability.
5. If a person found with this trait, then he can be used as a key communicator to promote developmental activities in a particular village.
6. Similarly, farmers with traits of empathy, sympathy, generosity can be engaged in trustworthy works like seed distribution *etc.*
7. The extension worker should get into deep study of knowing the capacities, types and traits of the personalities of his clients (farmers) with whom he has to work in order to plan and educate them. This also helps the extension worker to select a suitable farmer for a certain purpose
8. The study of personality of the people enables extension worker to judge and follow the method of guiding, by selecting suitable teaching method.
9. This enables the extension worker to properly plan the programmes to advise the farmers for desirable changes in rural communities

PERCEPTION AND MOTIVATION

Meanings

Perception is usually described as a response to stimulus. Perception is an active process, whereby sensory reactions are related to relevant past experiences of an individual when confronted with stimulus and more structured and meaningful picture is printed in the mind, which is finally perceived as the object. Our reaction to any situation is determined by the way we perceive it. Two different individuals may perceive the same object in two different ways. Therefore perception is very personal thing.

Definitions of Perception

Perception is the process of organizing and interpretation of sensory data in terms of one's previous experience and present needs - Ruch Perception is the process by which impressions, opinions and feelings about an object is formed by means of a sensory operation - Kuppu Swamy Perception is the process of assimilating experiences and relating them to previous experiences, attaching meaning or value to them and ordering them in to organized patterns of knowledge and feeling - Mc David

Motivation

Motivation is the process of initiating a conscious and purposeful action. Motive means an urge (drive or force) or combination of urges, to induce conscious or purposeful action. Motivation is a goal directed and need satisfying behavior. It explains why people do the things they do. It influences a person to do a thing in a certain way. Motive is something (a need or desire) that causes a person to act.

Definitions

Motivation may be defined as goal seeking or goal directed behavior or activity. Behavior is a function of the person, which is interaction within a situation Motivation is also defined as a stage of the organism in which bodily energy is mobilized and selectively directed towards parts of the environment

This definition breaks in to two parts:

1. Mobilization of bodily energy or drive and
2. Direction

Factors for the Motivation or motivated behavior are:

1. Environmental conditions
2. Internal urge
3. Incentive awards

Theories of Motivation

Maslow's' Classification of Needs

According to the famous psychologist Abraham Maslow (1954), the needs in order of importance to individuals *i.e.* the fulfillment of needs starts from the first order and if first order needs are fulfilled the individual thinks of second order and so on as given below

1. **Physiological needs**: These are called first order needs. These needs are necessary for survival of the individual *e.g.* food, clothing, shelter *etc.* these are most important, if these needs are fulfilled then only the individual steps in to second order needs
2. **Safety needs**: These are second order needs. These needs are for the security of the individual from physical, physiological, economic and social viewpoints *e.g.* protection from danger, threatening *etc.*
3. **Social needs:** These are third order needs. Man is a social animal and has an inherent desire to be with others in some form *e.g.* friendship, company *etc.* These are also called response needs.
4. **Esteem needs**: These are fourth order needs. The individual likes to get appreciation and recognition from others in the society *e.g.* power, status, prestige *etc.*
5. **Self-actualization**: These are called last order or fifth order or highest order needs. It is self-realization and knowing self or the ultimate purpose of human being *e.g.* what is human being, how he should live, what is his purpose of living *etc.*

Importance of Motivation in Agricultural Extension

1. Society composes of individuals. Everyone is different from the other and the degree of difference is based on the relative importance each one of us assigns to the different motivating forces and to the various interests we have, which impel us to act as we do
2. The job of the extension worker is to understand the basic wants or incentives of the people with whom he is working
3. He should show the leaner how to satisfy these basic wants by learning new things (*i.e.* by adopting new methods or practices)
4. The extension worker should find the personal goals of the learner and tie with his teaching goals
5. When people are shown how learning a subject will enable them to gratify (satisfy) a desire or realize a need, they are being motivated to learn
6. A great impetus (momentum or speed) is given to learning when the learner can see that what he learns will be of immediate value to him in making his own life more satisfying
7. A desire or a want therefore, is the most definite and dynamic of motives, from the educational point of view
8. It is possible for extension workers to motivate people to satisfy the four categories of basic needs of security, new experience, response and recognition through the increase of income, provision of new knowledge, skills *etc.*, encouraging farmers to work in groups and awarding prizes, certificates *etc.* to the winners in crop competitions respectively.

INTELLIGENCE

Among the millions of species that exist on the earth, the human being is said to be superior and exclusive (separate) because of its reasoning of distinguishing between right and wrong. The ability to adopt to the environment with and to master situations, understanding, ability to command and capacity to carry on difficult tasks by learning and putting the past experience to the most beneficent use. This quality, which we describe as intelligence is found in different degrees in different human beings

Definitions

Intelligence is the ability of an individual to make profitable use of past experience – Thorndike Intelligence is the ability demanded in the solution of problems, which require the comprehension, and the use of symbols – Grprett Intelligence is the ability of an individual to adjust himself to the conditions that arise in his environment – Brown Intelligence is the ability to adopt oneself to judge well, understand well, reason (think) well and act well – Binet Intelligence is the organization of abilities to learn a group of facts with alertness and accuracy to exercise mental control and display flexibility in seeking the solution of problem – Skinner Intelligence is the global or aggregate capacity of an individual to think rationally, act purposefully, and to deal effectively with her/his environment (Wechsler). Intelligence is the capacity to learn and adjust to relatively new and changing conditions (Wagnon). Intelligence is the ability to adapt, to shape and select environment to accomplish one's goals and those of one's society and culture (Sternberg). Intelligence is the capacity to do something useful in the society in which we live. Intelligence is the ability to respond successfully to new situations and the capacity to learn from one's past experiences (Gardner).

Three types of intelligence:

1. **Abstract intelligence**:- Abstract intelligence it is the ability to understand and deal with verbal and mathematical symbols. Out of the three, this is the one that receive greatest weight. Professional people are high in abstract intelligence. For instance, if a farmer knows about iron deficiency in sugarcane, if same mineral deficiency occurs in other crop, he can find it.
2. **Mechanical intelligence**:- Mechanical intelligence it is the ability to understand and deal with things, objects *etc.* and more concerned with the skills of individuals. Industrialists and building traders' are high in this intelligence. For instance, a farmer who is having mechanical skill to perform spraying operation does fast and effectively than those with less mechanical skill.
3. **Social intelligence**:- Social intelligence It is the ability to understand and deal with persons, to apply psychological principles of human relationship. Salesperson, politician, diplomat possess this intelligence.

Binet Test of General Intelligence

He noted that the intellectual capacity increase with age, the intelligent person would be less susceptible to distraction, more like to adopt the situation to achieve a goal and likely to criticize his own work.

Mental Maturity

If you give a child same test twice with enough time between two tests, you will find that his scores would improve as he grew older. He would certainly do better at the age of 18th than at the age of 4 years. However, a time would come when his ability to answer the test questions would no longer improve. A person does not stop learning even he reaches to mental maturity. Generally mental maturity reached somewhere between 14 th to 18th age. The generally accepted figure is 15th year.

Mental Age

It is a degree of intelligence exhibited by an individual in relation to other of his age group. Binet and Simson have development a scale called as mental age scale for measuring brightness and dullness among the individuals. Binet testing of school children classified individuals into levels of mental development, which is called as mental age. (M. A.)

Binet and Simson scale was applied as follows: A child who passes the entire eight year old test is regarded as 8 year old mentally, however his actual age (chronological age) may be 6 year. So he is mentally 2 years in advance. Another child with mental age of 8 may be 11 years old means in this case he is retarded in three years. So, brightness or dullness in terms of number of year will be advance of retardation.

Intelligence Quotient (I.Q.): William Sterm says that you can get a number that would show how intelligent a person. This is by dividing his mental age by his chronological age. The I.Q. is obtained by dividing mental age by chronological age and multiplying by 100. Formula of I.Q. is Mental Age Intelligence Quotient (I.Q.) = X 100 Chronological Age

$$\text{Intelligence Quotient (I.Q.)} = \frac{\text{Mental Age}}{\text{Chronological Age}} \times 100$$

The concept of mental maturity raises an interesting problem. *e.g.* Think of 15 years old with a mental age of 15. Clearly, his I.Q. is 100 or his mental age is still presumable 15 year. Therefore his I.Q. sunk from 100 to 75 I.Q. we must take the denominator age as 15. For avoiding these difficulties Wechsler-Bellevue adult Intelligence scale is used. Here final score is not calculated from mental age values of the tests passed by the person being examined, but from the total number of points made by him on all tests.

The Level of Intelligence and Categorization

Sl.No.	Categories	$I.Q = \frac{M.A.}{C.A.} \times 100$
1	Genius	170 and above
2	Very Brilliant	160-170
3	Brilliant	150-160
4	Very Superior	140-150
5	Superior	130-140
6	Very Bright	120-130
7	Bright	110-120
8	Average (Superior)	100-110
9	Average (Inferior)	90-100
10	Dull	80-90
11	Dull (Inferior)	70-80
12	Mental defective	65-70
13	Feeble minded	45-65
14	Imbeciles	25-45
15	Idiots	25 and above

Factors affecting Intelligence

G. Brown a psychologist pointed out that, there are numerous factors which directly or indirectly affect the intelligence or abilities of the individual and which makeup the behaviour pattern of the individual.

1. **Heredity and environment**: heredity provides the physical body to be developed with certain inherent capabilities while environment provides maturation and training of the organism. Newman concludes that the variations in I.Q. or intelligence were determined about 68 per cent by heredity and 32 per cent by environment. It means that 68 per cent of intelligence of the individual comes through heredity and 32 per cent by environment
2. **Age**: Brightness or dullness in childhood tends to remain bright or dull throughout his life. The intelligence is maximum at 20 years and remains relatively stable if health and other factors do not interfere, until around 70 years when it rapidly decreases due to decline in physical efficiency
3. **Health and physical development:** Health and physical development are directly related to mental activity. Physical and physiological defects result in sub-normal intelligence or less intelligence

4. **Race:** As it is race has no influence over the intelligence but certain races which are socio-economically and culturally week show marginal effect on intelligence
5. **Sex:** Not much difference is noticed as per the sex of the individual. According to Crow and Crow males are slightly superior than females in questions that involve mathematical material and scientific concepts or in performance of certain scientific tasks (work related to science) and girls excel that deal more directly with the humanities (languages, literature, philosophy, fine arts, history *etc.*)
6. **Social and economic conditions**: if these conditions are good then physical development and mental development will also be fairly good and intelligence will be better
7. **Culture**: Culture influences the intelligence of the individuals to a degree. It determines the attitude and abilities of the individuals.

Importance of Intelligence in Extension Work

1. Intelligence does not follow a set of similar types of pattern but depends largely on the complexity of demand of their environment and the kind of training they receive
2. Intelligence remain constant when the conditions remain constant *i.e.* health, types of education and situation
3. In all, the differences in intelligence can be treated to either heredity or environment since individual is a product of both
4. Gifted persons with higher intelligence can be better utilized by offering broader opportunities and with programmes for their accelerated growth
5. It is easy to identify the mentally retarded people or people with less intelligence and problem men (persons with less intelligence due to physiological defects) in rural society and such people should be given special attention while training them in agricultural technologies
6. An extension worker can increase his effectiveness by using appropriate techniques for teaching farmers with different levels of intelligence and thereby smooth introduction of the programs of change

Chapter 4

Communication Skills and Personality Development

COMMUNICATION SKILLS

Berlo (1960) points out there are five verbal communication skills. Two of these are encoding skills, writing and speaking while two of them are decoding skills, reading and listening. The fifth is crucial to both encoding and decoding *i.e.*; thought or reasoning. There are other encoding skills also, such as painting, drawing and gesturing. What we say about writing and speaking can be generalized to the other encoding skills as well.

From extension education's point of view, communication skills also include as extension worker's ability to plan a communication strategy vis-à-vis a given objective, his ability to write, speak, prepare and use a technical aid, treat a message to suit a situation and to analyze feedback.

As source-encoders, our communication skills determine our communication behavior in three ways:

i) They affect our ability to comprehend or understand what others are saying and expecting.

ii) They affect our ability to analyse our purpose and intentions.

They affect our ability to encode messages that express precisely what we intend to express.

What is Communication?

The word 'communication' comes from the Latin word communis, meaning common. 'Communication is a process by which two or more people exchange ideas, facts, feelings, or impressions in way that each gains a common understanding of meaning, intent and use of message.

Principles of Communication

The principle of communication helps to develop models and theories of communication. (Dipak De)

1. C.O.I.K (Clear Only If Known).
2. Leaky-pipe.
3. Least-effort.
4. KISS (Keep It Short Stupid/Keep It Simple and Straight Forward).
5. Seeing is believing.
6. One should express rather impress.
7. Brevity is the soul of wit and essence of good communication.
8. Good listening habits.
9. Ideas per minute are more important than words per minute.
10. Gesture and posture are more important than words per minute.
11. Come-on-Educate.
12. We may hear but not really listen. We may see but do not observe. We may read the lines but fail to read between the lines.
13. The poor speaker quits when he gets tired, but good speaker quit just before the audience get tired.
14. The amateur worries about what he is going to put in his speech. The expert worries about what should take out.

Structural Communication

Structural communication was developed in the 1960s by John G. Bennett and his research team to simulate the structure and quality of a small group tutorial through automated means. It provides access to high level learning for many students without much supervision. A communication consists of both information and structure. An effective two-way communication involves mutual verification of understanding through comparisons of structure. In structural communication, information is provided that participants have to organize in explicit ways.

Structural Communication is an instructional approach that provides a simulated dialogue between an author of instructional materials and the students. It has been called "an interactive technique for communicating understanding". Understanding is "inferred if a student shows the ability to use knowledge appropriately in different contexts and to organize knowledge elements in accordance with specified organizing principles" (Egan, 1972, p. 66). The technique was designed to encourage creative thinking in learners, allowing them to develop an understanding of a topic, not simply to memorize facts. Furthermore, Structural Communication was designed to promote learning for social action. The distinctions between the learning of knowledge and the learning for social action are evident in the components Bennett and Hodgson designed into the Structural Communication technique. The seven components of a Structural Communication unit are described below.

1. Intention

The opening statement, which defines what is to be studied, provides an overview, possibly an "advance organizer", and sometimes a rationale. It is used to provide a context for the content of the study unit and a focus for the study.

2. Presentation

The material, experience, exercise, case study, *etc.* which supplies the essential facts and concepts of the domain being studied. This may be an existing text, a video, a case study, a simulation, or real-life experience, depending on the overall strategy of the exercise and any sort of computer-based instruction, including simulations.

3. Response Array

A randomized array of items which summarize the key parts, concepts or principles from the knowledge base, that is being used and studied in the exercise. Often it resembles a "key point summary" of the Presentation.

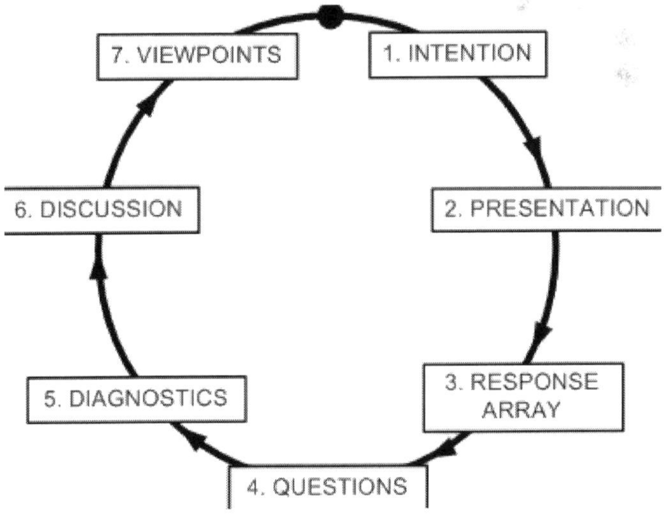

4. Questions

A set of problems for solution, which are designed to present the "intellectual challenge" that is an essential part of the Structural Communication methodology. These problems are interrelated and are open-ended to allow multiple responses and viewpoints. The purpose of the investigation section is for the learner to interact with the subject matter. To respond to a question, a student selects a number of items from the Response Array that, taken together, summarise what must be brought to bear on it.

5. Diagnostics

These are sets of logical tests of inclusion and exclusion. Relative to any given question, the author ascribes values of "essential", "relevant", "irrelevant" and "misleading" to the items of the Response Array and builds his tests around them. The student is not shown what these tests are but is given a set of comments based on his response.

6. Comments

The Comments are constructive statements that discuss in depth the rationale for including or excluding certain items. They can accommodate alternative interpretations and also serve to further the discourse represented in the Presentation.

7. Viewpoints

An outline of the author's, and other alternative viewpoints; this may review some aspects stated in the Intention, make explicit some biases or standpoints held dear by the author, draw attention to other views in the literature, *etc.* Ideally, the viewpoint section plays a final, interactive role between author and learner.

An additional aspect of the Structural Communication study unit is the assessment of the learner's responses to the questions posed by the study unit.

Elements and their Characteristics

Key Elements of Communication Process

From the different definitions it is clear that the communication process is called the interchange or the transfer of message, feelings and information between two persons or more than two persons. For completing this process some elements are necessary. But how many are these elements and which are these elements? On this issue many scientists conflict among themselves. The great thinker and philosopher Aristotle has analyzed these elements, observing the common talking among them that there are three elements of communication– (i) Communicator, (ii) Message, (iii) Receiver or Audience. Verifying this point Hareld Lasswell has analysed these elements of communication process that "Which person, what, to whom, by which channel, under what circumstance and how effectively speaks" Lasswell has told two more factors–(i) Channel (ii) communication-effectiveness, along with the three elements. Considering it Shannan C. and Weaver (1949) have told that there are five elements. *i.e.*,

Source → Communicator → Signal → Receiver → Destination Person

↑ ↓

← ← Feedback ← ←

Process of Communication

A process is any phenomenon which shows a continuous change in time or any continuous operation or treatment. In this context communication is a process wherein the act of communication is constantly changing as the communication proceeds. Further, communication is not an intermittent but a continuous function of human beings. It is not an occasional act in which human being choose to engage, rather its an essential continuum in their existence. Berlo (1960) views communication as a dynamic, ongoing, ever-changing, continuous act. He says that the process has no beginning, end or fixed sequence of events. It is not static, it is dynamic. The ingredients within the process interact, each affecting the other. In other words, we cannot really freeze communication at any one point but must consider it a by-product of interacting individuals. A message from one individual creates a response from the other, which then becomes the message for the first individual in a continuous process. Communication is the process of transmitting meaning between individuals. This process has been going on ever since the dawn of history. It is vital for the progress of society. Early man communicated through symbols and gestures and later on the spoken word, in the form of language, was used for communication. As technology developed, written words and other media were used, in addition to the symbols, gestures and spoken words. Every symbol and gesture had a meaning in the primitive times and even today they carry meaning and have an important place in the communication process.

Communication is as old as human history; many indicators prove that effective communication is the main factor enhancing civilization through history. This is why communication is, considered multi-culture phenomena; Ancient-Greeks, Ancient Egyptians, Ancient-Chinese, Arabs, Europeans and Americans have got their impact on communication. On the other hand, the study of human communication is interdisciplinary. It began with the mathematically theory of communication by Claude E. Shanon and Warren Weaver in 1949, scientists considered as strictly mathematical. Their aim was to measure the amount of information, in the mess ages, that is transmitted through the media the telephone; however, as years passed by, inter-disciplinary approaches to study human communication came up. They rely on psychology, sociology, speech communication, political, journalism, anthropology management, education, marketing and philosophy. Thus, we can say that every discipline concerned with human behaviour must deal with communication.

Meaning of Communication

Schramm defines communication as "a tool that makes societies possible and distinguish human from other societies". Berelson and Steiner define communication as the transmission of· information, ideas, emotions, skills through the use of symbols, words, pictures, figures, and graph. Rogers says, "Communication is the process of transmitting/ideas, information, and attitudes from the source to a receiver for the purpose of influencing with intent". Kar defines communication as "all those planned or unplanned processes through which one person influences behaviour of others."

A more comprehensive suggested definition to define communication would be: "a process of transmitting ideas, information, attitudes (images which we have formulated for ourselves) by the use of symbols, words, pictures, figures from the source (who is the originator of the message) to a receiver, for the purpose of influencing with intent", So communication is considered as a process through which senders and receivers of messages interact in a given social context.

We can conclude that communication is a process used to timely and properly exchange information between a sender and a receiver to achieve a desired goal.

- **Process:** It suggests that the components of interaction are dynamic in nature. They cannot be regarded as unchanging elements in time and space. This simply means that no single aspect of communication can be meaningfully understood apart from the other elements
- **Interaction:** It is the process of linking between senders and receivers of the message. The process specifies interaction or linkages between or among countless factors; so that the changes in any set of forces affect the operation of all other processes to produce a total effect. The concept of interaction is central to an understanding of the concept of process in communication. Communication is an attempt to bridge the gap between two individuals through producing and receiving messages which have meaning for both.
- **Social Context:** Human communication is, to a great extent. Influenced by the social context in which it occurs. The context or the situation that consists of a set of rules which govern the origin, flow and effect of the messages.
- **Overview:** Communication is a process whereby information is encoded and imparted by a sender to a receiver via a channel/medium. The receiver then decodes the message and gives the sender a feedback. Communication requires that all parties have an area of communicative commonality. There are auditory means, such as speaking, singing and sometimes tone of voice, and nonverbal, physical means, such as body language, sign language, paralanguage, touch, eye contact, by using writing

Types of Communication

1. Verbal Communication

It includes listening speaking reading and writing

2. Non Verbal Communication

Communication without language is known as nonverbal. It is also known as gestural communication. This includes gestures, facial expressions, *etc.*, which make communication more effective.

3. Intra Personal Communication

It occurs within ourselves and continuous as long as we are alive.

4. Small Group Communication

In Mutual conversation between five to ten people. It may be formal or informal.

5. Mass Communication

It is a communication system in which an identical message is originated by an institutional organization and sent to a large number of receiver through public channels (TV, Radio, Newspaper, Magasines, Film, Internet)

6. Dyadic Communication

In dyadic communication, two individuals communicate with each other.

7. Interpersonal Communication

Interpersonal communication most commonly occurs in face to face situations, where we can see, hear and even touch the other person or persons.

8. Public Speaking

One person addresses a large audience. Public speaking is mostly one way from the speaker to audience. Messages are conveyed by the speaker and received by the audience.

Verbal Communication

Verbal communication is one way for people to communicate face-to-face. Some of the key components of verbal communication are sound, words, speaking, and language. Speaking can be looked at in two major areas: interpersonal and public speaking. The basis of communication is the interaction between people. Verbal communication is one way for people to communicate face-to-face. Some of the key components of verbal communication are sound, words, speaking, and language. At birth, most people have vocal cords, which produce sounds. As a child grows it learns how to form these sounds into words. Some words may be imitative of natural sounds, but others may come from expressions of emotion, such as laughter or crying. Words alone have no meaning. Only people can put meaning into words. As meaning is assigned to words, language develops, which leads to the development of speaking.

Speaking can be looked at in two major areas: interpersonal and public speaking. Since the majority of speaking is an interpersonal process, to communicate effectively we must not simply clean up our language, but learn to relate to people. In interpersonal speaking, etiquette is very important. To be an effective communicator one must speak in a manner that is not offending to the receiver. Etiquette also plays an important role in an area that has developed in most all business settings: hierarchical communication. In business today, hierarchical communication is of utmost importance to all members involved. The other major area of speaking is public speaking. From the origin of time, it has been obvious that some people are just better public speakers than others. Because of this, today a good speaker can earn a living by speaking to people in a public setting. Some of the major areas of public speaking are speaking to persuade, speaking to inform, and speaking to inspire or motivate.

Non-Verbal Communication

Nonverbal communication (NVC) is usually understood as the process of communication through sending and receiving wordless messages. NVC can be communicated through gesture and touch (Haptic communication), by body language or posture, by facial expression and eye contact. NVC can be communicated through object communication such as clothing, hairstyles or even architecture, symbols and infographics. Speech contains nonverbal elements known as paralanguage, including voice quality, emotion and speaking style, as well as prosodic features such as rhythm, intonation and stress. Likewise, written texts have nonverbal elements such as handwriting style, spatial arrangement of words, or the use of emoticons. However, much of the study of nonverbal communication has focused on face-to-face interaction, where it can be classified into three principal areas: environmental conditions where communication takes place, the physical characteristics of the communicators, and behaviors of communicators during interaction.

Verbal vs Oral Communication

Scholars in this field usually use a strict sense of the term "verbal", meaning "of or concerned with words," and do not use "verbal communication" as a synonym for oral or spoken communication. Thus, vocal sounds that are not considered to be words, such as a grunt, or singing a wordless note, are nonverbal. Sign languages and writing are generally understood as forms of verbal communication, as both make use of words although like speech, both may contain paralinguistic elements and often occur alongside nonverbal messages. Nonverbal communication can occur through any sensory channel sight, sound, smell, touch or taste. NVC is important as:

"When we speak (or listen), our attention is focused on words rather than body language. But our judgment includes both. An audience is simultaneously processing both verbal and nonverbal cues. Body movements are not usually positive or negative in and of themselves; rather, the situation and the message will determine the appraisal."

Different Non-verbal communication

Arbitrariness

While much nonverbal communication is based on arbitrary symbols, which differ from culture to culture, a large proportion is also to some extent iconic and may be universally understood. Paul Ekman's influential 1960s studies of facial expression determined that expressions of anger, disgust, fear, joy, sadness and surprise are universal. Uniforms have both a functional and a communicative purpose. This man's clothes identify him as male and a police officer; his badges and shoulder sleeve insignia give information about his job and rank. Elements such as physique, height, weight, hair, skin color, gender, odors, and clothing send nonverbal messages during interaction.

Physical Environment

Environmental factors such as furniture, architectural style, interior decorating, lighting conditions, colors, temperature, noise, and music affect the behavior of communicators during interaction. The furniture itself can be seen as a nonverbal message.

Proxemics

Proxemics is the study of how people use and perceive the physical space around them. The space between the sender and the receiver of a message influences the way the message is interpreted. The perception and use of space varies significantly across cultures and different settings within cultures. Space in nonverbal communication may be divided into four main categories: intimate, social, personal, and public space. The distance between communicators will also depend on sex, status, and social role. The term territoriality is still used in the study of proxemics to explain human behavior regarding personal space. Hargie and Dickson identify 4 such territories:

1. **Primary territory**: this refers to an area that is associated with someone who has exclusive use of it. For example, a house that others cannot enter without the owner's permission.
2. **Secondary territory**: unlike the previous type, there is no "right" to occupancy, but people may still feel some degree of ownership of a particular space. For example, someone may sit in the same seat on train every day and feel aggrieved if someone else sits there.
3. **Public territory**: this refers to an area that is available to all, but only for a set period, such as a parking space or a seat in a library. Although people have only a limited claim over that space, they often exceed that claim. For example, it was found that people take longer to leave a parking space when someone is waiting to take that space.

4. **Interaction territory**: this is space created by others when they are interacting. For example, when a group is talking to each other on a footpath, others will walk around the group rather than disturb it.

Chronemics

Chronemics is the study of the use of time in nonverbal communication. The way we perceive time, structure our time and react to time is a powerful communication tool, and helps set the stage for communication. Time perceptions include punctuality and willingness to wait, the speed of speech and how long people are willing to listen. The timing and frequency of an action as well as the tempo and rhythm of communications within an interaction contributes to the interpretation of nonverbal messages. Gudykunst and Ting-Toomey (1988) identified 2 dominant time patterns:

- ✰ **Monochronic time schedule (M-time)**: Time is seen as being very important and it is characterized by a linear pattern where the emphasis is on the use of time schedules and appointments.
- ✰ **Polychronic time schedule (P-time):** Personal involvement is more important than schedules where the emphasis lies on personal relationships rather than keeping appointments on time.

Movement and Body Position

Information about the relationship and affect of these two skaters is communicated by their body posture, eye gaze and physical contact. Kinesics is the study of body movements, facial expressions, and gestures. It was developed by anthropologist Ray L. Birdwhistell in the 1950s. Kinesic behaviors include mutual gaze, smiling, facial warmth or pleasantness, childlike behaviors, direct body orientation, and the like. Birdwhistell proposed the term kineme to describe a minimal unit of visual expression, in analogy to a phoneme which is a minimal unit of sound.

Posture

Posture can be used to determine a participant's degree of attention or involvement, the difference in status between communicators, and the level of fondness a person has for the other communicator. Studies investigating the impact of posture on interpersonal relationships suggest that mirror-image congruent postures, where one person's left side is parallel to the other's right side, leads to favorable perception of communicators and positive speech; a person who displays a forward lean or a decrease in a backwards lean also signify positive sentiment during communication. Posture is understood through such indicators as direction of lean, body orientation, arm position, and body openness.

Facial Expression

The face is most prominent part in our body for sending non-linguistic symbols. It tells others what kind of person we are. It reflects our personality. Our face expresses many feelings and attitudes by contracting and relaxing facial muscles. For example, by smiling we show our happiness to others.

Some more common facial expressions and their usual meanings are:

- ☆ Arching the eyebrows – anger or seriousness
- ☆ Raising the eyebrows – disbelief
- ☆ Biting of lips – uncertainty
- ☆ Both the mouth and the eyes open – surprise/fear

Body Posture

The body posture refers to the way one stands or sits. It plays a very important role in non-linguistic communication. Depending on the posture, we can say whether he is interested in talking to you. For example, when a person stand learning against table with both his hands on the table, it shows that he is tired or not willing to participate in the act of communication.

Gestures

Gestures are another important sign of no-verbal communication. Gestures are very common in our life and they are the natural movements of our body, especially, head, shoulders and hands. Some of the gestures and their meanings are giver below:

A handing head or a hand signifies boredom, restlessness, or tiredness. It may also indicate frustration or disgust.

- ☆ Picking something, biting nails or sucking fingers show that the person is uneasy and nervous.
- ☆ Raising a hand in a classroom, a meeting or a seminar, or at a gathering means that the person wants the attention of the speaker.
- ☆ Pointing index finger means the indication of a particular direction. It may also indicate authority and domination.
- ☆ Covering the mouth may indicate that the person is either unsure of what he is saying or suppressing his feelings.
- ☆ If a person rubs his eyes, it may mean that he is in doubt or trying to deceive.
- ☆ Patting on the back of someone means encouraging that person.
- ☆ Hands on hips generally may indicate an aggressive attitude or readiness and ability to do a job.

Eye Contact

Eyes play a very important role in communication. It is generally understood that eyes reveal one's nature/character. A confident and honest person has a direct eye contact with others. People like thieves and other convicts do not look directly into other's eyes.

Listening

Listening is both a physiological as well as a cognitive process, a sort of qualitative add-on to hearing. Listening also helps the cultivation of smooth interpersonal relationships with fellow workers, which is essential for efficient functioning of the organization.

Listening Skills

Research studies and surveys have shown that most of the adults are poor listeners. A basic reason for this is that we practice it seldom though few of us have been taught how to listen; listening efficiency can actually be raised by merely giving some attention to it. A prime cause of poor listening is the difference between the think speed and a person's rate of speech. This refers to the difference between the average rate of speech-about 125 to 175 words a minute and brain's capacity to think words, which is at the rate of 500 to 1000 words a minute or in some case 5000. Thus, our brain works too fast and the relatively slow input of these 125 to 175 words a minute leaves the mind much room to think about other things. When we are engaged in conversation we find so many times that the listener is miles away.

Guidelines for Listening Effectively

Active and effective Listening is must to become an effective speaker. By Listening carefully to his audience, a communicator can gauge how his message is being received. By Listening patiently to their comments and questions, he can tell how his points are being understood. The listeners can see the following techniques/ideas to improve their basic Listening. A communicator can put them to use to improve his overall Listening ability.

1. **Stop talking**: You cannot listen if you are talking.
2. **Put the talker at ease:** Help a person feel free to talk. This is often called a permissive environment. In this environment speaker feels to express opinions, feelings ideas and attitudes.
3. **Show a talker that you want to listen:** Look and act interested. Do not do any other (reading mail) while someone talks. Listen to understand rather than to oppose.
4. **Remove distraction:** Do not tap or shuffle papers. If there is some noise or distraction from outside it will be better to shut the door.
5. **Give feedback:** When you are communicator, immediate feedback is very important. If there is a misunderstanding, the best time to rectify it is immediately, to avoid future problems.
6. **Listen between the lines:** Very often, we may say one thing while we really mean something quite different. Others may also do like this. So, it

is important to remember this advice. Do not listen to what I say, listen to what I mean.

7. **Listen for purpose motive:** Every speaker hopefully has a purpose, but it may not related to the content of the speech. In order to determine the speaker's motive, free your mind of traditional evaluative thoughts and ask yourself the question what is the speaker's purpose.
8. **Listen for attitudes:** Our behavior is a reflection of our attitudes, and our attitudes are shaped by our motives. To understand a person's motives, you must listen carefully or expressions about other people, classes, groups or ideas.
9. **Listen to non-verbal language:** This language is expressed through eye contact and facial expressions such as smiles, frowns, raised eye brows, moving versus steady eyes, a tense versus a relaxed face, and look of approval or disapproval *etc.*
10. **Avoid negative feedback:** When people become emotionally upset they tend to interrupt, argue with, or criticize the speaker. It makes the speaker defensive and he/she can hide his real opinions, feelings, ideas and attitudes.
11. **Emphasize with the talker:** Perhaps the best trait of a good listener is that of empathy. Being able to put ourselves in other's positions and sincerely trying to see things from their point of view is truly an ability teacher and trainers should cultivate.
12. **Be patient:** allow plenty of time. Do not interrupt a talker. Do not start for the door or walk away.
13. **Hold your temper:** an angry of disturbed person takes the wrong meaning from words. An individual is better able to deal with the situation if he fully understands the opposing position.
14. **Go easy on argument and criticism:** This puts people on the defensive, and they may ' clean-up ' or become angry. Do not argue. Even if you win, you lose. Ask questions: This encourages a talker and shows that you are listening. It helps to develop points further.
15. **Stop talking:** This first last, because all other guides depend on it. You cannot do an effective listening job while you are talking.

Note Taking

Studies find that note taking helps students' focus attention, promotes more thorough elaboration of ideas, and encourages efforts to relate ideas and organize materials. In short, note taking helps students to process information more deeply. As an instructor, you can do a number of things to help your students take and use their notes more effectively. Here are some tips:

Outline your Lecture/Use Transition Phrases

Be overt in the organization of your lecture, both orally and visually. For example, write your lecture outline on the board. Second, since studies show that students usually record what the instructor has written, be discriminating in your use of the board or transparencies. Third, during the lecture, refer to your outline to highlight shifts in topic. Last, use signaling phrases and transition statements such as "this is important," "you'll want to remember," "these differ in three important ways," "the second point is, "or "next,........"

Use a Framework

If appropriate to your subject matter, give students a framework or schema for how to organize information. Most information can be organized into one of two frameworks:

Sequence or Classification

Sequence is used to explain change, influence, or phases. For example, if you're presenting information that illustrates time, space, or a process it's most likely a sequence; "stages of development" suggests a process that can be illustrated with arrows.

WRITING SKILLS

You cannot make people understand a subject unless you understand that subject yourself. - William Jennings Bryan, researches conducted in the area of effective writing have mainly established four important steps *viz.*, plan, write, trim and check.

1. Planning

Good planning always helps the writer in visualizing receivers and identifying their interests and felt needs. Actually a good writer plans his/her writing like an architect who first drafts his plan and designs every detail and thus prepares a blue print of the building before its construction. Before writing, purpose of writing, message one wants to tell and characteristics of audiences/farmers/farm women should be very clear. Then rearrange ideas in order: Bait your reader *i.e.* catches their interest by drawing their attention by giving a good title. While writing, it is very important for writers to follow the 'Principle of empathy' *i.e.* putting himself in the reader's place and start judging the writing. Before you put words on paper that will bait, hold, sell and tell readers you must answer six questions.

a) **Who?** - Your Readers: Who are the people you want to reach?

   You cannot help people solve their problems until you know who they are and what their problems are? Know what your readers know and what they want to know. Visualize their educational, social, and· economic level, their interests, attitudes and beliefs.

b) **Where? :** Where do the intended readers live? Where will they use the publication? -rural area, urban areas, worldwide, continental, in school, at home, in the library, regional, national, provincial, local, in the office, in an enterprise *etc.*?

c) **What? :** Your facts: Have something to say and be sure of your facts? List facts that help readers solve a problem. Find a slant, a central theme to develop. Weigh facts. Is information timely and of local interest? Does it meet a need? It is practiced: Can people use it, can they afford to use it?

d) **Why? :** Your reasons for writing the publication. What is your purpose? What do you want your information to accomplish? Do you want to stimulate interest in a programme, or are you trying to show people how to do something?

e) **When? :** When is the material going to be used? Will it be of value only for a short time? For many years? It is needed at once? If not, when?

f) **How? :** Your outline for writing facts. Think through - how you are going to package your facts, how are you going to present them? First make a rough outline of facts listed.

2. *Writing*

After preparing a writing plan you should start putting down your ideas.

While Writing

1. **Make it brief:** Try to say what you want to say as briefly as you can. Of course, do not miss an)! of the important things you want to say.
2. **Write as you talk:** Simple writing is like talking with your readers.
3. **When you talk:** to a farmer, you bring your language down to his level, so that he understands you and you understand him. You thus create a common denominator between yourself and him. Simple writing is much the same.
4. **Write in clean terms:** Award may mean different things to different people. Your reader attaches a meaning to a word, prompted by his own experience. Use words which bring only one meaning you want him to have. You cannot go wrong, if you stick to simple words.
5. **Use concrete words:** Use concrete word, which create a picture in the readers mind. Concrete words always make meaning clear.
6. **Use personal sentences:** Like questions, exclamations sentences starting with verbs, spoken sentences directed to your reader. Sentence pattern should vary and sentence length on an average, should be about 17-19 words.
7. **Use familiar words:** Use familiar, every day words and the reader will feel that your writing is specially addressed to him. Technical words should be translated to their language without losing exact meaning, if possible.

8. **Say things accurately:** Use words that are precise and say exactly what you want to say. Check your facts and include only those which make the reader understand the subject. Don't generalize, give concrete examples, whenever possible which make him understand easily what you want to say.
9. **Make it short tales:** Use mostly short sentences. Let each sentence talk about one idea. Use a noun in each sentence. Use a strong verb too. Then your writing will carry your thoughts forward and smoothly. Do not make the sentence all of the same length; use a longer sentence now and then.
10. **Add human interest:** Write about people. Include names. Identify people.
11. **Address it to the reader:** Write directly to the reader. He will then feel your writing is specially meant for him.
12. **Make your writing friendly:** Do not sound bossy in your writing. Make yours a friendly approach, use appeals, persuade rather than force.
13. **Make it interesting:** Do not make your writing a drab account of facts. Adopt a style that will be interesting to read. See how others do it. You can adopt a style of writing that appeals to you most. There is nothing wrong in copying someone else way of writing, till you are able to, through experience, develop your own.
14. **Sift, sort and sell facts:** Screen out irrelevant ideas. Group facts under main heading, this helps you sort facts. Keep related ideas together, put first thing first. Rearrange ideas in the order your reader will take. Motivate your readers to read, believe and use your information must entertain as well as inform, must be clever as well as clear.

3. *Trimming of Written Material*

- ☆ Actually there is no such thing as good writing, there is only good rewriting.
- ☆ Never write more than required to convey the message.
- ☆ Say the same thing in few words. Change passive verbs to active so your reader can understand you're writing more easily.
- ☆ Weed out unnecessary ideas, introductions, irrelevant sentences or parts of sentences, words and conductance like and, as, while, because, with a semi colon, if possible.
- ☆ Weed out articles - a, an, the. You do not need them half the time *e.g.* (The) making (of) silage is one of the best way of preserving (the) roughage.
- ☆ Avoid vague adjectives and adverbs - those, indefinite words that express degree, such as substantial (substantially), appreciable (appreciably), considerable (considerably). These words clog your message; they give your reader no standard of comparison *e.g.* 'This is a subject of considerable importance'. Tell the same message with 'this is important'.

- ☆ Trim words but not the meaning.
- ☆ When you prune you polish.

4. *Check and Recheck Writing*

- ☆ Check and recheck while reading its average sentence length and word length to find ' How easy'.
- ☆ Figure percentage of personal words and personal sentences to know 'How interesting'.
- ☆ Technical information must also be rechecked.

Difference between Technical Writing and General Writing:

Sl.No.	*Technical Writing*	*General Writing*
1.	Technical writing focuses on factual and straight forward content.	General writing focuses on general subjects of writing.
2.	Technical writing aims to impart information, instructions and explanations.	General writing aims at amusement.
3.	Technical writing is based on objective tone.	General writing is based on subjective tone or personal style.
4.	It follows standard structured format.	It follows artistic unstructured format.
5.	It is usually based on a formal way of writing.	It may be formal or informal way of writing.
6.	The voice used in technical writing is 3rd person.	The voice used in general writing is 1st person.
7.	In technical writing specialized vocabulary is required.	In General writing evocative vocabulary is required.
8.	Technical writing has its specific set of audience or readers.	General writing has no specific set of audience or readers.
9.	Technical writing is archival.	General writing is non-archival.
10.	It is accurate and more precise in manner.	It is decorative and bombastic in manner.
11.	Technical writing is related to any specific area or domain.	General domain is related to any life in general.
12.	It focuses on more technical words specific to the area.	It focuses on more common words suitable to the general topic.
13.	For example writing any articles on Geeks for Geeks related to computer science field comes under technical writing.	For example writing any prose or story which will be published in a local magazine comes under general writing.

ORAL PRESENTATION SKILL

What is the Purpose of your Presentation?

Why are you giving a presentation? Have to explain, to convince, to entertain, to justify? The purpose of your presentation will determine how you structure it, what information you will include, what format you will adopt, the type and form of your visual aids, *etc.*

Who is your Audience?

Your audience will most likely include fellow classmates and your tutors or lecturers.

- ☆ What will they know of your subject?
- ☆ What background information do you need to provide?
- ☆ What technical terms or complex matters will you need to explain and what can you assume they already know?
- ☆ How can you keep them interested?
- ☆ What questions might they ask?
- ☆ Will your presentation be marked? If so, what are the marking criteria?

Preparing

Introduce your talk with a brief overview of the points you will cover, locating the topic in its wider context and clearly stating your argument or thesis. Describe what the presentation is about and how it will develop.

For example:

- ☆ 'I'm going to talk about.'
- ☆ 'This morning I want to explain.'
- ☆ 'The points I will focus on are first of all. Then. This will lead to. And finally.'

Tips: Display the outline of your talk in key points on an OHT or slide.

Try to begin with an attention grabber to capture your audience's interest. Some ideas are:

- ☆ a startling fact
- ☆ a cartoon
- ☆ a pertinent question
- ☆ an interesting statistic
- ☆ a dramatic visual aid

Tips: Avoid telling a joke unless you know it is going to work. If it falls flat, so will your confidence.

The Body

Decide the main ideas you want to get across and group them into logical and coherent 'chunks' of information.

a) Keep focused on what's important.
b) Ensure key ideas stand out.
c) Keep relevant and to the point. Omit trivial details and waffle.
d) Provide the necessary evidence to support your conclusions.
e) Use clear examples to illustrate your points.
f) Use 'connective devices' to move from point to point, to help your audience follow your thread of ideas, *e.g.*
   - 'The next point I want to make is.'
   - 'From this we can see that.'
   - 'As a consequence.' 'In spite of this, the results showed that.'
   - 'An example of this was when.'
   - 'On the other hand, it is also true that.'

Practicing

Being prepared means rehearsing your presentation as many times as it takes to get it right. Even experienced presenters advocate plenty of run-throughs. Use family, friends, or a mirror. This will increase your confidence and allow you to feel more relaxed and sure that everything will work on the day. You will know that your visual aids will fit in with your text, that you won't be shuffling around in a heap of notes, and that you will not 'dry up'. Also, most important, you'll get the timing right.

Delivery

- Make sure you know what type of equipment will be in the room and how to use it.
- Do not read from pages of notes. Use your visual aids for cues and have brief back-up notes on cards or light cardboard about A5 size or smaller. (Don't try to memorize your talk).
- Use appropriate language — *e.g.* get the right level of formality/ informality.
- If you think you are going to stumble over a particular word or phrase, use a different one if you can. Otherwise, rehearse it enough times so that it trips off the tongue.
- Don't use too much 'filler speech' — too many 'ums' and 'ahs' will irritate your audience?
- Sound and look enthusiastic. If you want your audience to be interested in your talk, then you must show interest and enthusiasm yourself.
- Pitch your voice at the back row, not the front row.

Body Language

Where will you stand in relation to the equipment you are using? If you are using a screen, make sure you don't stand in front of it.

- Try to adopt a relaxed posture but not so relaxed that you look sloppy or unprofessional.
- Avoid nervous gestures such as waving hands around, clicking pens, or brandishing a pointer.
- Make eye contact with your audience. Try to avoid gazing over people's heads or looking down at your notes too much. If you find it difficult to make eye contact, concentrate on a couple of friendly faces for a few seconds at a time.

Field Diary

The field diary is the basic document which contains all the data collected. Facts and interpretations are to be written separately and conversations to be noted in the vernacular languages. Two kinds of field notes are to be observed

- Taking notes on the spot
- Writing detailed diary

Daily Field Diary

It is instructed to write the field diary everyday in which

- Detailed notes of observations
- Field notes on discussion and conversations
- Details of Interpersonal communication, gossips, fold history's similar events
- Specific details of special events

Specific reference about the main field of study for instance, the details about the process of message diffusion, various kinds of channels of communication were described in detail.

Field Diary for Research Process Documentation

The most useful tool for a process documenter is a field diary. The documenter uses a field diary to record his or her observations and thoughts in an orderly fashion. Recording observations and impressions enables a researcher to pick up clues about how the system is operating.

A field diary should help the investigator understand the systems physical and social setting. It should help describe who, what, why, where, when and how. Who refers to the people or system being studied? What concerns the information gathered. Why, where, when and how provide important details about the observation.

There is no special format in writing a field diary but entries should be written daily in chronological order so information is not forgotten or changed because of forgetfulness.

Laboratory Record

Guidelines for Keeping a Laboratory Record

The following is a general description of how to keep a proper laboratory notebook. Requirements for different teaching, research, clinical, or industrial labs will most likely vary. Some institutions/labs will require less stringent record keeping, others will hold you to a very strict protocol. A well kept notebook provides a reliable reference for writing up materials and methods and results for a study. It is a legally valid record that preserves your rights or those of an employer or academic investigator to your discoveries. A comprehensive notebook permits one to reproduce any part of a methodology completely and accurately.

Choosing a Notebook

For most purposes you may select a bound notebook, quadrille-ruled. A teaching lab may require tear-out duplicate pages for making carbon copies. An engineering or industrial research/development lab will likely require a specific type notebook with pre-numbered pages and places for date and investigator's and supervisor's signatures on each page. Pads of tear-out graph paper or spiral bound notebooks without pre numbered pages are not acceptable.

Preparing the Notebook

Please use a ball point pen for all entries, so that the marks will not smear nor will they be erasable. Put your name, a telephone number and/or address, and project name or course number on the outside front cover of the record. Put that same information on the first page inside, or on the inside front cover. If your notebook does not include a pre-labeled table of contents section, then reserve the next several pages for a table of contents by labeling the top of each page as Table of Contents and numbering each page.

What to Enter

Above all, it is critical that you enter all procedures and data directly into your notebook in a timely manner, that is, while you are conducting the actual work. Your entries must be sufficiently detailed so that you or someone else could conduct any procedure with only the notebook as a guide. Few students (and not that many researchers for that matter) record sufficiently detailed and organized information. The most logical organization of notebook entries is chronological. If a proper chronological record is kept and co-signed by a coworker or supervisor, it is a legally valid record. Such a record is necessary if you or your employer is to keep your rights to your discoveries.

Depending on requirements set by a teacher, supervisor, company, or whatever, you may not have to confine your notebook entries to lab notes only. On the other hand a student might record your class lecture notes, lab lecture notes, ideas, questions, library research notes, and notes that are part of any pre-lab preparation. The bare minimum entries for an academic lab course, for each lab study, should include title of the lab study; introduction and objectives; detailed procedures and data (recorded in the lab itself); summary.

Making Entries

Someone else may need to consult your notebook sometime, so please make your entries clear and legible. When you make your first entries of the day, start by entering the date, writing out the month or abbreviation for the month (*e.g.*, 5 Apr '04, or April 5, 2004, but not 4/5/04). The use of numerals only can cause confusion. For example, in Europe the day comes before the month. Thus April 5, 2004 would be written as 5/4/04.

Depending on how your notebook is designed you may choose whether or not to use the backs of pages. If you leave them blank, put a corner-to-corner line through them to void all blank spaces. Some people use the backs for rough calculations, then void remaining blank space. You might also decide to save space (and trees) and use both sides of each page. Obviously you cannot use both sides with notebooks that are designed to make duplicate copies. In situations where you turn in duplicate copies to a supervisor, you obviously must start each new set of entries on a new page. If you make a mistake, put a line through the mistake and write the new information next to it. Never erase or obliterate an entry. When you finish a page, put a corner-to corner line through any blank parts that could still be used for data entry.

The Summary

When you have finished a project, summarize what you have accomplished. You don't have to draw conclusions, just indicate what sort of data or observations you collected, samples you saved (and where and how you saved them), or any other relevant information that wraps up the study. For a continuing study keep the summary extremely brief. In fact, if the notes are well organized and it is obvious where the study left off, you need write nothing more than "To be continued." Summaries help maintain continuity. They indicate where the work left off and how it might resume.

Repeated Procedures

So far you have been advised to record each step you perform in the laboratory, regardless of whether the procedure is published somewhere. However, once you carry out a procedure, you can refer to that part of your notebook, and only note changes you make. For example, the first time you prepare a sequencing gel you should write down the exact formulation, how you mix the gel, how long you let it cure, *etc.* The next time, just refer to the name of the procedure and the appropriate page(s) of your notebook.

Loose Materials

Suppose you enter raw data into a computer and have a printout with 400 pieces of data or, suppose you generate a graph using a software program. You might even have a silver-stained gel that you wish to refer to frequently, or a fluorescence photomicrograph that sums up your results nicely. Some investigators prefer to attach such materials to the notebook itself, but too many such items make a sloppy notebook and can stress the binding. Loose data should be kept in a separate folder or notebook, with location noted in the book.

Table of Contents

Record all entries in the table of contents as you go along. You can organize it anyway you like but it is advisable to include multiple levels in a table of contents, that is, indicate where a new study starts and include subheadings for specific parts of a study, methods, sets of data, *etc.* The idea is to enable someone (such a supervisor, grader, or yourself a year from now) to find anything quickly. List each set of entries with dates and page numbers.

Notebook Checklist

As you record your activities in the laboratory, ask yourself, "Did I."

- ☆ Keep up with the table of contents?
- ☆ Date each page?
- ☆ Number each page consecutively?
- ☆ Use continuation notes when necessary?
- ☆ Properly void all blank pages or portions of pages (front and back)?
- ☆ Enter all information directly into the notebook?
- ☆ Properly introduce and summarize each experiment?
- ☆ Include complete details of all first-time procedures?
- ☆ Include calculations?

Indexing

An index is a reference list for locating a particular document in the file. Indexing helps:

1. Easy location of files
2. Ensures cross-reference
3. Improves the efficiency of the records administration and reduces costs

A good Indexing system should be

a. Simple to operate and use,

b. Flexible in expansion,
c. Suitable for the organisation.

Types of Indexing

Types of indexes are also known as methods/systems of Indexing. Each office uses the Indexing method that suits and aids its filing system. The following are some of the types of Indexing systems.

1. **Page Index** - A page index consists of a page for each letter of the alphabet fitted with a tab showing the letter, and on each letter is written the names beginning with that letter and quoting the relevant page number. Bound Book Index, Loose leaf Index, Vowel Index are some of the forms of Page Index.
2. **Loose or Vertical card Index-** It is used to overcome the drawbacks of the Page Index. A card index consists of a number of small size cards (12 cms x 7 cms) each concerned with one item of the index. Heading is on the top edge of the card.
3. **Visible Card Index** - Cards, under this system, are laid flat in transparent covers in a shallow tray or in a metal frame. Each card is fitted into a metal hinge so that it overlaps the one before it in such a way that a narrow strip at the bottom containing the name or title remains visible.

Automatic Card Index: Visible Books and Staggered Card Index are some of its variations.

Advantages

1. Occupies less space,
2. Though costs more saving in time and labour offsets it,
3. Additional information can be added without disturbing the order,
4. Speedy reference,

Disadvantages

1. More costly,
2. Training necessary.

Yet its advantages make the user's ignore its drawbacks.

4. **Strip Index-** Strip index is designed for maintaining a list of the names, addresses and telephone numbers of the correspondents. It consists of a frame into which strips of stiff papers can be fitted in any required order. Frames containing these strips may be fixed on the wall or arranged on a rotary stand. Strips can be protected from, exposure/damage with removable transparent celluloid or plastic window sheets. The advantages and drawbacks of this type are similar to those of Visible Card Indexing.

5. **Wheel Indexing -** It is a sort of modern method of visible indexing. Here cards are arranged about the circumference of a wheel which may be portable or set in a cabinet or desk. A single wheel can hold as many as 5,000 cards and as many as six such wheels can be set up within the easy reach of a clerk.

Footnotes

Most students will have to reference an assignment at some stage of their studies. Understanding the correct way of references, this will help students to avoid plagiarism. The footnote system of citation is used in some discipline areas, *e.g.* history, fashion, art and design. Footnotes must be used to reference and acknowledge the sources of your information. This unit will take you through how to reference using footnotes.

Inserting Footnotes

How to insert footnotes using Word XP

a. Place cursor where the footnote is to be inserted
b. From the menu bar, select Insert
c. Select reference>footnote and select options
d. The (next) numeric footnote will automatically be inserted and also placed at the bottom of the page.

How to insert footnotes using Word 2000

a. Place cursor where the footnote is to be inserted
b. From the menu bar, select Insert>footnote and select options
c. The (next) numeric footnote will automatically be inserted and also placed at the bottom of the page

Bibliographic Procedures

All literature cited in the text should: be included in the list of references giving complete details. Standard style of literature citation should be followed.

A. Citation in Text

During preparation of citation, the style of reference in the text varies from journal to journal and therefore consults the guidelines of the journal and goes to references cited for noting style. In case of one author, give last name/surname first, year of publication within parentheses, *e.g.* Singh (2002). In case of two authors, give first the last name of first author followed by last name of second author, year of publication *e.g.* Singh and Grover (2002). In case of more than two authors give last name of first author followed by *et al.,* and year of publication *e.g.* Singh *et al.* (2002). In case the author is referred at more than one place in the same year, number references as 'a' and 'b' *e.g.* Singh (2002Sa). Singh (2002b).

B. Citation in Journal

This must include the name of author(s), followed by year of publication within parentheses or stops, title of the paper/article, name of the journal/periodical, volume and number in brackets, and page number. In case of male authors, initials are used and in case female authors give full name *e.g.* Singh, J. and Grover, Indlj. 2002.

Academic achievement and entrepreneurial performance of undergraduate students of CCS Haryana Agricultural University, CCSHAU J. Research. 88(3), 416-421.

Note: In some journals the name of journal is written in italics.

C. Citation from Book

a. Khetarpaul, Neelam and Grover, Indu.2000. A Consumer Guide for Home Makers. Agrotech Publishing Academy, Udaipur, p 40.
b. Khetarpaul, Neelam; Katyal, Sudha and Grover, Indu. 2001.nfant Health and Nutrition. Agrotech Publishing Academy. p 298.

   (In this example, total pages of the book have been referred).

Citation from an Edited Book

1. Grover, Indu., Agarwal, Deepti and Grover, O.K. 2000. Gender perception of voters towards women and governance. In: Singh Raj (ed.). New Panchayati Raj: A functional analysis, Anmol Publications Ltd. pp 66-77.

   (pp 66-77 indicates that chapter is within these pages).

Citation of Seminar/Conference Proceedings

1. Sinhg, Saroj., Grover, Indu and Grover, O.K. 2001. Participatory role and adoption of buffalo management practices by rural women in Haryana (India). In: National Conference on Extension Education for Sustainable Devel,?pment, Mumbai, 29-31 December, 2000.Proceedings edited by O.S. Verma, AK Singh and J.P. Yadav, New Delhi, ISEE. pp 201-222.
2. Rana, Manju. 1992. Effectiveness of training methods among farm women. Thesis, Ph.D., CCS Haryana Agricultural University, Hisar, p 156.

Citation from Institutional Publication

a. Department of Science and Technology, New Delhi. 1997. General information on research and development funding schemes of Central Government departments/agencies, New Delhi. GOI, Ministry of Science and Technology, Deptt. of Science and Technology, p 135.

Citing Government Publications

☆ India. 2002. Ministry of Human Resource Development. Annual Report. New Delhi. p 270.

- ☆ India. 2001. Ministry of Agriculture, Department of Agriculture and Cooperation, Directorate of Economics and Statistics. Indian Agriculture in Brief. (30th ed.), Delhi. Controller df Publication. p 389.

Abbreviations for Names of Journals

The names of journals should be abbreviated according to international standards/rules. Most journals give abbreviated forms on top/bottom side of each page. Some examples are as follows:

- ☆ Curr. Sci. ·
- ☆ Ind. J. Ext. Edu. ·
- ☆ Maha. J. Extn. Edu. ·
- ☆ Ind. J. Genet and Plant · Breeding
- ☆ Plant Physiol.
- ☆ Plant Dis. Rep.

READING

Reading Skills

Communication is sharing ideas in common. Sharing ideas involves receiving information and giving information. Listening and reading are the most important methods or receiving information. All that we speak and write is primarily dependent on what we have heard or read. However, commensurate importance is not attached to those skills and deliberate attempts to understand and improve these skills are seldom made.

Reading

It is matter of compulsion that we read as much as possible to become good speakers and good writers. Reading is also a part of our professional responsibilities as teachers, scientists or reporters. Reading is the key that unlocks the doors to all other subjects and to other doors beyond, throughout life. It is, therefore, necessary that we improve our reading skills. Reading is a complex mental process involving many different skills. In this process our minds translate printed symbols into ideas which the writer is trying to communicate. Inspite of the complex visual and mental activity the aim of reading is simple. We read so that we may understand. Our mind has to register meaning of the printed symbols. This implies that comprehension is important in reading. The other basic skill is speed. It is the rate at which you can cover printed material through reading. Reading speed can be measured as number of words per minute. The average reader covers about 250 words per minute. Very fast readers read 500 to 600 words per minute. Occasionally there is a person who reads at the phenomenal rate of 1000 words per minute. However, the capacity to read fast is valueless without maintaining levels of comprehension. Both speed and comprehension needs to be developed. Reading skills enable readers to turn writing into meaning and achieve the goals of independence, comprehension, and fluency.

Kinds

Here are some kinds of reading skills:

- Word attack skills let the reader figure out new words.
- Comprehension skills help the reader predict the next word, phrase, or sentence quickly enough to speed recognition.
- Fluency skills help the readers see larger segments, phrases, and groups of words as wholes.
- Critical reading skills help the reader see the relationship of ideas and use these in reading with meaning and fluency.

Independence

Independence is the ability to function in a given area without depending upon another's help. Independence in reading and writing is the ability to read and write anything one can say or understand in his or her language without depending upon another's help.

Comprehension of General and Technical Articles

Comprehension is the ability to grasp something mentally and the capacity to understand ideas and facts.

Comprehension Skills

Reading requires understanding, or comprehending, the meaning of print. Readers must develop certain skills that will help them comprehend what they read and use this as an aid to reading. Comprehensibility in writing is related to comprehension in reading. Comprehension skills are the ability to use context and prior knowledge to aid reading and to make sense of what one reads and hears

Discussion

Comprehension is based on:

- knowledge that reading makes sense
- readers' prior knowledge
- information presented in the text, and
- the use of context to assist recognition of words and meaning.

Examples

Here are some examples of reading comprehension skills:

Understand that print conveys meaning. Using context as an aid to reading

Comprehensibility: Comprehensibility is the ability to be understood; intelligible.

Fluency: Fluency is the ability to read, speak, or write easily, smoothly, and expressively.

Fluency Skills

Fluency should be the aim of every reading and writing lesson. It should increase as learners' progress from beginning to advanced readers and writers. Fluency enables learners to read and write with more understanding. They gain this skill through practice and observation (Gudschinsky 1973). Fluency skills are the ability to see larger segment and phrases as wholes as an aid to reading and writing more quickly.

Examples

Here are some examples of fluency skills:

- ✰ Immediately recognizing letters and frequent clusters of letters.
- ✰ Learning frequent words by sight
- ✰ Seeing phrases as wholes
- ✰ Using prediction skills within the phrase or clause

A fluent reader is one who reads and understands what he or she is reading quickly and with a minimum of effort.

Characteristics

Here are some characteristics of a fluent reader. A fluent reader

- ✰ has automatic decoding skills
- ✰ has built up a vocabulary of sight words
- ✰ has good comprehension skills
- ✰ has good word attack skills
- ✰ reads smoothly and with expression, and
- ✰ self-corrects (recognizes when something doesn't sound right).

Decoding Skills

Decoding skills are the skills necessary to analyze and interpret correctly the spoken or graphic symbols of a familiar language.

Also known as: word attack skills

WRITING

Precise Writing

The word 'precise' is derived from French and it means 'precise' a good précis is a straight forward condensed statement of the salient features, ideas or information contained in the original passage and it is written in a clear and concise way in the writing own words. A precise is a summary and precise-writing means summarizing. Precise writing is an exercise in compression. A precise is the gist or main theme of a passage expressed in as few words as possible. It should be lucid, succinct, and full (*i.e.* including all essential points), so that anyone on reading it may be able to grasp the main points and general effect of the passage summarized.

Skills Required

The aim of précis is to present to a busy reader, the main ideas of original passage as concisely and clearly as possible, in a readable form. In this sense a précis is like an original piece of writing. Precise writing demands power of judgment and evaluation as the passage usually contains important ideas and a number of unnecessary words.

Method of Procedure

1. Reading

(a) First read the passage through carefully, but not too slowly, to get a general idea of its meaning. If one reading is not sufficient to give you this clearly, read it over again, and yet again. The more you read it, the more familiar will it become to you, and the clearer will be (i) its subject, and (ii) what is said about that subject. Ask yourself, 'What is it I am reading? What does the author mean? What is his subject? What is he saying about it? Can I put in a few words the pit of what he says?"

(b) Usually you are required to supply a title for your précis. This is a good stage at which to do this. Think of some word, phrase or short sentence that will sum up briefly the main subject of the passage. For example, look at Exercise 155, No.20, in which the first sentence gives the subject, all the rest of the passage being an expansion and illustration of it: "Hospitality is a virtue for which the natives of the East in general are highly and deservedly admired". This at once suggests the short title of "Eastern Hospitality". But you will not always find such convenient key-sentences in the passage you have to summarise.

(c) Further reading is now necessary to ensure that you understand the details of the passage as well as its main purport. Take it now sentence by sentence, and word by word. If the meanings of any words are not clear, look them up in a dictionary. Detailed study of this king is necessary, because a phrase, a sentence, or even a single word, may be of prime importance, and the misunderstanding of it may cause you to miss the whole point of the passage.

(d) You should now be in a position to decide what parts of the passage are essential and what parts are comparatively unimportant and so can be omitted without any loss. This process of selection is not so easy as some people think. Beginers select; but they often select in a haphazard or mechanical way. It requires some practice to be able to say, "This is essential to the meaning of the passage, and that is only incidental and unimportant."

2. **Writing**

   (a) Rough Drafts – You should now be ready to attempt the writing of the precise; but be sure of the limits within which it must be compressed. If the number of words is given you, this is easy; but if you are told to reduce the passage to say, a third of its length, count the number of words in the passage and divide by three. You may use fewer words that the number prescribed, but in no case may you exceed the limit.

It is not likely that your first attempt will be a complete success. The draft will probably be too long. In fact you may have to write out several drafts before you find how to express the gist of the passage fully within the limits set. A good deal of patience and revision will be required before you get it right. It is a good plan to write the first draft without having the actual words of the original passages before one's eyes.

Summarizing

A summary is a short and concise representation of the main points, ideas, concepts, facts or statements of a text written in your own words. Unlike a paraphrase, which is generally of a similar length to the original text, a summary is much shorter.

When either summarizing or paraphrasing, you should not alter:

- the author's original idea(s)
- the degree of certainty with which the ideas are expressed.

How to Summarize?

To create/write a good summary, you should read the article or text a number of times to develop a clear understanding of:

- the author's ideas and intentions
- the meaning and details
- the force with which the ideas are expressed.

Referencing

Summaries need to be referenced. Whether you have summarized or paraphrased an author's words, thoughts, ideas *etc.*, a citation to the original source of the words, thoughts, ideas *etc.* must be provided.

Write a Summary

Use the following steps to write a summary.

Step 1

- **Write notes in point form using keywords;** this will make it easier to express the ideas in your own words.

Step 2

- **Write the summary directly from your notes** without re-reading the passage.

Step 3

- **Refer back to the original** to ensure that your summary is a true reflection of the writer's ideas.

Tips

- Topic sentences provide a quick outline of the main idea(s) presented in a paragraph.
- When summarizing a chapter or article, the introduction and conclusion should provide a good overview of the content.

Precise Writing/Abstracting/Summarizing

Precise Writing/Abstracting/Summarizing is the one of the most useful skills for both the student and the professional for reporting purpose. It is the art of shortening a document to extract the maximum amount of information, then conveying this information to a reader in the minimum number of words. But, it is necessary at the same time to be very careful not to lose or distort the original meaning. The goal is to preserve the core essence of the original report in a manner which is both clear and concise.

Precise Writing

Precise is a French word and concerned with the English word precise. A precise is a summary and precise writing means summarizing. This is an exercise in compression. It is the gist or main theme of a passage expressed in as few words as possible. It should be lucid, succinct and full (including all the essential points so that anyone, on reading it, may be able to grasp the main points and general effect of the passage summarized.

- It is not paraphrasing. A paraphrase should reproduce not only the substance of a passage, but also all its details. It will, therefore, be at least as long as, and probably longer than, the original.
- A precise should not contain more than a third of the number of words in the original passage.
- It is an exercise to grasp the gist of the passage and to read the material with full concentration. So summarizing is an excellent training in concentration of attention. It teaches us to read with the mind, as well as with the eye, on the page.

- It is a good exercise in writing a composition.
- It teaches us to express our thoughts clearly, concisely and effectively.
- It corrects our common tendency to be vague, disorderly and diffused in our thinking and written expressions.
- It is an exercise to choose our words carefully and to construct our sentences with an eye to fullness combined with brevity and to put our matter in a strictly logical order.
- In other words, it is an ability to grasp quickly and accurately what is read, or heard and to reproduce it clearly and concisely.

Method or Procedure

- Reading carefully for comprehending clearly its main theme or general meaning.
- Think of some word, phrase or short sentence that will sum up briefly the main subject of the passage. Sometimes this is supplied by what we call key sentence.
- Key sentence may be found at the beginning or the end of the passage.
- We are required to supply a title for the precise.
- Sift the essential from the inessentials. Leave out superfluous details. Omit repetitions, illustrations and examples. Important ideas should find place in the precise.
- Jot down brief notes, title and the essential details.

Final Draft

- The precise should be all in your own words.
- It must be a connected whole. The ideas must be joined together in such a way to read continuously.
- The precise must be complete and self-contained.
- It must convey its message fully and clearly without requiring any reference to the original to complete its meaning.
- The language should be simple and it should be grammatically correct.
- A suitable title should be provided to the passage.

Abstract Writing

An abstract is a very important portion of an article. Abstract is best placed before the text of the manuscript so that the reader might comprehend the essence of the report. The formulation of an abstract into a short paragraph forces the author to express precisely the most important information of the technical report. The abstract should be informative, give a succinct condensation of the article, and complete in itself and intelligible without reference to the text, figures or tables.

It generally appears at the beginning and written in the past tense. Always begin the abstract with rationale and objective statements; never jump directly into the materials and methods. It should include:

- ✰ Principal objectives and scope of investigation.
- ✰ Methodology/technique(s) employed/but not mega emphasis on techniques but on results.
- ✰ The main and important findings of the experiment.
- ✰ The contribution of this particular piece of work towards new knowledge

The abstract is not needed in a short report though it is required in a long report. The busy technocrat or an over busy executive may read only the abstract and refer to the relevant parts of the report only if it is absolutely necessary. The abstract is placed on the separate page or immediately after the title.

Abstract and Summary

Abstract is generally restricted to 250 words or even less, but a summary may be somewhat longer. If a summary is the precise of a report, an abstract is a precise of the summary. Long reports, sometimes has a summary as well as an abstract; in short reports the abstract tends to coincide with the summary. A **synopsis** is a summary with some particulars, especially of the results, given in greater detail than in a summary.

Extract, Summary

Extract is one or more portion of a document selected to represent the whole (an excerpt). Summary is a brief restatement within the document (usually at the end) of its salient findings and conclusions and is intended to complete the orientation of a reader who has studied the preceding text.

Synoptic

A Synoptic is a concise first publication in a directly usable form of key results selected, from an available but previously unpublished paper. It differs from an abstract (which it contains) in that it is often a combination of text, tables, and figures, and may contain the equivalent of 2000 words.

Individual and Group Presentations

The practice of showing and explaining the content of a topic to an audience or learner is known as presentation. There are as many forms as the life situations. A presentation is a process of offering for consideration or display. It can be a social introduction, as of a person at court, or a demonstration, lecture, or welcoming speech, or a manner or style of speaking, instructing, or putting oneself forward. The manner of presenting, especially the organization of visual details to create an overall impression is a presentation. It may be the formal introduction of a person, as into society or at court. In the business world, there are sales presentations, informational and motivational presentations, first encounters, interviews, briefings, status reports, image-building, and the inevitable training sessions. Presentations can also be categorized as vocational and non-vocational. In addition, they are expository or persuasive. They can also be impromptu, extemporaneous, written, or memorizes. But it's more important to focus on their purpose. There are three basic purposes for giving oral presentations:

1. To Inform
2. To Persuade
3. To Build Goodwill

Types

1. Informative Presentations

Include talks, seminars, proposals, workshops, conferences, and meetings where the presenter or presenters share their expertise and information is exchanged.

2. Persuasive Presentations

Sometimes called transactional, are often motivational. Persuasive Presentations are designed and delivered to come out with a specific goal in mind.

3. Goodwill Presentations

Goodwill presentations, which often take the form of after dinner speeches, are often designed to be entertaining-for example by sharing video highlights of a football season or anecdotes based on some past experiences.

4. Multipurpose Presentations

Presentations usually have more than one purpose. A presentation to employees may be announced as an informative session on new regulations but in reality may also be an all-out effort to persuade workers to buy into the new rules.

5. Sales

Sales techniques are complex and require two essential items for success in sales presentations which are knowing and understanding your audience, and building rapport.

6. Political

Presentations in the political arena are primarily grouped in the persuasive category. To be effective they must include lots of information and also build goodwill.

7. Motivational Presentations

Motivation is another form of persuasion, but one that somehow takes on a more fervent, highly charged tone.

Interviews

A Job Interview is yet another presentation form, one where the presenter should make an effort to identify his or her immediate audience, but also take great pains to know as much as possible about the larger audience.

Audiences can be classified into four basic categories:

1. Captives
2. Pragmatists
3. Socially Motivated
4. Committed

When an individual stands up to deliver a presentation before an audience, it is essential that the audience know who the presenter is, why they are there, what specifically they expect to get from your presentation, and how they will react to your message. You should try to gather as much background information as possible before your presentation. There will be times, especially with presentations that are open to the public, when you will only be able to guess.

It has been found that audiences believe presenters who use visuals are more professional and credible than presenters who merely speak. Some research indicates that meetings and presentations reinforced with visuals help participants reach decisions and consensus in less time.

A presentation program, such as Microsoft PowerPoint, is often used to generate the presentation content. The development of information technology has made the use of these solutions possible, but Liquid Crystal Display (LCD) Projectors as well as overhead transparencies are still used widely for the purpose of presentations.

Helpful Presentation Tips

- ☆ Easy Format
- ☆ Organize Your Thoughts
- ☆ Know Your Material
- ☆ Prep Time
- ☆ Outline Your Presentation

Three Ingredients of Group Presentations

Clarity

Clarity means clearness of purpose, thought or style. Developing clarity within your group will help you develop a clear message for your audience.

Clarity of Purpose

Just as your presentation will have a clear purpose, your group should create a **Charter Statement** that explicitly captures the group's desired outcome. This Charter Statement becomes the test of everything that will go into the presentation and help guide the efforts of the team. For example, if your group agrees that your general purpose is to sell your product, and, more specifically, you know that the key decision maker in the audience is doubtful about cutting checks to companies like yours, build that into your Charter Statement. The purpose of our presentation is to sell our Product to ABC Company by overcoming the objections of the company's Purchasing Officer through clear examples of how our Product provides a fast return on investment. The Charter Statement will come in handy when you have a team member who may want to go off track to tell personal anecdotes that don't pass the test of the group's charter.

Clarity of Roles

Personalities come into play when groups meet to develop presentations. The struggles for position and ego can quickly deplete the group's momentum, resulting in hurt feelings and, potentially, a weaker presentation. Providing clarity to group roles helps to establish expectations and keep the entire group moving towards a common objective, leading thus to a great group presentation. Identify the roles your group needs during message development. For example, to ensure that team members are meeting assignments, select a **Project Manager**. This person is not the boss of the presentation, but rather will focus on schedule and assignments. Other roles could include a **Gap Analyst** who is responsible for identifying gaps in content and support materials (handouts, graphics, *etc.*), which in turn could work closely with other roles within the group like the **Chief Researcher**. Capitalize on the unique personalities within your group to develop roles that work well for all, but be sure to discuss the roles openly so they are clear to everyone.

Clarity of Message

Instead of writing speeches for each *individual speaker*, try creating one *master presentation*, a unified narrative, and *then* decide who speaks to which points, and when. This is a shift from the traditional segmented method of group presentations where often group members are directed to give five minutes of talking and then are left to develop content independently. In a master presentation, each speaker may weave in and out at various points during the presentation. When done well, this fluid dynamic can hold an audience's attention better by offering a regular change in speaker's voices and presence.

Control

Group presentations face unique logistical challenges beyond just developing presentation content.

Control in Introduction

The way of introducing a group is noticed by the audience, so plan those introductions with your presentation. Your presentation may be part of a larger event that includes an master of ceremonies who will introduce the team. If so, be sure that you provide pertinent information to the emcee that will allow her/him to generate interest in your presentation even before you begin speaking. If your group is responsible for making its own introductions, however, you will need to decide if you will introduce your group members in the beginning, or when they first speak.

Control in Transitions

Decide how you are going to hand off from one speaker to the next. In the master presentation approach, you may want to consider simply have speakers pick up a narrative right where the previous speaker left off. If you use the more traditional segmented approach, each speaker may cue the subsequent speakers by identifying them and their subject matter. For example: and speaking of quality control, no one is more qualified as Mr. X, who is going to tell us about how this team will deliver a quality project for you.

Control vis-à-vis Time and Space

Multiple speakers translate to occupying more physical space, and the potential to gobble up more time with introductions and transitions. If you will be presenting in a small room, consider where each speaker needs to be positioned to quickly reach the speaking area, and whether they will sit or stand when not speaking. Your presentation must fit within your allotted time, so you will need to time your group's presentation, including equipment set up, introductions, and transitions.

Commitment

Commitment from each group member is going to give your presentation the best content and flair that will impress your audience.

Commit to a Schedule

Once you know the date of your presentation, create a schedule that includes specific milestones, such as presentation draft due and final rehearsal. Having a specific schedule allows members either to agree to the group's expectations or to offer dates that better fit their personal schedules. One can also assign specific responsibilities to the scheduled milestones; for example, who is responsible for bringing the handouts, projector, and laptop to the presentation?

Commit to Rehearsing

If you find group members who lack the commitment to rehearse, consider finding group members who will commit. Rehearsing is one of the most important steps for presentation success. Have your team members agree from day one that they will make themselves available to practice with the group. If you find group members who lack the commitment to rehearse, consider finding group members who will commit. Practice makes perfect, and no rehearsal means your group does not know what will happen to the content, timing, or quality of the presentation.

Commit to Answering your Audiences Questions

Always be ready to pepper your group with questions after your formal presentation is over. Your presentation is not over yet. Handling those questions is as important as the presentation itself. A well-done presentation means nothing if presenters fumble questions so badly that they appear incompetent. Have each member develop a list of potential questions and then, as a group, review the list. Discuss who will be responsible for handling which types of questions.

Suggestions for an Effective Group Presentation

First, appoint one of the speakers to be in charge. If this is not done and something unexpected occurs, no one has the responsibility for the group which may lose credibility because of uncertainty in how to respond. In addition, when a question is asked, the speaker-in-charge immediately refers it to the person with the appropriate expertise; thus there is no uncertain pause because no one knows who will answer the question.

Second, each speaker should know what the other speakers are going to say. This knowledge will help them avoid duplication of material and they can make appropriate references to another speaker's content if it applies. This knowledge helps insure continuity among the different speakers.

Third, the last words of each speaker should synchronize with what the next speaker will cover. This adds unity to the entire presentation and gives the new speaker a smooth opening to his or her material. An example might be, "Now Susan will cover the financial aspects of our proposal and help you understand the benefits this will give you."

Fourth, if possible the strongest speaker should end the group presentation. The ending is the most important part of the presentation; people remember best what you say last. You want to have an ending speaker who can show passion and enthusiasm for the topic.

Finally, a dress rehearsal is essential. The group members need to get a feel of the complete message to see how they can best contribute. In addition, with several people speaking it is hard to gauge the time the speakers will take. This dress rehearsal allows the group to time the presentation and to make adjustments in each portion in order to be under the time limit; the more people involved the more unpredictable the total time. With everyone hearing the other speeches, each speaker can give feedback to the others that will improve the quality of the whole presentation. Extraneous and redundant material can be eliminated as well.

Impromptu Presentation and Extempore

Impromptu Presentation

A speech and/or event that involves an eight-minute speech (with up to three of these eight minutes available for use as preparation time) is known as impromptu speaking. The preparation time is known as prep time, or simply preps.

This event varies depending upon the situation; and, the time allowed also varies at different levels:

- ☆ The speaker at the college level is granted 7 minutes to divide, as he or she sees fit.
- ☆ Another variation exists in which the speaker must speak for five minutes and half of a minute is given for preparation time. The speaker is given a slip of paper, which provides three choices for his speech. The topics can be abstract or concrete nouns, people, political events, quotations or proverbs.
- ☆ There is another variation where the speaker is given an envelope with slips in it, each with a quotation. They have to draw 3 slips of paper, choose one, and then put two back. They are then given 5 minutes to prepare a 5 minutes speech on the subject.

Types of Impromptu

OHSSL Impromptu (Ohio Impromptu)

In this type, the speaker is provided with seven minutes of time to divide at his own discretion between prepping and speaking. All topics are quotations and generally random, but occasionally may be random. It is also more factual and less humor-based, resembling extempore. Other structures can also be used. Double analysis is normally used by experienced impromptu speakers. The ratio usually being used by experienced speakers to divide the preparatory and speaking time is 1:6.

Editorial Impromptu

It is an attempt to return impromptu to impromptu speaking. The students are given a short editorial (ideally 3 to 5 paragraphs) to which they will develop a response. The students are allowed nine minutes to divide between preparation and speech time. At least five minutes must be used for speaking. Limited notes, prepared in the round, are permitted. The speech should involve the development of an argument in response to the idea generated or opinion shared in a given editorial.

High School Competition

In high school speech and debate competitions, thirty seconds are given to the competitor to select a topic from a set of topics (usually three). During those 30 seconds, the competitor must also compose a speech of five minutes with a 30 second grace period.

Structure of the Speech

An impromptu speech is simply a mini-speech and therefore it has: an Opening, a Body and a Conclusion.

Opening

- ☆ Open by addressing the Chairman or Master of Ceremonies, *e.g.* Mister/ Madam Chairman
- ☆ Brief introduction/opening sentence attention getting

Body

- ☆ Cover the main points (Try to find one or two central themes)
- ☆ Use clearly worded simple sentences and tries to link the themes

Conclusion

- ☆ Be brief and look for an elegant closing that links back to the opening
- ☆ End by acknowledging the Chairman or Master of Ceremonies

A general outline for impromptu speeches is as follows:

1. Introduction/Roadmap (1 minute)
2. First section (1 minute)
3. Second section (1 minute)
4. Third section (1 minute)
5. Conclusion (1 minute)

Techniques to Deliver a Better Impromptu Speech

Give yourself Time to Prepare

- ☆ Take deep breaths
- ☆ Rise slowly from your chair and walk slowly to the lectern (or stand behind and away from your chair)
- ☆ Use this time to collect your thoughts and decide on the purpose and plan of your speech
- ☆ Think about the opening sentence.
- ☆ Always remember that never start the speech immediately.

Feel Confident

- ☆ Look around at audience and smile. Stand tall.
- ☆ Don't slump, don't fidget, don't grasp the lectern, don't put hands in your pockets
- ☆ Speak and act in a confident manner

Slow Delivery

- ☆ Gives you time to think ahead
- ☆ The audience has time to absorb and react to what you are saying
- ☆ Helps you reduce umms and ahhs

Focus

- ✰ Keep the focus on the subject while talking
- ✰ Don't think of any negatives (eg being unprepared)
- ✰ Talk directly to the audience and adapt to their feedback
- ✰ Maintain good eye contact with the audience
- ✰ Be brief and to the point
- ✰ Don't ramble or say too much on the subject
- ✰ Speak at the audience's level

SPEAKING

The process of speaking to a group of people in a structured, deliberate manner intended to inform, influence, or entertain the listeners is known as public speaking. It is closely allied to "presenting", although the latter has more of a commercial connotation. In public speaking, there are five basic elements, often expressed as "who is saying what to whom using what medium with what effects?" The purpose of public speaking ranges from simply transmitting information, to motivating people to act, to simply telling a story. Good orators should be able to change the emotions of their listeners, not just inform them. Interpersonal communication and public speaking have several components that embrace such things as motivational speaking, leadership/personal development, business, customer service, large group communication, and mass communication. Public speaking can be a powerful tool to use for purposes such as motivation, influence, persuasion, informing, translation, or simply entertaining. A confident speaker is more likely to use this as excitement and create effective speech, thus increasing their overall ethos. Professional public speakers often engage in on-going training and education to refine their craft. This may include seeking guidance to improve their speaking skills such as learning better story-telling techniques, for example, or learning how to effectively use humor as a communication tool as well as continuous research in their topic area of focus.

Importance of Good Public Speaking

Good public speaking skills also works at the time of the interview. It is compulsory to face an interview before getting a new job. Interviewer will not only judge your personality but also your communication and fluency in general. In this technology- savvy world, your presentation skills and public speaking skills are also counted along with your basic qualification.

It is not that we only require public speaking skills to speak in front of audience or to give lectures. Public speaking skills are an asset that is worth millions. These skills are counted not only in our professional life but are also useful when a person is college go-getter.

Tips for Public Speaking

- ☆ Know the needs of your audience and match your contents to their needs.
- ☆ Know your material thoroughly.
- ☆ Put what you have to say in a logical sequence.
- ☆ Ensure your speech will be captivating to your audience as well as worth their time and attention.
- ☆ Practice and rehearse your speech at home or where you can be at ease and comfortable, in front of a mirror, your family, friends or colleagues.
- ☆ Use a tape-recorder and listen to yourself. Videotape your presentation and analyze it.
- ☆ Know what your strong and weak points are. Emphasize your strong points during your presentation.
- ☆ When you are presenting in front of an audience, you are performing as an actor is on stage. How you are being perceived is very important. Dress appropriately for the occasion.
- ☆ Be solemn if your topic is serious. Present the desired image to your audience.
- ☆ Look pleasant, enthusiastic, confident, proud, but not arrogant.
- ☆ Remain calm. Appear relaxed, even if you feel nervous.
- ☆ Speak slowly, enunciate clearly, and show appropriate emotion and feeling relating to your topic.
- ☆ Establish rapport with your audience.
- ☆ Speak to the person farthest away from you to ensure your voice is loud enough to project to the back of the room. Vary the tone of your voice and dramatize if necessary. If a microphone is available, adjust and adapt your voice accordingly.
- ☆ When using audio-visual aids to enhance your presentation, be sure all necessary equipment is set up and in good working order prior to the presentation. If possible, have an emergency backup system readily available.
- ☆ Have handouts ready and give them out at the appropriate time. Tell audience ahead of time that you will be giving out an outline of your presentation so that they will not waste time taking unnecessary notes during your presentation.
- ☆ Know when to STOP talking. Use a timer or the microwave oven clock to time your presentation when preparing it at home. Just as you don't use unnecessary words in your written paper, you don't bore your audience with repetitious or unnecessary words in your oral presentation. To end your presentation, summarize your main points in the same way as you normally do in the CONCLUSION of a written paper.

☆ Have the written portion of your assignment or report ready for your instructor if required.

Group Discussion

Group

A group means many people are working together to achieve some targets. The performance of a group depends on the collaborative work of the individuals. The team of three or more than three people is considered as a group.

Discussion

Discussion means exchanging ideas between two or more than two people, which is generally a face-to-face interaction. It is a process of talking between people to reach a specific decision. It is also defined as a conversation about a specific topic. The result or end product of group discussion can be a particular decision, enhanced knowledge, actions, an argument, doubt-clearing, disagreement, *etc.*

Why Group Discussion?

Over the last years, GD is used to test the soft skills of the candidates. It is used as a quick method to shortlist a few candidates from the massive strength. The written exam tests the knowledge and aptitude. Organizations check the social skills, technical, and communication skills so that a candidate can work effectively without any obstacle in between his/her skills.

The GD or Group Discussion is used as an exam for many organizations and schools. It tests the personality skills, communication skills, teamwork, confidence, problem-solving skills, social skills, behavior, and leadership abilities of a candidate. It is also an essential part of short-listing candidates for a job or admission to a university.

An intelligent candidate with good knowledge and skills sometimes lacks communication skills, which can be a lack of confidence, hesitation in speaking, unable to explain his/her point of view, *etc.* Due to this, Group Discussion nowadays has become an essential part of most of the organization and institutes.

Knowledge and technical skills are essential things for a candidate, but in an organization, he/she might have to face the meetings, discussing technical issues, handling new employees, *etc.* Here, a lack of communication skills becomes the disadvantage of that candidate. Hence, organizations and institutes ensure that the candidate must possess all the qualities before joining.

Importance of Group Discussion

Increases Confidence

The GD allows the participants to speak in public, which increases their confidence to speak without any hesitation.

Focuses on Deep Thinking

It tests the knowledge of a candidate. The participants get only a few minutes (3 to 5) for the preparation. In such a small duration, they have to make a list of points to speak. The participants in a group listen and understand other participants points of view, making them think deeply about the topic.

Improves Communication Skills

GD helps the students to represent their views and ask the questions. It not only increases the confidence of a candidate but also enhances their communication skills.

Removes Hesitation of Speaking

Some candidates usually have hesitation in speaking in public. At starting, such candidates might face difficulty while speaking. But, after two or three Group Discussion, it improves. Group Discussion helps such candidates to speak loudly and express their views on the topic. It further removes their hesitation in speaking.

Team Work

Group Discussion is also based on teamwork. In a company, it is very important to work as a team for the assigned project. Due to this, the quality of working together with team members is checked in the Group Discussion. The panelist also judges the cooperation of participants in the group. It allows participants to share their views with other participants in the group.

Behavior

It helps to understand the attitude and behavior of the candidates towards other participants in a group.

Listening Skills

GD improves the listening skills of the candidates. For example, the panelist may ask any candidate in a group to summarize the discussion. A candidate can only summarize if he/she has good listening skills. Due to this, every candidate presents in the discussion listens carefully. It further improves the listening skills of the candidates.

Diversity in the Ideas

The Group Discussion involves sharing ideas with other participants in the group. Every participant shares his/her thoughts in the group, which makes the diversities in the ideas.

Recruitments

Group Discussion is widely used for the recruitment process. It helps panelist to select a few candidates from the group. It also helps them to judge that the candidate is suitable for the organization or not.

Types of Group Discussion

- ☆ Topic-Based Group Discussion
- ☆ Case Study Based Group Discussion

Topic-Based Group Discussion

There are few candidates in a group, where each group is given any topic. The group members discuss the topic, which is called a Group Discussion.

The topic-based Group Discussion can be further classified as:

1. Controversial Topics
2. Knowledge-Based topics
3. Abstract Topics
4. Conceptual topics

Let's discuss the above topics in detail.

Controversial Topics

The discussion on controversial topics becomes a debate. Such topics are given to judge the participants' temper and how they can handle the discussion without losing their calm. It shows that how a candidate can represent his/her views without arguing with other participants.

The example of the controversial topics can be Reservation System, Religion equality, *etc.*

Knowledge-Based Topics

The participants should have a proper understanding of the topic before proceeding for a discussion. The information should be enough to convince the panelists. The essential thing is to be confident. Do not initiate the discussion if you are not aware of the topic.

The examples of Knowledge-Based topics can be 'CAT vs. GATE' and 'Government jobs vs. Private jobs'.

Abstract Topics

The abstract topics test the creativity and thinking of a candidate. It also tests the communication skills. The example of Abstracts topics can be 'A walk to remember'.

Conceptual Topics

The conceptual topics are widely used today by companies. It tests the knowledge, logical, and aptitude skills of a candidate. The topic for the discussion is randomly selected, giving panelists a space to judge the participants' creative thinking and ability to handle the situation.

Case Study Based Group Discussion

The Case study based Group Discussion is generally followed by the MBA Institutes, such as IIM, *etc.* In such a discussion, a problem is given, and the participants are asked to resolve them. The preparation time is also higher as compared to other discussions. The panelists test the teamwork and decision-making skills of the participants.

The participants need to active and updated with the things around them. It also tests the observation capability of the participants.

The example of Case study based topics can be 'Discussion between boss and manager'.

Rules of Group Discussion

There are some rules of a Group Discussion. The rules of Group Discussion are listed below:

- **Prepare well for the topic.** The quick trick is to note all the important points during the preparation time. Do not write brief paragraphs. Try to write short points and explain them briefly in the discussion.
- **Be confident about your content.** It would help if you always were confident about your content. If you are not confident, you may hesitate while speaking, which will create a negative impression.
- **Participants Introduction.** Most participants only focus on their content. You should be aware during the introduction of all the participants in your group. To get attention, you can call out the person's name when the discussion gets diverted.
- **Body Language.** Always take care of your body language. Do not focus only on the content.
- **Leadership.** Try to take the lead in the group if you are confident about the topic and the knowledge.
- **Avoid false starts or commitments.** Do not initiate the discussion if you are not sure about your content. Do not present any wrong information.
- **Follow your Domain.** Every company has its own culture. Prepare your content according to the company requirements. For example, industrial companies will focus more on technical talks, while marketing companies seek the combination of creative and oriented content.
- **Do not fight.** In the Group Discussion, some candidates often get aggressive. Do not fight during the discussion. Most of the companies require sensitive and polite candidates rather than aggressive ones.

Organizing Seminars and Conferences

The popularity of conferences and seminars results from the distinct benefits they offer. Attendees pre-qualify themselves as prospects; if they weren't interested in the issue being discussed, they wouldn't come. You get to talk with them face-to-face, both formally during presentations and informally during breaks, meals, and receptions. As a speaker, you are automatically classified as an expert. Outside speakers can add credibility and drawing power. Because seminars can run from a few hours to several days, they can be adapted to a wide array of subject matter and contexts. In combination with outbound telemarketing, the offer of a quality seminar is an effective tool for multiplying the number of qualified prospects for your products and services.

Benefits of Seminar/Conference

It is better to address the benefits of a seminar/conference before going through the first steps of conference organization. A conference is the broadest of the three kinds of meetings (Seminar, Workshop, and Conference) can provide, among others, the opportunity to:

- Meet students, researchers, professors *etc.* from other linguistics departments of your own or other countries
- Become aware of other cultures and ideas related to linguistics
- Familiarize yourself with other research work, projects *etc.*
- Promote your association, university/department within the field
- Attract new and active students in your association, university/department
- Build connections and create a network of co-workers and institutions
- Travel and get to know the hosting country (geographically, culture, religion *etc.*)
- Promote your own research and get feedback
- Gain and train management skills
- Trained interpersonal skills
- Learn how things are behind the scenes
- Become an active member of the academic community

Seminars can be of the following Types

- A half-day seminar on your products and services.
- A one-day or two-day conference, featuring multiple tracks and noted speakers, run by a professional conference organizer on highly technical issues, attracting participants from across the country.
- A series of half-day or one-day seminars offered at several locations across the country to teach the fundamentals of a specific application to those just beginning to work in that area.

Optimizing the Seminars Success

The subject for a business conference must:

Be Educational

People come to conferences to learn. Entertainment is a plus, but promotion of your business is acceptable only if it is heavily disguised. There has to be something worth learning.

Be Urgent

There must be a need to know now. A conference on "New Genomic Tools" is not likely to draw as many people as one entitled Speeding Your Drugs to Markets. The term "urgent" here means that there is value to learning something now rather than later.

Provide Solutions

The conference must offer attendees solutions or progress toward solutions, because people attend conferences in order to solve problems better. You should test seminar content against these criteria. A brainstorming session with your staff can usually provide you with a good assortment of subjects to pick from that will ensure good attendance.

Length

Next you must decide on the conference's length. You can determine length by asking yourself the following questions:

a) **How much time and support do I have to set it up?** A two-day conference requires much more work than a half day seminar. You must adjust the length to the level of effort you can put into setting it up.

b) **How much content do I have to offer?** There is simply not enough material for a lengthy conference on some subjects.

c) **Is my audience willing to pay?** A two-day conference gets expensive, and most firms expect to be paid for them. The audience, in turn, is faced with not just a fee but a substantial opportunity cost in time invested away from other duties. A longer conference must therefore deal with a particularly weighty, urgent, or complicated subject. Generally you must also have a longer mailing list to find people with budgets and schedules that will allow them to attend.

d) **How many speakers can I attract?** Many conferences use only one speaker, but the speaker had better be good. Using three or four speakers a day diversifies your risk. Some of the time the audience will be listening to some exceptional speakers, even if not all are of equal quality.

e) **How geographically scattered is my market?** Few people will drive more than an hour to attend a half day seminar. Assuming that very few of the intended clientele respond to your invitations (an optimistic target, unless outbound telemarketing is utilized), you will need to invite two hundred prospective buyers of your services to attract a twenty-person audience.

Timing

Timing will also affect attendance. Most are scheduled to avoid summer vacations and end-of-the-year holidays. Longer conferences tend to be scheduled for either the beginning or end of the week, allowing participants to travel on the weekend and perhaps enjoy a short vacation at the city where the conference is being held.

Location

Possible locations for your seminar or conference include your own offices, a hotel, a corporate or privately operated conference center, or space borrowed or leased from another organization. If your office has adequate meeting space and is close to the market you are trying to reach, holding the conference there provides you the opportunity to show the place off and introduce other members of the firm. The location serves as a subtle advertisement for your firm.

The Invitation

Because the quality of your invitation or flyer greatly influences attendance, it deserves special attention. For a one- or two-day conference, the flyer should include seven things:

- ☆ **The conference theme** - The reader should be able to determine almost instantaneously what the conference is about. The wording should be as compelling as possible.
- ☆ **A concise description stressing benefits** - Don't leave it to the readers to figure out what they will get out of attending. If you tone down your flyer, be sure you don't obscure the benefits in the process.
- ☆ **A description of who should attend** - This inclusive list indicates who the conference is designed for and allows the reader to say, "This is meant for me!"
- ☆ **A review of the schedule and content** - In this portion, list what will occur: From the morning through the afternoon of each day, noting speakers and subjects followed by two or three bullets stressing the benefits of each.
- ☆ **A concise speaker description.** Include credentials, background and experience.
- ☆ **Logistics of time, place, fee, payment, registration, and accommodations.** This should include directions to the site, if not included with confirmation letters.

- ☆ **A sign-up sheet**. This can be a tear sheet or a response card to fax or mail in. Make sure to ask qualifying questions, including seminar topic interest, and product interest and timing.

The Speakers

People who organize seminars for the first time often see themselves or others in their firm as primary speakers. You can often do better by including outsiders and sometimes by using them exclusively.

Something to take Home

For-profit conferences universally provide participants with printed material summarizing what they have learned. If your event is free, you must decide whether to do so. Materials you can use for this purpose include copies of exhibits used by speakers, white papers or reprints of articles on the subject, and speaker bios. A high quality package of materials adds to the professional appearance of a seminar and gives those who attend something they can save with your name on it.

Post-Seminar Follow-up

Within a week of a conference's completion you should follow up with letters to everyone you have met, thanking them for attending. If you have promised someone information, you can put it in the letter, or, if appropriate, call. As in all forms of marketing, follow-up has a major impact on lead generation and separates the serious marketer from the casual one.

Chapter 5

Entrepreneurship Development and Business Communication

ENTREPRENEURSHIP

Entrepreneur plays a vital role in the economic development of a nation. Economic development refers to the income levels and its stability. This chiefly depends on its entrepreneurs. An Entrepreneur is a kind of individual who aims at achieving his goals and who possesses knowledge, skills, inventiveness, drive and spirit of innovation. An entrepreneur always finds opportunities and transforms opportunities into achievement and gains economic benefits. Entrepreneurship involves process of manufacture, innovation in production, usage of new materials, and market etc. An entrepreneur does have that mental attitude to foresee risk and uncertainty with a view to achieve a specific motive.

Concept of Entrepreneur

The word "Entrepreneur" is derived from the French verb 'entrepredre' which means 'to undertake'. It means Between Takers. Entrepreneur is another name of Risk Taker. An entrepreneur is an individual who takes moderate risks and brings innovation. Entrepreneur is a person who organizes/manages the risks in his/her enterprise. "Entrepreneur is an individual who takes risks and starts something new"

During the early 16th century the Frenchmen who organized and led military expeditions were referred to as 'Entrepreneurs'. During the early 18th century French economist Richard Cantillon used the term entrepreneur for business. Since that time the word entrepreneur means one who takes the risk of initiating a new organization or introducing a new thought, product or service to society.

Definition

An entrepreneur is an individual who takes moderate risks and brings innovation. An entrepreneur is a person who has possession of a new enterprise, venture or idea and assumes significant accountability for the inherent risks and the outcome. An entrepreneur is "one who undertakes an enterprise, especially a contractor, acting as an intermediary between capital and labour."

According to J.B. Say, "An Entrepreneur is the economic agent who unites all means of production; land of one, the labour of another and the capital of yet another and thus produces a product. By selling the product in the market he pays rent of land, wages to labour, interest on capital and what remains is his profit". Thus, an entrepreneur can be considered as a person who possesses the initiative skill and drives to set up a business or enterprise on his own and ultimately achieves his desired goals. He brings social change. He looks for opportunities, and finds his own way mainly for economic gains. An entrepreneur is generally a highly calculative individual who always undertakes risks in order to achieve his goals.

According to Joseph Schumepeter, "An entrepreneur in an advanced economy is an individual who introduces something new in the economy, a method of production not yet tested by experience in the branch of manufacture concerned, a product with which consumers are not yet familiar, a new source of raw material or of new market and the like".

Concept of Entrepreneurship

The term 'entrepreneurship' is associated with the term 'Entrepreneur'. Though, they apparently look same, conceptually they are different. Entrepreneurship is the process of finding opportunities in the market place, planning for the resources required to convert these opportunities into success and to achieve long term gains. It involves creating capital by exploiting resources in new ways to initiate and operate an enterprise. Entrepreneurship can be defined by describing what entrepreneurs do. For example: "Entrepreneurs use individual inventiveness, and take on calculated risk to create new business ventures by raising capital to apply inventive new ideas that resolve problems, meet challenges, or satisfy the needs of a clearly distinct market." But the entrepreneurship is not limited to business and profit: "Entrepreneurship involves bringing about change to achieve some benefit. This benefit may be financial but it also involves the satisfaction of knowing you have changed something for the better."

"Entrepreneurship is essentially the act of creation requiring the ability to recognize an opportunity, shape a goal, and take advantage of a situation. Entrepreneurs plan, persuade, raise resources, and give birth to new ventures."

According to Cole "Entrepreneurship is the purposeful activity of an individual or a group of associated individuals undertaken to initiate, maintain and aggrandize profit by production or distribution of economic goods and services".

According to Higgins "Entrepreneurship is meant the function of foreseeing investment and production opportunities, organizing an enterprise to undertake a new production process, raising capital, hiring labour, arranging the supply of raw materials, finding site, introducing a new technique, discovering new resources or raw materials and selecting top managers for day to day operations of the enterprise".

While John Kunkel considered entrepreneurship as a function of social, political and economic structure, Max Weber treated it as a function of religious beliefs. The concept of entrepreneurship involves four key elements.

They are:

a) Organising
b) Risk-bearing
c) Vision and
d) Innovating.

These four elements are interrelated and form a continuous process in business.

1. Organising

Organising involves mobilization of resources and utilization of them to initiate, maintain or enhance profit by the production or supply of goods or services. Entrepreneurs also mobilize other factors of production such as land, labour and capital.

2. Risk-Bearing

Starting a new enterprise always involves risk. The enterprise may earn profit or incur loss. The entrepreneurs should be bold enough to assume this risk.

3. Vision

Entrepreneurial vision encompasses the relentless pursuit of operational excellence, innovative technology and responsiveness to the needs of the market.

4. Innovation

Innovation refers to introduction of something new in the economy, ex. a new product or a new method of production. According to Schumpeter, economic development can take place only through creative innovations and changes. The innovations may be related to new products, new technology, new sources of raw material, new market or new organisation.

Functions of Entrepreneurs

An entrepreneur always seeks opportunity. He also manages and coordinates production and services. He has to start with a small scale enterprise. He not only finds the business opportunities but also utilizes the other resources like 5 Ms-man, money, machine, materials and methods. However, the main functions of the entrepreneurs are as under:

Idea Generation

This is the most essential function of the entrepreneur. Idea generation is possible through the vision, insight, observation, experience, education, training and exposure of the entrepreneur. It specifically implies proper selection of product and identification of project. Ideas can be generated through surrounding environment and market survey.

Determination of Objectives

Another function of the entrepreneur is to determine and lay down the objectives of the business. Entrepreneur should be quite clear regarding the following things:

- The nature of business
- The type of business

This indicates whether the enterprise belongs to the category of a manufacturing concern or a service-oriented unit or a trading business. Thus, the entrepreneurs can attempt to venture in accordance with the objectives determined by him.

Raising of Funds

Fund raising is the most essential function of an entrepreneur. All the actions of a business depend upon the financial support and its apposite management. It is the sole accountability of the entrepreneur to raise funds from both, internal and external agencies. In this regard, he should be aware of the different sources of funds and the official procedure to raise funds. He should be well aware of different government sponsored schemes such as PMRY, SGSY, REGP, *etc.*

Procurement of Raw Materials

Another important function of the entrepreneur is to procure raw materials. Entrepreneur has to find out the cheap and regular sources of supply of raw materials, which will enable to reduce the cost of production.

Procurement of Machinery

The next function of the entrepreneurs is to procure the machineries and equipments for establishing the new venture. While procuring the machineries, he should specify the following details:

- The details of technology
- Capacity of the Installed machines
- Details of the manufacturers and suppliers
- After-sales service

- Warranty period of the machineries and instruments
- All these details are to be minutely observed by the entrepreneurs.

Market Research

Another important function of the entrepreneur is market research and analysis of product. Market research involves the systematic data collection regarding the product which the entrepreneur wants to manufacture. It has to be done indefatigably in order to know the details of the product, *i.e.* the demand for the product, the supply of the product, the cost of the product, the customers, *etc.* while initiating an enterprise.

Determination of Form of Enterprise

Entrepreneur has to consider the nature of the product, level of investment, nature of activities, types of product, quality of product, quality of human resources, *etc.* to decide the form of enterprise. The principal forms of ownership organizations are sole proprietorship, joint venture, Joint Stock Company and cooperative society. Determination of ownership right is crucial on the part of the entrepreneur to obtain legal title to resources.

Recruitment of Manpower

Entrepreneur has to perform the following activities while undertaking this function:

- Estimating manpower need of the organization
- Forming selection procedure
- Devising scheme of reimbursement
- Forming the rules of training and development

Implementation of the Project

Entrepreneur has to work on the implementation of the project. The acknowledged project is to be implemented in a time-bound manner. All the activities from the initial stage to the final stage are to be accomplished by him in accordance with the implementation schedule to manage cost and time, as well as to manage competition. Thus, implementation of the project is an important function of the entrepreneur. To conclude with, all these functions of the entrepreneur can precisely be put into the following categories:

- Innovation
- Risk bearing
- Organization
- Management

Meaning of Entrepreneurship

In the above definitions, entrepreneurship refers to the functions performed by entrepreneurs in establishing an enterprise. Just as management is regarded as what managers do, entrepreneurship is regarded as what entrepreneurs do. Thus, entrepreneurship is the act of being an entrepreneur. It is the process of giving birth to a new enterprise. Thus, entrepreneurship means creating something new, organizing, coordinating and undertaking risk, and handling economic uncertainty. The meaning of entrepreneurship varies from country to country. While some consider entrepreneurs primarily-as innovators, certain others think of them as managers of an enterprise, some as bearers of risks and others as mobilizes and allocators of capital. In India, an entrepreneur can be defined as a person or a group of persons who are responsible for the existence of a new business enterprise.

Characteristics of Entrepreneurship

Entrepreneurship is characterized by the following features:

1. Economic and Dynamic Activity

Entrepreneurship is an economic activity because it involves the creation and operation of an enterprise with a view to creating value or wealth by ensuring optimum utilisation of scarce resources. Since this value creation activity is performed continuously in the midst of uncertain business environment, therefore, entrepreneurship is regarded as a dynamic force.

2. Related to Innovation

Entrepreneurship involves a continuous search for new ideas. Entrepreneurship compels an individual to continuously evaluate the existing modes of business operations so that more efficient and effective systems can be evolved and adopted. In other words, entrepreneurship is a continuous effort for synergy (optimization of performance) in organizations.

3. Profit Potential

"Profit potential is the likely level of return or compensation to the entrepreneur for taking on the risk of developing an idea into an actual business venture." Without profit potential, the efforts of entrepreneurs would remain only an abstract and a theoretical leisure activity.

4. Risk Bearing

The essence of entrepreneurship is the 'willingness to assume risk' arising out of the creation and implementation of new ideas. New ideas are always tentative and their results may not be instantaneous and positive.

Entrepreneurial Process

Entrepreneurship is a process, a journey, not the destination; a means, not an end. All the successful entrepreneurs like Bill Gates (Microsoft), Warren Buffet

(Hathaway), Gordon Moore (Intel) Steve Jobs (Apple Computers), Jack Welch (GE) GD Birla, Jamshedji Tata and others all went through this process.

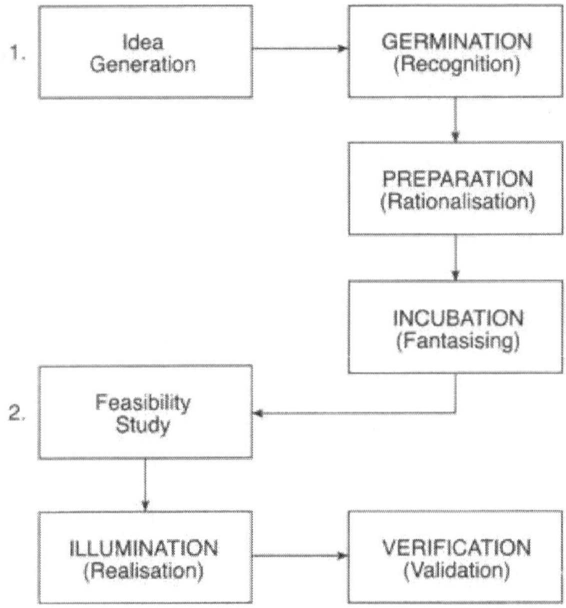

The Entrepreneurial Process.

1. Idea Generation

To generate an idea, the entrepreneurial process has to pass through three stages:

a. Germination

This is like seeding process, not like planting seed. It is more like the natural seeding. Most creative ideas can be linked to an individual's interest or curiosity about a specific problem or area of study.

b. Preparation

Once the seed of interest curiosity has taken the shape of a focused idea, creative people start a search for answers to the problems. Inventors will go on for setting up laboratories; designers will think of engineering new product ideas and marketers will study consumer buying habits.

c. Incubation

This is a stage where the entrepreneurial process enters the subconscious intellectualization. The sub-conscious mind joins the unrelated ideas so as to find a resolution.

2. Feasibility Study

Feasibility study is done to see if the idea can be commercially viable.

It passes through two steps:

a. Illumination

After the generation of idea, this is the stage when the idea is thought of as a realistic creation. The stage of idea blossoming is critical because ideas by themselves have no meaning.

b. Verification

This is the last thing to verify the idea as realistic and useful for application. Verification is concerned about practicality to implement an idea and explore its usefulness to the society and the entrepreneur.

Importance of Entrepreneurship

1. Development of Managerial Capabilities

The biggest significance of entrepreneurship lies in the fact that it helps in identifying and developing managerial capabilities of entrepreneurs. An entrepreneur studies a problem, identifies its alternatives, compares the alternatives in terms of cost and benefits implications, and finally chooses the best alternative.

2. Creation of Organisations

Entrepreneurship results into creation of organisations when entrepreneurs assemble and coordinate physical, human and financial resources and direct them towards achievement of objectives through managerial skills.

3. Improving Standards of Living

By creating productive organisations, entrepreneurship helps in making a wide variety of goods and services available to the society which results into higher standards of living for the people. Possession of luxury cars, computers, mobile phones, rapid growth of shopping malls, *etc.* are pointers to the rising living standards of people, and all this is due to the efforts of entrepreneurs.

4. Means of Economic Development

Entrepreneurship involves creation and use of innovative ideas, maximisation of output from given resources, development of managerial skills, *etc.*, and all these factors are so essential for the economic development of a country.

Factors Affecting Entrepreneurship

Entrepreneurship is a complex phenomenon influenced by the interplay of a wide variety of factors.

Types of Entrepreneurs

Depending upon the level of willingness to create innovative ideas, there can be the following types of entrepreneurs:

1. Innovative Entrepreneurs

These entrepreneurs have the ability to think newer, better and more economical ideas of business organisation and management. They are the business leaders and contributors to the economic development of a country. Inventions like the introduction of a small car 'Nano' by Ratan Tata, organised retailing by Kishore Biyani, making mobile phones available to the common may by Anil Ambani are the works of innovative entrepreneurs.

2. Imitating Entrepreneurs

These entrepreneurs are people who follow the path shown by innovative entrepreneurs. They imitate innovative entrepreneurs because the environment in which they operate is such that it does not permit them to have creative and innovative ideas on their own. In our country also, a large number of such entrepreneurs are found in every field of business activity and they fulfill their need for achievement by imitating the ideas introduced by innovative entrepreneurs. Development of small shopping complexes is the work of imitating entrepreneurs. All the small car manufacturers now are the imitating entrepreneurs.

3. Fabian Entrepreneurs

The dictionary meaning of the term 'fabian' is 'a person seeking victory by delay rather than by a decisive battle'. Fabian entrepreneurs are those individuals who do not show initiative in visualising and implementing new ideas and innovations wait for some development which would motivate them to initiate unless there is an imminent threat to their very existence.

4. Drone Entrepreneurs

The dictionary meaning of the term 'drone' is 'a person who lives on the labor of others'. Drone entrepreneurs are those individuals who are satisfied with the existing mode and speed of business activity and show no inclination in gaining market leadership. In other words, drone entrepreneurs are die-hard conservatives and even ready to suffer the loss of business.

5. Social Entrepreneur

Social entrepreneurs drive social innovation and transformation in various fields including education, health, human rights, workers' rights, environment and enterprise development. They undertake poverty alleviation objectives with the zeal of an entrepreneur, business practices and dare to overcome traditional practices and to innovate. Dr Mohammed Yunus of Bangladesh who started Gramin Bank is a case of social entrepreneur.

Difference between Entrepreneur and Manager

An entrepreneur is involved with the start-up process whereas a manager with running the business

Over a long period of time an entrepreneur assumes financial, material and psychological risks whereas a manager does not have to bear risks

An entrepreneur is driven by perception of opportunity whereas a manager by the resources he currently possesses

An entrepreneur initiates change whereas a manager follows rules and procedures

An entrepreneur is his own boss whereas a manager is a hired employee

An entrepreneur gets uncertain rewards whereas a manager gets fixed rewards and salary

Scope for Entrepreneurship Development in Agriculture

Technologies those reduce the expenditure of production and increase the benefit of the farmers will unlock new opportunities for Agri-entrepreneurship. New technologies that are trouble-free, time saving and relieve farmers from hard work will also provide opportunity for entrepreneurship in agriculture.

On-Farm Activities

Depending upon the geological location and resources accessibility, the probable areas of entrepreneurship in agriculture are:

1. **Agro produce processing units**: These units do not manufacture any novel product. They simply process the agriculture produce. *e.g.*, Rice mills, Dal mills, Decorticating mills *etc.*
2. **Agro produce manufacturing units:** These units produce completely innovative products based on the agricultural produce as the key raw material. *e.g.*, Sugar factories, Bakery, Strawboard units *etc.*
3. **Agro-inputs manufacturing units:** These units manufacture goods either for automation of agriculture or for increasing production. *E.g.*, Fertilizer manufacturing plants, insecticides manufacture units, food processing units, agricultural equipment *etc.*
4. **Agro-service centers:** These include the workshops and service centers for repairing and serving the agricultural equipment used in agriculture.

Off-Farm Vocations

Entrepreneurship development is also beneficial in various off-farm activities such as cloth stitching, knitting, needlework, cloth printing, carpeting, dari making, wrapping and synthetic bag making, agarbatti making, candle making, rope making, case making, bamboo-work, distilled water making, grease extraction, chalk making, biogas mechanic, electric cabling, mason, woodwork, black smithy, solar mechanic, electrician, auto mechanic, welding, pottery, and other rural crafts. The possible areas of entrepreneurship in allied activities of agriculture

This includes the activities like, Dairy technology, Horticulture, Goat rearing, Rabbit rearing, Floriculture, Fisheries, Shrimp farming, Poultry farming, Sheep rearing, Vegetable cultivation, Nursery farming, Grafting, Farm forestry, *etc.*

SWOT

Strengths, Weakness, Opportunities and Threats (SWOT) Analysis

It has always been important for any enterprise to know and understand how it fits in and interacts with the surrounding environment on both an internal (office/factory/shop environment) and external level (how the business operates with the outside world). Examining the environment proves fruitful to develop a strategy for both long and short term plans. The process of liberalization and economic reforms, of course, create remarkable opportunities for development of numerous industries but it also put up new challenges against the industries. The 21st century, would guarantee a bright future for industries provided that a conscious and sincere effort is made to avoid all hurdles. The entrepreneurs have to give more importance to the quality of their production and hence, sufficient attention must be there on research and growth. Innovation is very much necessary to maintain constant progress. And for that, an entrepreneur has to analyze the existing situation. The analysis of the current situation is done by conducting (1) Internal Analysis and (2) External Analysis, jointly known as SWOT analysis that refers to identifying the strengths, weaknesses, opportunities and threats of an organization. SWOT Analysis is basically a tool, often used by organizations in planning its future. This tool can be explained in a better way as under:

S - Strengths

W - Weaknesses

O - Opportunities

T - Threats

The Internal Analysis of the organization includes the organizational position with respect to different functional areas like Production, Finance, Marketing and Distribution and so on. This will reveal its strength and weakness. The External Analysis will include the industry performance, competitive activity and a review of the growth and decline of the user industries.

Analyzing the Business

The most influential way to evaluate the business is to perform a SWOT analysis of the company. It is important to know that the SWOT analysis is only based upon information that is known by the assessors, but SWOT is still a powerful tool considering its immediate benefits. It's quite obvious that recognizing the Strengths and Weaknesses before tackling the Opportunities and Threats is the best way to approach the analysis. There is no fixed way of doing a SWOT analysis, but it should be done in the most comfortable way. The objective of SWOT should be to plan a strategy for the future to improve your company's overall performance.

Strengths and Weakness

Every business must assess its strengths and weaknesses sporadically. It can be done by the management itself or by some outside expertise. The sole aim of SWOT analysis should be to overcome their weakness gradually and transform it into its strength later on.

Some of the strengths of an organization are:

- ☆ Accessibility of required infrastructure
- ☆ Sufficient manufacturing capability
- ☆ Skillful manpower
- ☆ Quality assurance and quality control
- ☆ Manufacture at low cost
- ☆ Facilities for product development
- ☆ Good locality
- ☆ Wide network
- ☆ Motivated employees
- ☆ Latest equipment installed
- ☆ Brand name at National/Regional/Local level
- ☆ Consistency in gaining profits
- ☆ Good financial position
- ☆ Efficient administration
- ☆ Human resource development
- ☆ Excellent transportation
- ☆ Healthy competition

Some of the Weaknesses of an organization are

- ☆ Increasing expenditure of operations
- ☆ Union pressures
- ☆ Low level of employees
- ☆ Lack of raw material
- ☆ Lack of capital
- ☆ Weak credit in market
- ☆ Problem of utilization of capacity
- ☆ Out-of-date technology
- ☆ Poor planning

- Insufficient infrastructure
- Less number of trained technicians
- Poor managerial expertise
- Unorganized work culture
- Ineffective co-ordination
- Brand name not recognized on a National/Regional/Local level
- Poor location for business
- Too much waste

Recognizing these weaknesses require one to be honest and practical. This is an important part as to realize what needs to be done to minimize weakness in the future.

Opportunities

If the business is to achieve its goals, an entrepreneur has to know the parts of the environment to be monitored. A business has to monitor environmental forces like customers, competitors, distribution channels, suppliers, *etc.* as they can affect its profits. A marketing intelligence system should be developed to track important developments. For each trend or development, management needs to identify the hidden opportunities and threats. An opportunity is an area of need in which a company can perform profitably.

Some of the **opportunities** of an organization are:

- Rising population
- Increase in disposable earnings
- Skilled workforce
- trouble-free accessibility of money
- Appropriate technology
- Positive government policies
- Availability of market information, distribution outlets and media.
- Existence of good cultural environment.
- Opponent going bankrupt

Threats

Some developments in the external environment characterize threats. An environmental threat is a challenge posed by an unfavourable trend or development.

Some of the **threats** of an organization are:

- Scarcity of power, water, fuel
- Negative response by the market
- Recession

- Cut throat competition
- Political unsteadiness
- End of season approaching
- Technical obsolescence
- Financial crisis in the market
- Ever-increasing interest rates
- Resource crisis
- Difficulty in retaining technical expertise
- Rising cost of wages
- Change in customers' tastes and preferences
- Existing product becoming old-fashioned

On completion of a SWOT Analysis, an organization can know 'where it stands'. Therefore, it can proceed to develop specific objectives and goals for the future.

Advantages of SWOT Analysis

The advantages for conducting a SWOT Analysis are mentioned as under:

- An entrepreneur can come to know the demand and can accordingly choose respective field, keeping their strengths in mind.
- It certainly helps in the development of innovative technology.
- An entrepreneur can make an analysis as to develop a completely new product, or to just bring slight change in the existing product.
- Improvements can be brought about in the practice of production.
- New goals can be set through SWOT on the basis of previous findings.
- Short term task and long term task can be planned.
- Opportunities and threats can be recognized.
- Batter plans can be formalized for effective communication and implementation.
- Major functions and minor-functions can be set out in achieving the objectives of the firm.
- Expansion of the range of services and benefits.
- The social, political, legal, economic, technological factors can be analyzed
- An analysis of the competitor's strategies can be made and their policies can also be known.

Case Study

To understand SWOT analysis better, let us go through a case study of **SWOT Analysis** conducted in a **five star Hotel in Bangalore:**

The **strengths** of this five star Hotel are

- ☆ It has got more than 95 years of experience
- ☆ It has hotels in different places in India
- ☆ It has many heritage properties which adds to its charm and represents Indian culture
- ☆ It has a few low budgeted hotels, which caters to the needs of middle class tourists
- ☆ It has hotels not only in big cities but also in small cities which are of tourist importance

The **weaknesses** are

- ☆ It has tough competition from rivals, and hence their concentration is diverted.
- ☆ It has very few low budget hotels, which are of need today.
- ☆ It has more than one hotel in one city, which is a waste of investment.

Its **opportunities** are

- ☆ There is availability of government support in the form of land and financing of projects.
- ☆ There is an increase in the tourist inflow in India, India is expecting to receive 2.4 million tourists, *i.e.*, there will be lot of business.

Its **threats** are

- ☆ Its competitors are also located nearby the hotel.
- ☆ It has to face the threats of various resorts, which people are looking for now-a-days.
- ☆ It also has threats from smaller hotels, which tourists prefer, because of cheaper rates and almost the same kind of hospitality.

Therefore, this hotel has to overcome its threats and weaknesses and concentrate more on its strengths and opportunities available.

Achievement Motivation

Achievement motivation is affect in connection with evaluated performance in which competition with a standard of excellence was paramount. (McClelland, Atkinson, Clark, and Lowell, 1953) Intentional actions are usually considered the prototype of all acts of will. Theoretically, a complete intentional action is conceived of as follows: Its first phase is a motivation process, either a brief or a protracted vigorous struggle of motives; a second phase is an act of choice, decision, or intention, terminating this struggle; the third phase is the consummator intentional action itself, following either immediately or after an interval short or long. (Lewin, 1926, reprinted in 1999) Achievement motivation can, therefore, be defined as the striving to increase or to keep as high as possible, one's own capabilities in all activities in which a standard of excellence is thought to apply and where the execution of such activities can, therefore either succeed or fail. (Heckhausen, 1967)

a. nAch reflects S's social status,
b. S's social status also exerts certain specifiable effects on their task preferences and performances, and
c. Therefore the complex interrelationships between nAch and performance can be rendered more comprehensible by viewing nAch, aspiration, and performance to constitute aspects of an individual's more or less integrated role-related activity. ...

The concepts of "motive" and "role" are obviously not mutually exclusive. Rather they represent different levels of analysis. (Klinger and McNelly, 1969) In the case of an individual, it is not sufficient to say that an individual is at the X percentile in achievement motivation: but alternatively, his profile of the six dimensions identified in the present study [status with experts, acquisitiveness, achievement via independence, status with peers, competitiveness, and concern for excellence], possibly together with others, might be used more precisely to identify the combination of characteristics determining the unique direction of his motivation to achieve. (Jackson, Ahmed, and Heapy, 1976)

Achievement motivation consists of a varied and complex set of assumptions, assessments, predictions, inferences, values, standards, and affective reactions that may be irrational, inaccurate, and contradictory. (Dweck and Elliott, 1983, p. 644) A Sample Evolution Achievement behavior is best defined as behavior on skill tasks or, at least, on tasks where individuals believe or feel that their competence affects outcomes. Achievement behavior is distinguished from other forms of behavior by its purpose: the goal of achievement behavior is to be or feel competent or incompetent (Maeher and Nicholls, 1980).

Government Policy and Programmes

Throughout the world, micro and small enterprises (MSEs) are considered as an important constituent of the national economies, as they significantly contribute to expansion of employment. Considering the importance of micro and small enterprises, the Central and State Governments have been implementing several schemes for promotion of these enterprises such as:

Prime Minister's Employment Generation Programme (PMEGP)

Government of India has approved a new credit linked subsidy programme known as Prime Minister's Employment Generation Programme (PMEGP). It has been done by merging the two schemes that were in process till 31.03.2008 specifically Prime Minister's Rojgar Yojana (PMRY) and Rural Employment Generation Programme (REGP) that were meant for creation of employment opportunities through establishment of small enterprises in rural and metropolitan areas. PMEGP will serve as a central sector scheme to which shall be administered and monitored directly by the Ministry of Micro, Small and Medium Enterprises (MoMSME). The Scheme shall be implemented and governed by Khadi and Village Industries Commission (KVIC), which is a statutory organization under the administrative control of the Ministry of MSME as the single nodal agency at the National level. Whereas at the State level, the State KVIC Directorates, State Khadi and Village Industries Boards (KVIBs) and District Industries Centres (DICs) as well as banks will implement the Scheme. The Government subsidy under the Scheme will be routed by KVIC through the recognized Banks to the entrepreneurs in their Bank accounts.

Objectives of Prime Minister's Employment Generation Programme (PMEGP)

The main objective of the scheme is to produce more and more employment opportunities in both, rural as well as urban areas of the country. For this purpose, setting up of new self-employment ventures/projects/micro enterprises shall also be done. It also aims at bringing together a widely isolated traditional rural and urban youth seeking employment and providing them self-employment opportunities.

Through the scheme, the government wishes to provide a sustainable employment to a large segment of traditional rural and urban unemployed youth in the nation, and thereby to help them. It is also an attempt to increase the earning capability of rural and urban youth.

Conditions of Eligibility for Beneficiaries

- The age should be above 18 years.
- There will be no limit of income for setting up projects under PMEGP.
- The beneficiaries should have studied at least up to VIII standard for setting up of project costing more than Rs.10 lakh in the manufacturing sector and above Rs. 5 lakh in the business/service sector.

The assistance under the Scheme is provided only in case of new projects sanctioned particularly under the PMEGP.

- Self Help Groups along with those belonging to BPL are also eligible for assistance under PMEGP provided that they must not have availed benefits under any other Scheme
- The institutions should be registered under Societies Registration Act,1860.
- Production Co-operative Societies and Charitable Trusts are also eligible for the scheme

Market Development Assistance Scheme for Micro/Small Manufacturing Enterprises/Small and Micro Exporters

The scheme offers financial support by manufacturing Small and Micro Enterprises in International Trade Fairs/Exhibitions under MSME India stall. It also offers funding for sector specific market studies done by Industry Associations/ Export Promotion Councils/Federation of Indian Export Organization. SSI Associations that initiate anti-dumping cases are covered under the scheme.

Objectives of Market Development Assistance Scheme for Micro/Small Manufacturing Enterprises/Small and Micro Exporters

- ☆ The scheme aims at encouraging Small and Micro exporters for tapping and developing markets in foreign country.
- ☆ Through this scheme, government wants to motivate the representatives of small/micro manufacturing Enterprises to participate under MSME India stall at International Trade Fairs/Exhibitions for further exposure.
- ☆ Further the scheme also aims at popularizing the implementation of Bar Coding on a large scale.

Schemes for Women Entrepreneurs

A. **Mahila Udyami Yojana (MUY):** IDBI has set up special fund under this scheme with corpus fund of Rs.5Crore to provide seed capital assistance to the women entrepreneurs intending to initiate projects in SSI sector. This scheme is implemented by SIDBI. 51 per cent of equity should be managed by women.

B. **Stree Sakthi Package (SBI):** Under this scheme, EDPS are exclusively designed and conducted for women entrepreneurs. Moreover, an amount of Rs.25000/- is provided without collateral security.

C. **Priya Darshini Yojana:** It is implemented by Bank of India. Financial assistance is provided to women entrepreneurs.

   - ☆ Maximum loan provided in the scheme is up to 2 lakhs for term loan and 1 lakh for working capital.
   - ☆ Assets acquired with finance are hypothecated as security.
   - ☆ Repayment period is 3-5 years.
   - ☆ Margin money is 20 per cent depending on type of activity.
   - ☆ Besides these programmes other schemes such as PMRY, SEPUP (Self Employment Programme for urban Poor.), SGSY, IRDP and Rashtriya Mahila Kosh help women entrepreneurs in providing financial assistance.
   - ☆ In conclusion Government incentives available are of two types of subsidies.

- ☆ Investment subsidy to establish more enterprises. It includes various subsidies such as capital investment subsidy, transport subsidy, power generator subsidy, and social subsidy to women entrepreneurs.
- ☆ Other subsidies include Export/import subsidy, tax subsidy, excise subsidy/duty exemption, and capital subsidy for technology up-gradation.

Impact of Economic Reforms on Agribusiness/Agri-enterprises

Agriculture plays a vital role in the Indian economy. Over 70 per cent of the rural households depend on agriculture. Agriculture is an important sector of Indian economy as it contributes about 17 per cent to the total GDP and provides employment to over 60 per cent of the population. Indian agriculture has registered impressive growth over last few decades. The food grain production has increased from 51 million tones (MT) in 1950-51 to 250MT during 2011-12 highest ever since independence.

Some salient facts about Agricultural scenario

1. Agriculture is the largest provider of livelihood in rural India
2. It contributes 25 percent to India's GDP
3. It is still dependent primarily on the monsoons
4. The growth in agricultural production has been stagnant for the past several years.
5. The drought in north and western parts in FY09 created shortages in supply of food grains.

The following points highlight the thirteen major impacts of liberalisation on Indian agriculture.

They are:

1. Raising the Production of Food grains,
2. Increasing Trend in Horticultural Output,
3. Diversification of Agriculture,
4. Increase in Floricultural Output,
5. Agricultural Exports,
6. Food Processing,
7. Rising Productivity of Agricultural Resources and Others.

Raising the Production of Food Grains

India has been experiencing the increase in the production of food-grains particularly after the inception of new agricultural strategy (*i.e.*, Green revolution). Annual growth rate of 2.08 per cent was recorded during seventies. The decade of nineties could not maintain this pace and annual growth rate has fallen to 1.7 per cent. It is felt that if the country maintains 4 per cent growth rate in agricultural production, then after meeting its domestic demand, the country can export the surplus amount of food grains to the foreign countries.

Increasing Trend in Horticultural Output

The diversity of physiographic, climate and soil characteristics enables India to grow a large variety of horticultural crops which includes fruits, vegetables, spices, cashewnut, coconut, cocoa, arecanut, root and tuber crops, medicinal and aromatic plants *etc.* India is the largest producer of fruits, and second largest producer of vegetables.

Diversification of Agriculture

Agriculture is not only meeting the demand for food-grains but also other needs of development. In recent years, agricultural sector has been diversified to produce commercial crops and horticultural crops *viz.*, fruits, vegetables, spices, cashew, arecanut, coconut and floricultural products like flowers, orchards *etc.* dairy and other animal husbandry products.

Increase in Floricultural Output

About 31,000 hectares of land spread over Karnataka, Tamil Nadu, Andhra Pradesh and West Bengal are under flower production. Since the inception of liberalisation, commercial farming of floricultural activities has been increasing gradually. The demand for Indian cut flower is increasing continuously in the international market.

Agricultural Exports

Another important emerging trend of agriculture is the increasing volume of agricultural exports. Agricultural exports are playing an important role in expanding economic activities along with generating employment opportunities. The Export-Import Policy (Exim) 2002-03 has provided ample opportunities for increasing the volume of agricultural exports.

Food Processing

Economic liberalisation has made ample scope for the development and expansion of food processing industry in India. The Government is also offering necessary incentives by exempting the industry from excise duty. In order to invite foreign capital into this industry the Government has permitted 51 per cent foreign equity partnership and also offered prompt approval of foreign technology transfer to the food processing industry of the country.

Rising Productivity of Agricultural Resources

Another impact of liberalisation has been felt that it boosted the productivity of agricultural resources. Improvement in the productivity of resources is being done through better allocation of resources and latest technology between different areas under present circumstances.

Developing Agriculture in Backward Areas

In the post-Green Revolution period, application of new agricultural strategy, research and technology was very much restricted in the production of two main crops *i.e.* wheat and rice. But under the liberalisation wave, with the growing demand for agricultural exports, many new areas of agricultural operations have become favourable and lucrative. In backward areas, having no irrigation system, dry land farming is becoming popular. The other activities like horticulture, floriculture, animal husbandry, fishery *etc.* have been encouraged.

More Employment Opportunities, in Agricultural Sector

As a result of Green revolution and mechanisation there is considerable fall in employment opportunities in the rural areas. Even special employment programmes could not serve the purpose. But increasing potentiality of the agricultural sector as emerged from the liberalisation/globalisation wave has set up new trends in horticultural, floricultural and animal product and has created ample opportunities and scope for employment of huge number of population. This allied sector being labour intensive can provide better solution to the unemployment problem specially rural India.

Growing Volume of Subsidies

The volume of subsidies granted to agriculture, in respect of fertiliser, irrigation and electricity charges *etc.* has been increasing in our country. Out of these total amounts of about 75 per cent is allotted in the area of fertiliser and food-grains. Under the present era of liberalisation, although there is a move to reduce the volume of subsidies in the budget but political compulsions have prevented the government to undertake that move.

Growing Trend of Investment in Agriculture

Agricultural sector is experiencing a growing trend in the volume of its investment during the post liberalisation period. The share of public sector investment has declined from 28.2 per cent in 1993-94 to 21.0 per cent in 1998-99. On the other hand, the share of private sector investment in agriculture has increased considerably from 71.8 per cent in 1993-94 to 79.0 per cent in 1998- 99.

Institutionalization of Agricultural Credit

The wave of liberalisation has encouraged the institutional agricultural credit. In the initial stage of post-independence period, Indian farmers were depending too much on unorganised sources of agricultural credit, *i.e.* on village money lenders, landlords, traders *etc.* who charge exorbitantly higher rate of interest.

Entrepreneurial Development Process

Entrepreneurship development is the means of enhancing the knowledge and skill of entrepreneurs through several classroom coaching and programs, and training. The main point of the development process is to strengthen and increase the number of entrepreneurs.

Process of Entrepreneurship Development

The below-mentioned steps will illustrate how to build an effective entrepreneurship development program for an entrepreneur to organize and launch the new ventures.

- **Discover** – Any new process begins with fresh ideas and objectives, wherein the entrepreneur recognizes and analyzes business possibilities. The analyzing of opportunities is a risky task, and an entrepreneur looks out for inputs from other persons, including channel partners, employees, technical people, consumers, *etc.* to reach an ideal business opportunity.
- **Evaluation** – The evaluation of an opportunity can be done by asking several questions to oneself. For instance, questions like whether it is worth taking a chance and investing in the idea, will it attract the consumer, what are the competitive advantages and the risk linked with it are asked. A reasonable and sensible entrepreneur will also analyze his skills and whether it matches his entrepreneurial objectives or not.
- **Developing a plan** – After the identification of an opportunity, an entrepreneur has to build a complete business plan. It is the most important step for new business as it sets a standard and the assessment criteria and sees if a company is working towards the set goals.
- **Resources** – The next step in the process of entrepreneurial development is resourcing. Here, the entrepreneur recognizes the source of finance and from where the human resource can be managed. In this step, the entrepreneur also tries to find investors for his new business.
- **Managing the company** – After the hiring process and funds are raised now it's time to start the operation to accomplish the desired goals. The entire entrepreneur will decide on the management structure that will be assigned to resolve the operational problems whenever it occurs.
- **Harvesting** – The last step in this process is harvesting, where an entrepreneur determines the future growth and development of the business. Here, real-time development is compared with the projected growth, and then the business security or the extension is initiated accordingly.

SKILL DEVELOPMENT

Business Leadership Skills

Effective leadership skills are necessary in any business in order to succeed. Whether you own the business, manage it, or are employees looking to make his or her way up the corporate ladder, the right leadership style is essential in order to reach your business goals. Listed below are six **business leadership skills** that every great leader should possess. Adaptability means reacting in an effective manner to changing circumstances. Everyone experiences challenge when trying to adapt to the new and unfamiliar, but true leaders are able to resolve those challenges with a solid plan of action. If there is one trait that every good business leader should possess in today's workforce, it is adaptability.

But more people are leaders than they realize. Leadership takes on many different faces; it's just a question of understanding how you lead and why.

Here are 10 key business leadership skills you'll need to succeed as a leader:

1. **Lead By Example.** You can't be an aloof leader, someone that's never around and incapable of getting your hands dirty. One of the best ways to lead is by example – pitching in where needed, lending a helping hand, and making sure that the work you do is clearly understood by your team.
2. **Passion.** A leader without passion isn't a leader. He's a paper pusher. Or a taskmaster. Or a government employee... Passion drives a lot, and you can inspire so much in others through your own passion and enthusiasm.
3. **Be Organized.** A disorganized leader isn't leading; he's chasing his own tail. Disorganization breeds nothing but more disorganization. If you're frazzled and messy, your team will be too. When you're organized you'll be much more productive and so will everyone else.
4. **Delegate.** You can't do everything. A great leader needs to be able to delegate effectively. The key to delegating successfully is giving employees ownership of the work you assign them.
5. **Take Ownership and Responsibility.** Although you've just delegated work and truly given your team ownership, you also have to take ownership and responsibility at all times. Your team has to know you'll be there for them through the good and the bad times.
6. **Communicate Effectively.** Everyone knows great leaders have to be great communicators. But there are certain points of communication that many people forget.
7. **Be Brave and Honest.** Cowardly leaders will shy away from any number of situations that crop up regularly when running a team. The project your team has worked on for 6 months just got shelved.
8. **Great Listener.** A huge part of being a great communicator is being a great listener. If all you want to do is talk, you're not a leader. Keeping people motivated means listening to them, asking them questions, understanding their issues.
9. **Know Your People.** You have to know your people. You don't have to be best friends or even socialize outside work, but you do have to know what makes them tick.
10. **Be a Follower.** Benjamin Disraeli said, *"I must follow the people. Am I not their leader?"* That sums up many of the other points so beautifully. Great

leaders are followers too. If you're a leader without following, you're a dictator. And as fun as that sounds... Being a leader-follower means finding value in your team, getting inspired by your team, encouraging your team to communicate, brainstorm and be open.

Developing Organisational Skills

Organizational skills are some of the most important proficiencies you can have as an employee. Being organized will allow you to meet deadlines, minimize stress and carry out your duties more efficiently. To learn how improving your organizational skills can affect your work performance, you need to know what this skill set entails. In this article, we look at different types of organizational skills, suggest what you can do to improve your skills and discuss how to highlight organizational skills on a resume or in a job interview.

What are Organizational Skills?

Organizational skills are skills that allow you to use your resources efficiently and effectively. Being organized means you manage your time, energy and workspace well and can accomplish all your assigned tasks successfully. Organizational skills can take different forms depending on your particular workplace and job title, but they typically involve maintaining an orderly workspace, meeting deadlines and communicating well with your team.

How to Highlight Organizational Skills

Here are suggestions for how to highlight your organizational skills:

On a Resume

Before you begin writing or editing your resume, consult the job description on the listing to find out what specific types of organizational skills your potential employer is searching for. Look for organizational keywords like *"multi-tasking," "managing resources"* and *"meeting deadlines."* Try to incorporate these phrases into the work history entries or skills list on your resume. You can also highlight your organizational skills in your short bio or objective statement by using keywords like *"task-oriented," "reliable"* and *"goal-setter."*

In a Job Interview

One of the best ways to highlight your organizational skills in a job interview is by sharing examples of how you remained organized while working your previous jobs.

Ways to Improve your Organizational Skills

Although many successful people feel that they work best under pressure, there is no doubt that honing your organisational skills makes your business life a lot easier. You need to be able to manage your time and resources to successfully juggle all the tasks in your diary. Organisational skills don't come naturally to everyone, but you may be relieved to hear that they can be developed.

1. *Avoid Fire Fighting*

Preparing for the day ahead means planning your schedule and deciding what tools you will need to complete your daily tasks.

2. *Plan your Time*

Whilst you are planning your day and preparing for each of the tasks ahead, schedule in how long you think each task will take. This allows you to break up your day into manageable chunks.

3. *Successful Planners make Lists*

Lists enable your mind to remain tidy. Busy people have ideas buzzing around in their heads all the time.

4. *Begin Tasks Early*

Those of you who enjoy working under pressure may need to feel a deadline snapping at your heels before you are able to settle down and complete a task.

5. *Decide when you Work Best*

It can help to be very honest with yourself about when you are most productive. This will differ for every individual. Some people work best when they get up early and begin working as the sun comes up, whilst others are more productive in the evening.

6. *File Smart*

We have all found ourselves in a situation where we're desperately trying to meet a tight deadline but cannot find the crucial file that we need to complete the work.

7. *Avoid Distractions*

Avoiding distractions is a great way of organising your time well. Having your email server open can create a long line of distractions if you check your emails each time you hear a pindrop.

8. *Make your Workspace work for you*

Do you find yourself sorting through piles of paper to locate the one document you need? Do you have to leave the room each time you need to use the photocopier or collect more stationery?

9. *Delegate*

Delegation doesn't come easily to many people, but it is an incredibly useful skill to learn if your time is precious. If you have a team around you then allow people to work to their strengths, they may be better at some tasks then you are.

Controlling

Controlling involves ensuring that performance does not deviate from standards. Controlling consists of three steps, which include:

1. Establishing performance standards,
2. Comparing actual performance against standards, and
3. Taking corrective action when necessary.

Performance standards are often stated in monetary terms such as revenue, costs, or profits but may also be stated in other terms, such as units produced, number of defective products, or levels of quality or customer service.

The measurement of performance can be done in several ways, depending on the performance standards, including financial statements, sales reports, production results, customer satisfaction, and formal performance appraisals. Managers at all levels engage in the managerial function of controlling to some degree.

The managerial function of controlling should not be confused with control in the behavioral or manipulative sense. This function does not imply that managers should attempt to control or to manipulate the personalities, values, attitudes, or emotions of their subordinates. Instead, this function of management concerns the manager's role in taking necessary actions to ensure that the work-related activities of subordinates are consistent with and contributing toward the accomplishment of organizational and departmental objectives.

Supervising

Supervisors are an important asset to any business or organization. Supervisory skills are essential for every employee since they are the foundation for moving up in your career to leadership positions.

In this Supervisory Skills post, we will be giving you guidelines on how to become a supervisor. If you currently work as a supervisor, you will find this guide a helpful tool to improve and excel in your job.

The Core Supervisory Skills

Here are the 5 Core Supervisory Skills that are essential to acquire:

1. **Leadership Skills:** Being a good leader is critical for a supervisor, it is the first step towards managing a team. Your team members or employees would rely on their leader for guidance and mentor which is vital for success.
2. **Time-Management Skills:** Time Management is the seed to thriving in any task, it is implemented in every workplace and for any position. As a supervisor, you should learn how to manage your time as well as scheduling tasks for your employees. Therefore, time planning is critical to the success of any job or project.

3. **Technical Skills:** When an employee asks for your technical help and knowledge, you should be able to deliver as their supervisor. Mentoring, and passing on your technical skills are of great importance as a supervisor.
4. **Communication Skills:** Your communicative skills should be perfected as they happen to help you to deliver the message or task to an employee properly, as well as setting an example to the team members; it also maintains your prestige as a supervisor.
5. **Judgmental Skills:** Judgement of a certain situation or deciding on a task is a skill acquired with experience, time, as well as trial and error.

Supervisory skills are easy to acquire and apply, yet it is still important to improve them. So, improve your Core Skills and become a better supervisor with excellent leadership and time management skills.

Problem Solving

Effective problem solving is a key management skill and a major factor in determining individual and organizational success. People with good problem solving skills adapt more quickly in times of rapid change and are generally high achievers, whether it is by putting things right when they go wrong, making the best use of resources or creating and exploiting opportunities. Whatever our personal and professional ambitions, we can improve our chances of achieving what we want by developing our problem solving skills. Each of us has an innate ability to solve problems. To develop this ability, we need a clear understanding of the skills and techniques involved, and practice in applying them in different situations.

To be a successful problem solver we must go through these stages:

- Recognizing and defining the problem
- Finding possible solutions
- Choosing the best solution
- Implementing the solution.

In this lesson the following objectives have been defined:

- Define problem-solving
- Differentiate between analytical and creative problem solving
- Recall various steps in the creative problem-solving process.

Monitoring and Evaluation

Monitoring and Evaluation (**M and E**) is used to assess the performance of projects, institutions and programmes set up by governments, international organisations and NGOs. Its goal is to improve current and future management of outputs, outcomes and impact. Monitoring is a continuous assessment of programmes based on early detailed information on the progress or delay of the ongoing assessed activities. An evaluation is an examination concerning the relevance, effectiveness, efficiency and impact of activities in the light of specified objectives.

Monitoring and evaluation processes can be managed by the donors financing the assessed activities, by an independent branch of the implementing organization, by the project managers or implementing team themselves and/or by a private company. The credibility and objectivity of monitoring and evaluation reports depend very much on the independence of the evaluators. Their expertise and independence is of major importance for the process to be successful.

Difference between Monitoring and Evaluation

In monitoring, the feedback and recommendation is inevitable to the project manager but in evaluation, this is not the case. The common ground for monitoring and evaluation is that they are both management tools. For monitoring, data and information collection for tracking progress according to the terms of reference is gathered periodically which is not the case in evaluations for which the data and information collection is happening during or in view of the evaluation. The monitoring is a short term assessment and does not take into consideration the outcomes and impact unlike the evaluation process which also assesses the outcomes and sometime longer term impact.

This impact assessment occurs sometimes after the end of a project, even though it is rare because of its cost and of the difficulty to determine whether the project is responsible for the observed results. Evaluation is a systematic and objective examination which is conducted on monthly and/or yearly basis, unlike Monitoring, which is a continuous assessment, providing stakeholders with early information. Monitoring checks on all the activities on the last [implementation stage] unlike Evaluation which entails information on whether the donated funds are well managed and that they are transparently spent.

Developing Managerial Skills

Developing management skills is perhaps one of the most critical activities facing any organization. Yet it can also be one of the most complicated, given the diversity of the average manager's work.

How do you go about developing the broad range of skills managers need to be effective? Try thinking about the 8 P's of effective management. In this mini-series on developing management skills we consider the topic from three crucial management perspectives:

Focusing – the ability to direct attention and effort onto what matters. This is the subject of this article – developing management skills.

Doing – the ability to get things done. This is the topic of our article effective management skills: Making it Happen.

Knowing – the ability to build your explicit and tacit knowledge, to inform how and why you manage. This topic is explored in our article leadership and professional development: be in the know.

Developing Management Skills – the 8 P's of Effective Management

Purpose

- We are all better motivated when we feel we're engaged in something of real significance.
- Look beyond the mundane and functional to see the bigger picture, then help others to do the same.
- Foster a sense of value and purpose amongst your team.
- Set a direction with a clear sense of purpose and value in what you intend to achieve as a team/organization.
- Build plans to deliver your purpose and explain them.

People

- The customer comes second, it's *people* who come first.
- Help the people you manage to flourish is a fundamental focus for any manager.
- Finding the right people is critical.
- Match an individual's strengths with the right role.
- Coach to develop your team's capabilities.
- Develop talent, help people realize their potential.
- Build a better team.
- Delegate.
- Handle conflict promptly and fairly.
- Build relationships with your team, your manager, peers, the wider organization and customers.

Place

- People spend a larger part of their waking lives at work.
- Make your workplace a great place to work. Create the conditions for a happy and productive workplace.
- Give people more choice and control over how they do their work.
- Remember technology is changing the nature of the workplace, so adapt to new conditions.
- Don't forget remote workers, just because they are remote!
- Design interesting and challenging jobs, in appropriate environments.
- Encourage dynamism and enthusiasm by being an example.
- Build trust and value team member contributions.

Processes

- ☆ Ensure your processes deliver value for your customers/clients.
- ☆ Deliver value simply and efficiently.
- ☆ Learn how to solve problems and make decisions effectively.
- ☆ Strive to enhance quality continually.
- ☆ Review processes from your customers perspective.
- ☆ Develop and sustain good customer relationships.
- ☆ Understand and review order generation (sales and marketing).
- ☆ Do the same for order fulfilment (the complete process from purchase to after sales service).

Projects

- ☆ Initiate projects to improve ways of working and introduce new practices.
- ☆ Manage change and improvement.
- ☆ Learn the basics of project management, as a minimum.
- ☆ Apply continuous improvement approaches.

Performance

- ☆ Management is about getting the right things done, and doing them well – it's all about performance!
- ☆ Motivate others to perform.
- ☆ Manage performance.
- ☆ Set goals.
- ☆ Monitor and control.
- ☆ Review progress.
- ☆ Provide timely feedback.

Profit

- ☆ Get the other areas right and profit (or surplus) will follow.
- ☆ Balance costs with income.
- ☆ Achieve performance within budget.
- ☆ Monitor trends to indicate if changes are having an impact.
- ☆ Improve financial performance.
- ☆ Remember profit can be measured in more than financial terms.

Personal Development

- ☆ The most important investment you can make is in your own personal development.
- ☆ Developing management skills on a regular, planned basis – it's not just a course or annual top-up!

- Plan your own personal development.
- Keep up to date.
- Use a variety of development methods – personal, professional, formal, informal.
- Develop your strengths.
- *Why not?*

Business Leadership Skills

What is Business Leadership?

The term business leadership refers to how individuals go about making decisions, working toward goals and exercising authority in a professional environment. Business leadership can take many different forms, but it usually involves a CEO or several higher-level employees guiding and inspiring the rest of the team. Every leadership team will adhere to a different leadership style. The goal of business leadership is to find the leadership model that works best for a particular company and its team of employees.

Skills for Business Leaders

Several specific skills are associated with strong leaders. Most of these involve your ability to accomplish tasks efficiently, influence others to perform well and consistently meet and exceed expectations. Here are some of the most common skills seen in successful business leaders:

- Self-motivation
- Organization
- Delegation
- Communication
- Responsibility
- Setting goals
- Integrity
- Interpersonal skills

How to Improve Business Leadership Skills

Find a Role Model or Mentor

Often, the first step in becoming a more effective leader is finding a strong leader to emulate. Seek out an accomplished leader in your network and talk to them about how they achieved success. Ask them about their personal and professional goals and the path that they followed to get to their current position. Using their wisdom, look for ways to implement their advice with your own team.

Seek Out Leadership Opportunities

The key to improving as a leader is often practice. Look for opportunities to hold leadership positions in your community. This could be at work, in your church or with another group in your personal network. Taking the lead in these settings can provide you with valuable experience and equip you for future opportunities.

Volunteer for more Responsibility

If there are not any leadership opportunities available at your workplace, you can still show your capabilities by volunteering to take on more responsibility. This could mean offering to assist with a project, tackle a new task or even take on a new title. If you show a willingness to serve, you are likely to earn the attention of your team's current leadership.

Communication

Why Communication is Important in Leadership?

It's critical. Good communication is a core leadership function and a key characteristic of a good leader. Effective communication and effective leadership are closely intertwined. As a leader, you need to be a skilled communicator in countless relationships at the organizational level, in communities and groups, and sometimes on a global scale. You need to think with clarity, express ideas, and share information with a multitude of audiences. You must learn to handle the rapid flows of information within the organization, and among customers, partners, and other stakeholders and influencers.

Facts about Communication for Leaders

1. *Authenticity Counts — a Lot*

Be honest and sincere. Find your own voice; quit using corporate-speak or sounding like someone you're not.

2. *Visibility is a Form of Communication*

If you want to communicate well, don't be out of sight. Don't be known only by your emails and official missives. Be present, visible, and available.

3. *Listening is a Powerful Skill*

Good communicators are also good listeners. When you listen well, you gain a clear understanding of another's perspective and knowledge.

Tips for Leaders to Communicate more Effectively

1. *Communicate Relentlessly*

Communicate information, thoughts, and ideas clearly - and frequently - in different media. Keep processes open and transparent, and find ways to help smooth the path of communication for your team or organization.

2. *Simplify and be Direct*

Say what you mean. Be direct. Don't hide behind complexity or pile on a ton of information. Simple communication can be smart communication.

3. *Listen and Encourage Input*

Pause. Be okay with silence. Encourage the other person to offer ideas and solutions before you give yours. Do 80 per cent of the listening and 20 per cent of the talking.

4. *Illustrate through Stories*

When you tell a good story, you give life to a vision, goal, or objective. Telling good stories creates trust, captures hearts and minds, and serves as a reminder of the vision.

5. *Affirm with Actions*

While effective leaders master the art and craft of language, speaking clearly, and presenting logical and compelling arguments, skilled leaders also know that communication goes beyond words.

Direction Skill

Directional skills, such as making goals and plans and then solving problems as they arise are imperative as they allow you to guide your organization toward the future. These skills become even more important with the passage of time as our global environment is becoming more volatile, uncertain, complex, and ambiguous (VUCA). Before you can guide your organization to greatness, you have to have a vision and then be able to execute it. The following tools and processes will aid you in your journey.

Plan, Do, Check, and Act

The *PDCA cycle* (**Plan, Do, Check, Act**) was developed by Dr. Walter Shewhart as a plan of action for creating processes and products. It is a four-step method that uses direction and control to execute, while also providing an iterative process for continuous improvement:

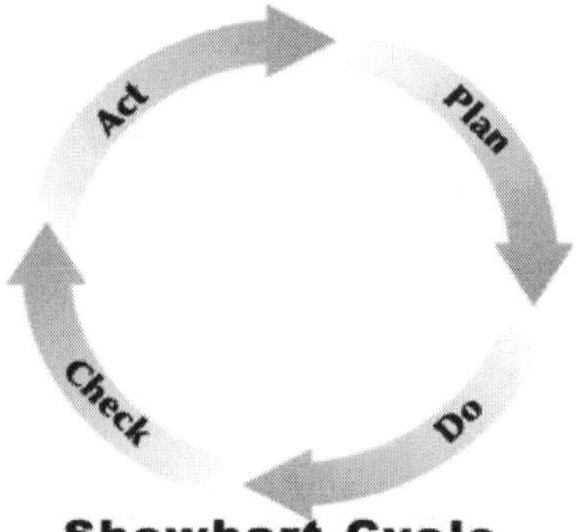

Shewhart Cycle

It is often called the *Shewhart Cycle* or *Deming wheel.* While the four steps of the cycle look easy, it actually takes a lot of work by all team members to complete the cycle correctly.

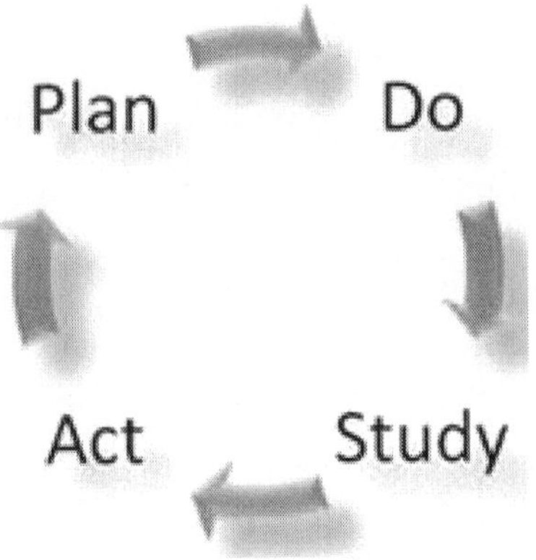

The PDCA cycle is often known as the *Deming Wheel* (*Smith, Hawkins, 2004*). Deming used a modified version — PDSA (Plan, Do, Study, Act) as he believed *study* (analysis) provided a better description than *check.* Good plans start with a brainstorming session that includes all the people involved with the project. This allows everyone to be part of the solution, in addition to gathering the best ideas.

Motivation Skills

Successfully motivating others is an important skill that you'll want to point out to potential employers. Employees can use motivation skills to increase their chances of getting positive results when interfacing with customers, subordinates, upper management, suppliers, donors, team members, funding sources and supervisors. For managers, who must inspire a whole team, the ability to motivate is usually a requirement.

What are Motivational Skills?

Motivational skills can be defined as actions or strategies that elicit a desired behavior or response from a stakeholder. These strategies and actions vary based on three major factors:

1. The motivator's style
2. The target audience
3. The personality of the person the motivator seeks to influence

Steps in the Motivational Process

Motivation is a process that requires careful strategy for success. These steps can help you to reach the desired outcome:

1. Assess the preferences and personality characteristics of the individual or group to be motivated. What motivates one group of people might not be the same as what motivates another group of people.

2. Define motivational strategies appropriate for that target.
3. Convey expectations for performance from the object of the motivation. Or, make it clear how the person can achieve the desired outcome.
4. Communicate benefits, rewards, or sanctions if expectations are or are not met.
5. Share feedback on progress or lack of progress toward desired outcomes.
6. Address problems or obstacles that are limiting success.
7. Provide rewards for desired outcomes.
8. Issue warnings before enacting sanctions.
9. Recognize people who have responded in the desired manner.

Problem Solving Skill

Problem solving then, is what you **should do** when you don't know what to do. However, we may all be more susceptible to panic, procrastination, pretending or ploughing-on than we would care to admit. To avoid these responses think through your own approach to problem solving. You may want to use some of the articles, tools and resources highlighted below to build your own repertoire.

For more answers to the question "what is problem solving?" see our articles and problem solving exercises. You will also find a detailed process with links to tools to help you solve problems in our article Seven Step Problem Solving Process or if you are short of time the essential steps are in our article 7 problem solving steps.

Á problem solving skill is perhaps normally seen as a reactive tool. That's to say, it's applied to existing problems, once these have been identified. However, problem solving can also be done pro-actively. The application of key questions to existing situations, even in the absence of obvious problems, can still yield benefits. Asking why and why not, is perhaps one key to innovation and improvement.

The seven step problem solving technique covers:

1. Finding the right problem to solve
2. Defining the problem
3. Analysing the problem
4. Developing possibilities
5. Selecting the best solution
6. Implementing
7. Evaluating and learning

Frustration and unrest with the current situation is often a good place to start. Use this below is to evaluate the current issues and to brainstorm any possibilities which could arise from those. The second table contains questions which might help you to identify future possibilities. Remember, the key to this problem solving skill is to ask: why not? Just as this same question is routinely asked by innovative experts:

- *Why not?*
- Two words that are filled with possibilities.
- They can turn a challenge into an opportunity.
- An obstacle into an inspiration.
- It's a question we ask ourselves at Toyota every day.
- Because we're continuously looking for new ways to improve what we do.
- By asking tough questions.
- Can we make a car that has zero emissions?
- Can we improve the economy of a community?
- Can we enrich the lives of people around us?

SUPPLY CHAIN MANAGEMENT

In commerce, supply chain management (SCM), the management of the flow of goods and services, between businesses and locations, and includes the movement and storage of raw materials, of work-in-process inventory, and of finished goods as well as end to end order fulfillment from point of origin to point of consumption. Interconnected, interrelated or interlinked networks, channels and node businesses combine in the provision of products and services required by end customers in a supply chain. Supply-chain management has been defined as the "design, planning, execution, control, and monitoring of supply-chain activities with the objective of creating net value, building a competitive infrastructure, leveraging worldwide logistics, synchronizing supply with demand and measuring performance globally". Marketing channels play an important role in supply-chain management. Current research in supply-chain management is concerned with topics related to sustainability and risk management, among others. An important concept discussed in SCM is supply chain resilience.

Function

Supply-chain management is a cross-functional approach that includes managing the movement of raw materials into an organization, certain aspects of the internal processing of materials into finished goods, and the movement of finished goods out of the organization and toward the end consumer. As organizations strive to focus on core competencies and become more flexible, they reduce ownership of raw materials sources and distribution channels. These functions are increasingly being outsourced to other firms that can perform the activities better or more cost effectively. The effect is to increase the number of organizations involved in satisfying customer demand, while reducing managerial control of daily logistics operations.

Importance

Organizations increasingly find that they must rely on effective supply chains, or networks, to compete in the global market and networked economy.[23] In Peter Drucker's (1998) new management paradigms, this concept of business relationships extends beyond traditional enterprise boundaries and seeks to organize entire business processes throughout a value chain of multiple companies.

In the 21st century, changes in the business environment have contributed to the development of supply-chain networks. First, as an outcome of globalization and the proliferation of multinational companies, joint ventures, strategic alliances, and business partnerships, significant success factors were identified, complementing the earlier "just-in-time", lean manufacturing, and agile manufacturing practices. Second, technological changes, particularly the dramatic fall in communication costs (a significant component of transaction costs), have led to changes in coordination among the members of the supply chain network.

Total Quality Management (TQM)

Total quality management (TQM) consists of organization-wide efforts to "install and make permanent climate where employees continuously improve their ability to provide on demand products and services that customers will find of particular value." "Total" emphasizes that departments in addition to production (for example sales and marketing, accounting and finance, engineering and design) are obligated to improve their operations; "management" emphasizes that executives are obligated to actively manage quality through funding, training, staffing, and goal setting. While there is no widely agreed-upon approach, TQM efforts typically draw heavily on the previously developed tools and techniques of quality control. TQM enjoyed widespread attention during the late 1980s and early 1990s before being overshadowed by ISO 9000, Lean manufacturing, and Six Sigma.

Total quality management (TQM) is a term that originated in the 1950s and is today used mainly in Japan. It is the equivalent of what other countries or organizations may call a company-wide quality management system, enterprise quality management system, or integrated quality management system, to name a few. The term 'total' means the entire organization all teams, departments and functions is involved in quality management. The 'system' refers to the managerial and technological methods to achieve quality requirements and business objectives throughout an entire organization. Although it may go by various names, Juran believes 'enterprise excellence' to be a more appropriate name for TQM.

Juran's Enterprise Excellence Management System has long been expressed as a company-wide business excellence and/or enterprise quality system. It embodies organizational beliefs and habits set forth in policies and processes to develop a culture capable or designing and delivering products and services that will exceed customer, regulatory, business and societal needs.

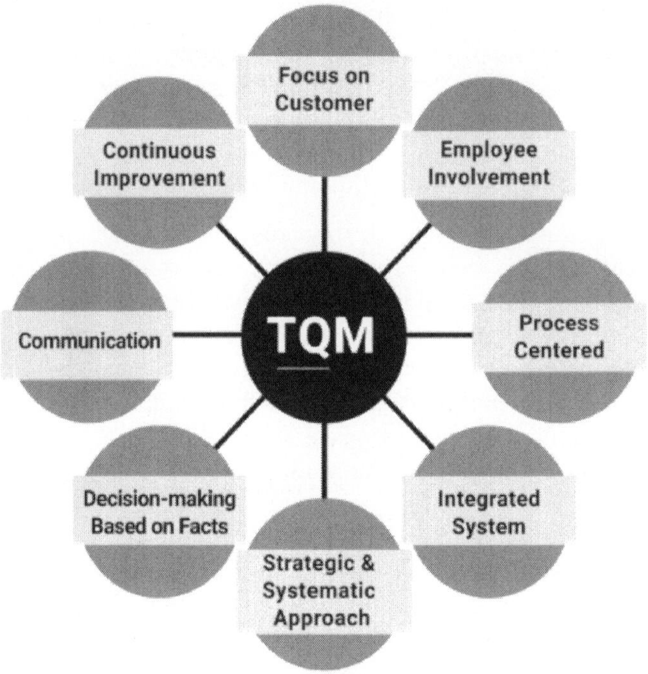

Key Principles of TQM

Total quality management has a number of key principles which when implemented together can move any organization towards business excellence.

Customer Focus

Central to all successful TQM systems is an understanding that quality is determined by the customer. No matter what measures you introduce to improve the quality of your products and services, the only way of knowing if they have been successful is customer feedback, whether in the form of reviews, return rates, or satisfaction surveys.

Employee Involvement

Every person in an organization from entry-level workers to management has a responsibility for the quality of products and services. However, employees can only be invested if they feel empowered to make their own decisions, something that depends on management creating the right workplace environment.

Centred on Process

A TQM system will fail without a clear focus on processes and process-led thinking. A process fault is ultimately the cause of most problems, which is why effective monitoring of every single step is an essential part of assessing, maintaining and improving quality.

Integrated System

An organization should have an integrated system that allows for effective total quality management. This may be a bespoke system, or one based on a quality standard such as ISO 9001, but it should be understood and applied across all functions and departments.

Strategic and Systematic Approach

Critical to quality management is the existence of a strategic plan that outlines how an organization intends to achieve its mission and business goals. It goes without saying that quality should be a core component of such a plan.

Decision-Making Based on Facts

Business performance can only be assessed using the available facts, such as sales data, revenue figures, and customer retention rates. The opinions of customers, employees and suppliers should never be used to inform decisions.

Communication

Effective communication is essential when an organization is implementing significant changes for the sake of business improvement. Every member of staff should be made aware of the strategy, the timescales involved, and the reasons for implementing it.

Continuous Improvement

Applying the principles of DMAIC and Lean Six Sigma will instill an organization with a culture of continuous improvement, driving all employees to constantly seek new ways to be more competitive and deliver high-quality products for all stakeholders.

Project Planning Formulation and Report Preparation

The Project Formulation phase, which may also be called the pre-investment phase can be divided into a number of stages. Different authors have divided this into different numbers of phases. One such division is:

1. Preliminary project formulation
2. Pre-feasibility study
3. Feasibility study
4. Support or functional studies
5. Detailed project report

1. Preliminary Project Formulation

The process of project development begins with the conception of a project idea. The identification of a project is in itself a vary important and difficult decision. What is the need? What are the objectives? What are he available alternatives of satisfying the need and meeting the objectives? What is the proposed investment? A host of similar questions has to be confirmed, while identifying the project ideas or identifying the opportunities for investment.

2. Pre-Feasibility Study

A pre-feasibility study as the name implies is a predecessor to detailed techno-economic feasibility study. It is a preliminary assessment of the project idea developed in preliminary project formulation that is before going into detailed feasibility study, which is a costly and time consuming task.

3. Feasibility Study

A project can be taken up for implementation, only if its feasibility offered. The objective of feasibility study is to define and analyse the critical demands that relate the objectives of the project, together with the alternative approaches to achieve such objectives. The outcomes of a feasibility study is an aspect of demand capacity, at a specified location, using a particular technology in relation to the defined materials and other inputs, at different investment and product costs and yielding a defined return on investment.

The feasibility study can be divided into two parts, the technical feasibility and the economic feasibility, though both are very much interrelated.

A) Technical Feasibility

The technical feasibility study comprises of ascertaining that the proposed project is technically sound, and will meet the technical requirements envisaged in the project. To assess the technical feasibility of the project, a very large amount of information about the machines, processes, materials, skills and their costs will be required. The various possible alternative processes and actions will have to be compared and evaluated for the selection of suitable plant and machinery.

B) Economic Feasibility

The economic feasibility study is a detailed analysis of the investment and returns. It needs a thorough understanding of the demand, market trends, details about the costs of resources, services, machines and equipment, and wages of the employees. Feasibility study should be able to justify the investment, location of the plant, and capacity of the plant. Feasibility study must include the estimates of costs, showing year wise finances required, foreign exchange required, tangible and intangible savings *etc.* It is the feasibility study on the basis of which the decision to go ahead with the project can be taken. It is not an end in itself, but only a means to arrive at an investment decision.

4. Support or Functional Studies

Support or functional studies are those activities of project formulation, which are confined to some specific elements or functions or a large investment project. These are conducted to generate information which is essential for the pre-feasibility and feasibility studies. Some examples of functional studies are summarized below:

- **Market Study** – To forecast the demand of the goods to be produced and their pattern.
- **Materials Study** – To determine the present and projected availability of various resources to be employed in the implementation of the project
- **Technology Selection Study** – selection of technology, evaluation of size of the plant and equipment, so as to arrive at optimum plant specifications.
- **Site Selection Studies** – In projects where the transport of materials and goods, is an important activity, location of the plant becomes critical.
- **Laboratory studies** – To determine the suitability of materials to be used in implementation of the project.
- **Pilot plant studies** – To determine the performance of actual and real life projects, experimental studies are conducted on pilot.

5. Detailed Project Report

Once the feasibility of the project is established, it can be implemented. The detailed project report is an exploded blue print of the project, which can be regarded as a tool for the implementation of the project. It will enlist all the steps, which need to be taken before the start of the project, and will have complete details of the activities and resources required for the execution of the projects.

Project Report

It is a concise copy of detailed analysis done for the project. An entrepreneur/ expert prepares the report before the investment in project is done. The report assesses the demand for proposed product/service, works out cost of investment and profitability on this investment. It acts as an instrument to convince investors to invest in the project

A project report gives information on the following

- Economic aspects – present market, scope for growth, justification for investment.
- Technical aspects – technology, machinery, and equipment needed.
- Financial aspects – Total investment needed entrepreneur's contribution, cost of capital and return on capital.
- Production aspects – Product details, justification for the choice of product, export worthiness.
- Managerial aspects – Qualifications, experience of people needed for managerial posts.

Contents of a Project Report

a. Objectives and scope of the report.
b. Product characteristics (product design, specifications, quality standards, uses and applications).
c. Market position and trends (current capacity for production, potential demand, export prospects, trends in import-export, price structure *etc.*).
d. Raw materials (types, quality, sources, price).
e. Manufacturing (process, production schedule, technique used.
f. Plant and machinery (types, infrastructure support, cost).
g. Land and building (Requirement, building construction schedule, choice of location, cost).
h. Financial implications (Capital structure, fixed and working capital investment, project cost, profitability).
i. Marketing channels (Trade practices, marketing and advertising strategy).
j. Personnel (Requirement of staff, skilled-unskilled labour, salary and wage payment, qualifications, experience)
k. The project report is submitted to financial institutions for grant of land and other financial concessions. Organisations like Small Industries Service Institute (SISI) and Small Industries Development Organisation (SIDO) help entrepreneurs to prepare project report.
l. The financial institutions ascertain from the report, whether the project can generate enough funds to repay the borrowings in stipulated time frame

Financing of Enterprise

What is Financing?

Financing is the process of providing funds for business activities, making purchases, or investing. Financial institutions, such as banks, are in the business of providing capital to businesses, consumers, and investors to help them achieve their goals. The use of financing is vital in any economic system, as it allows companies to purchase products out of their immediate reach. Put differently, financing is a way to leverage the time value of money (TVM) to put future expected money flows to use for projects started today. Financing also takes advantage of the fact that some individuals in an economy will have a surplus of money that they wish to put to work to generate returns, while others demand money to undertake investment (also with the hope of generating returns), creating a market for money.

Understanding Financing

There are two main types of financing available for companies: debt financing and equity financing. Debt is a loan that must be paid back often with interest, but it is typically cheaper than raising capital because of tax deduction considerations. Equity does not need to be paid back, but it relinquishes ownership stakes to the shareholder. Both debt and equity have their advantages and disadvantages. Most companies use a combination of both to finance operations.

Types of Financing

Equity Financing

"Equity" is another word for ownership in a company. For example, the owner of a grocery store chain needs to grow operations. Instead of debt, the owner would like to sell a 10 per cent stake in the company for Rs.100,000, valuing the firm at Rs.1 million. Companies like to sell equity because the investor bears all the risk; if the business fails, the investor gets nothing.

At the same time, giving up equity is giving up some control. Equity investors want to have a say in how the company is operated, especially in difficult times, and are often entitled to votes based on the number of shares held. So, in exchange for ownership, an investor gives his money to a company and receives some claim on future earnings.

Debt Financing

Most people are familiar with debt as a form of financing because they have car loans or mortgages. Debt is also a common form of financing for new businesses. Debt financing must be repaid, and lenders want to be paid a rate of interest in exchange for the use of their money. Some lenders require collateral. For example, assume the owner of the grocery store also decides that they need a new truck and must take out a loan for $40,000. The truck can serve as collateral against the loan, and the grocery store owner agrees to pay 8 per cent interest to the lender until the loan is paid off in five years. Debt is easier to obtain for small amounts of cash needed for specific assets, especially if the asset can be used as collateral. While debt must be paid back even in difficult times, the company retains ownership and control over business operations.

Loan Characteristics

- For early stage and established businesses, typically with a minimum of 3 employees.
- New/start-up business considered if viability can be demonstrated.
- Sole traders, partnerships or limited companies may apply.
- Charities and Social Enterprises may also apply.
- Can be used for any legitimate and legal business purpose including:

Working Capital

Purchase of equipment or machinery

Business Expansion

- Business assets considered as security if available.
- Enterprise Finance Guarantee Scheme will usually be considered.
- Personal guarantees from the director(s) (unsupported) are usually required for limited company lending.
- Maximum loan value Rs.250,000.
- Interest rate parameters 9 per cent – 15 per cent over Bank of England base rate, calculated on reducing balance.
- Arrangement fee 3 per cent
- Term of loans:
    - 3 year repayment for loans up to Rs. 25,000
    - Longer term considered for higher values (subject to maximum 7 years).

Application Process

- A completed Enterprise Loan Application Form should be accompanied by a Client Declaration form.
- Applications must be supported by a business plan (an extended *Executive Summary* might be considered acceptable for established businesses): financial forecasts should be included.
- Annual accounts for established businesses will be required along with *Management Accounts* for the current year.
- An initial enquiry with Finance for Enterprise Investment Manager is welcomed to help verify viability.

Opportunities for Agri-Entrepreneurship and Rural Enterprise

Many developing countries and economies in transition, particularly those with large rural communities, suffer from inadequate access to food and lack of employment. The problem is compounded by the dependence on outdated and inefficient technologies leading to poor productivity and slow economic growth. Agriculture-based industrial products account for half of all exports from developing countries, yet only 30 per cent of those exports involve processed goods compared to a figure of 98 per cent in the developed world. In this context, UNIDO aims to promote sustainable, inclusive business opportunities for the rural poor through agri-business and agro-value chain development. UNIDO's technical cooperation activities in this regard focus on adding value to agricultural commodities including non-food sectors at various points of the chain of economic transactions that links input providers, farmers, traders, processors, logistic providers, distributors and retailers. Through its technical assistance, UNIDO links resources and markets in the agribusiness value chains and strengthens forward and backward industrial linkages in order to leg up the economic transformation of countries, improve employment and income opportunities, and reinforce sustainable livelihoods.

Agro-industrial activities benefit a number of groups, including poor and marginalized rural populations, urban agro-industries and communities facing human security challenges or requiring urgent supplies of agricultural equipment and the rehabilitation of food industries. UNIDO promotes investment in agribusiness and value chain development; builds partnerships and linkages with strategic financing institutions; organizes various global forums and expert group meetings in related fields; and publishes specialized training manuals, guides and electronic media. To carry out its mandate in this area, UNIDO mobilizes expert services such as cluster development, conformity with quality and standards, rural energy, environmental management and cleaner production.

Agriculture sector provides food for 1.3 million people in India as well as supports over 60 percent of the population for their livelihood and plays a crucial role in sustainable growth of the country's economy. Creating more industries in the urban areas for migrating farmers cannot be a solution to the economic problems being faced by the rural poor. There is need to bring about transformational changes in the agriculture sector by making farmers to change their outlook toward farming. They need to consider their farms as their businesses. Entrepreneurship should be encouraged in agriculture as innovation would not just help farmers improve crop productivity and thus more profit but create new avenues of employment generation for rural youth as well. Tech entrepreneurship involves entrepreneurial skills, models and innovative ideas to solve problems in the farm sector and increase the profitability of the farming business with sustainable, community-oriented, directly marketed farm practices. It is not an opportunity but an imperative to achieve the Modi government's target of doubling farm income by 2022 through the integration of latest technologies and innovations.

Opportunities and Challenges for Rural Entrepreneurship in India

Rural entrepreneurship is defined in broader sense "as the enthusiastic willingness of a villager to organize his or her economics activity, whatever it may be (a business, a job, an investment *etc.*) with the help of appropriate technology and practices conceived for a sustainable living." Rural entrepreneurship has an important role to play in the development of Indian economy. Taking into account the fact that nearly 70 percent of the Indian population calls rural India it's home, adequate funding and support can provide a thriving entrepreneur atmosphere in these communities. So far as it is known that rural India as compared to the mainstream population is economically poor, younger, more isolated geographically, isolated from the main markets, culturally imbedded in tradition, less dynamic economically and experiencing depopulation. It is to be noticed that most of the entrepreneur development policies are tailored to meet the needs of the urban India. Such policies will have to be changed by establishing such centers which exclusively deal with the rural communities and provide consulting and research facilities. These centers can then provide a network between the isolated rural India and bridge the gap between these urban and rural communities. Most of the businesses in rural India are family owned thus providing a firm entrepreneurial base which can be exploited by the establishment of the rural entrepreneur centers. According to a recent study by the Rural Policy, rural areas who just need support to specify their thirst for the welfare of the public. These are the social entrepreneurs working in non-profit enterprises such as social justice organization, micro enterprises and business association.

Need for Agri-Entrepreneurship

After over six decades of independence and industrialization in our country, still large part of population remains under poverty line. Agriculture continues to be the backbone of rural society. As per this study, seventy percent of holdings are held by small and marginal farmers resulting in overcrowding on the agricultural land and diminishing farm produce. This also results in migration of farm worker in large numbers to the urban areas. In both the cases the population remains under poverty line. Agricultural work force has a share of seventy per cent in the total work force of the country. Cultivators who own farmland come to about sixty-eight percent of this work force while agricultural labor accounts for the remaining thirty two percent. These cultivators are increasing in numbers over the years but the large increase was among the agricultural labor, which went up from twenty percent of the rural work force to thirty two percent. One also needs to keep in mind that there is a continuous growth of population. Thus, the policy for rural entrepreneurship development has to tackle, the problems by providing other occupation option to the rural youths.

Opportunities of Agri-Entrepreneurship

1. **Support and Motivation to local people**: Rural entrepreneurs have a lot of support from the rural people. Rural village people always encourage and give the motivation to the entrepreneurs.
2. **Low establishment cost:** When compared to the urban areas, rural entrepreneurs' business establishment cost is very low. There is no need to construct or facilities huge infrastructure and buildings.
3. **Competitive advantages/Availability of labour :** In India seventy percent of the people are living in the village. Majority of the rural people are depending on the agriculture. The agriculture work is not available throughout the year. That is the reason why rural entrepreneurs have the competitive advantage in easily acquiring unskilled and semiskilled labor.
4. **Government policies and subsidies:** The government of India is continuously monitoring and introducing the new policies for encouraging the rural entrepreneurship. These policies are very flexible, innovative, liberalized and giving continues support to rural entrepreneurs. At the same time government has also announced huge subsidies for promoting the rural entrepreneurship.

5. **Availability of raw materials:** Most of the times the rural entrepreneurs are depending upon the farm based products as raw materials, which are available through-out the year. These raw materials are available in the rural area that is the reason there is no transportation cost and flotation cost.
6. **Cost of production:** Rural entrepreneurs cost of production is very low when compared to the urban industries. The factors of production are available with low cost, automatically the cost of production is also low. Because of this rural entrepreneurs can sell their goods and services with cheaper cost.
7. **Optimum utilization of produces:** Optimum utilization of farm produces is only possible through the rural entrepreneurship only. Most of the rural entrepreneurs depend upon the farm produces as raw materials.
8. **Employment generation for rural youth:** Rural entrepreneurs are providing hundred percent jobs for rural youth. If the rural entrepreneurs are succeeding in this activity the migration of the people to urban from rural will be immediately stopped to a maximum extent.
9. **Promotion cost:** There is no promotion cost for rural entrepreneurs; in fact the competition is very less. Particularly there is no need for advertising and other promotional activities for their products.
10. **Potential customer:** In this twenty-first century rural villagers are economically strong and also heavily populated. This heavy population can be converted as potential customers. That is the reason all the MNC's are concentrated in rural villages for their potentiality.
11. **Building the goodwill:** Rural entrepreneurs have a lot of scope building the goodwill. Most of the rural entrepreneurs have ethical values and also these people do not work for pure profits.

Challengers of Rural Entrepreneurship

Rural entrepreneurship has its own drawbacks. Policies such as keeping of land in protection when there is already an over production and pricing income are two of the greatest threads to rural entrepreneurship. Due to the remote access and unavailability of knowledgeable labor, commercial markets and managerial staff are hindered due to the remote locations. In order to alleviate the problems of rural entrepreneurship, under the government supported resources for these projects and select only the very best ideas that directly benefit not only the community but also can compete on a global scale. It is also vital for the success of the rural communities that the development of each rural project remain in the hand of the local agencies which in return cooperate with the government to oversee the leading factor that can help develop the rural areas.

Role of Government in Development of Rural Entrepreneurship

The different between rural and entrepreneur is only a matter of degree rather than the content. Many successful entrepreneurs are prospering in the cities who are hailing from rural areas. It is essential to have a balanced regional development of the country and to avoid the concentration of industry in one place. Rural areas must try for better utilization of human resources to improve the rural economy. Government has moral responsibility in designing, promoting, innovating rural entrepreneurial development programme for the up- liftmen of the rural economy on which the urban economy is built upon. Government should encourage the rural population and try to a bridge the gap between the extended and remote communities. But here, the government is not playing the role sincerely. Always, all the times government has only concentrate on the urban industrialization and urban economy development. This is the right time the governments are changing their policy, and come forward with the more beneficial policies, schemes and subsidies for development of rural entrepreneurship as well as economy.

References

"Block Panchayaths/Block Development Offices in Kerala". *Commissionerate of Rural Development, Kerala*. Retrieved 23 August 2020.

"Census 2011, West Bengal" (PDF). *censusindia.gov.in*. Retrieved 20 April 2020.

"First Five Year Plan". *Planning Commission*. Retrieved 10 September 2018.

"leadership". Oxford English Dictionary *(Online ed.). Oxford University Press.* (Subscription or participating institution membership required.)

"Modern leadership and management methods for development organizations". *Research Gate*. Retrieved 2018-11-07.

"Names of Blocks of Jharkhand". *Jharkhandi Baba. 21 October 2017. Archived from* the original *on 21 October 2017*. Retrieved 21 October 2017.

"Narcissistic Leadership (PDF Download Available)". *Research Gate*. Retrieved 2017-02-16.

"Rural development in West Bengal". *Department of Panchayat and Rural Development, Government of West Bengal homepage*. Retrieved 19 September 2019.

"Social Demography of Uttar Pradesh". *Government of Uttar Pradesh official portal*. Retrieved 22 August 2020.

"The Failure of the Community Development Programme in India". *Archived from* the original *on 12 July 2012*. Retrieved 6 April 2010.

"E-commerce will make the shopping mall a retail wasteland", ZDNet, January 17, 2013.

"Information and communication technologies for sustainable agriculture (2013)" (PDF). FAO. 20 May 2015, Retrieved 9 June 2016.

"Information in the Biosphere: Biological and Digital Worlds", Gillings, M. R., Hilbert, M., and Kemp, D. J. (2016), Trends in Ecology and Evolution, 31(3), 180–189; free access to the article http://escholarship.org/uc/item/38f4b791

"Introducing the Network of National ITS Associations!", Promotional web site, Retrieved 10 November 2016.

A. R. Desai (1969) Rural Sociology in India, Popular Prakashan, Bombay

Aamodt, M. G. (2010). I/O applications workbook: Industrial/organizational psychology an applied approach. Belmont, CA: Wadsworth.

Abraham, V (2013): "Missing Labour or Consistent „De-feminisation ?," Economic and Political Weekly, Vol 48, No 31, pp 99-108.

ACC (Administrative Committee on Coordination), United Nations (1984). Guiding principles for the design and of monitoring and evaluation in rural development projects and programmes. Rome: International Fund for Agricultural Development (IFAD).

Activities of Krishi Vigyan Kendras. MANAGE Extension Research Review 3 (1): 50–67.

Adams, M. E. (1982) Agricultural Extension in Developing Countries, Longman, UK

Adivi Reddy.A, (2001), Extension Education Seventh Edition, Sri Lakshmi press, Bapatla.

Agarwal J.C. (2007) Textbook on Essentials of Educational Technology Innovations in Teaching Learning, Second edition, Vikas Publishing House PVT LTD.

Aggarwal, A and N Kumar (2012): "Structural Change, Industrialization and Poverty Reduction: The Case of India," Development Papers 1206, Economic and Social Commission for Asia and the Pacific (ESCAP), United Nations, South and South-West Asia Office, New Delhi.

Agree, W.K., Ault, R.H. Emery, E. (Eds) (1979). Introduction to Mass Communication, New York, Harper and Row, Publishers.

Ahluwalia, Montek (2000), "Economic Performance of States in Post-reform period", Economic and Political Weekly, May 6

Albrecht, T.L. and Adelman, M.B, (1987) Communication Social support. Newbury Park, CA:Sage.

Alex, G., W. Zijp, and D. Byerlee (2002). Rural extension and advisory services: New directions. Rural Development Strategy Background Paper No. 9. Agriculture and Rural Development Department. Washington, D.C.: World Bank.

Alhaji Abubakar Aliyu, Rosmaini Bin HJ Tasmin (2012) "The Impact of Information and Communication Technology on Banks Performance and Customer Service

Delivery in the Banking Industry", International Journal of Latest Trends Fin. Eco. Sc, Vol-2, No. 1, March 2012, pp 80-90.

Ames, Daniel R.; Flynn, Francis J. (2007). "What breaks a leader: The curvilinear relation between assertiveness and leadership". Journal of Personality and Social Psychology. 92 (2): 307–324. doi:10.1037/0022-3514.92. 2.307. PMID 17279851.

Arvey, R. D.; Rotundo, M.; Johnson, W.; Zhang, Z.; McGue, M. (2006). "The determinants of leadership role occupancy: Genetic and personality factors". The Leadership Quarterly. 17: 1–20. CiteSeerX

Aycan, Z. (1987) Paternalism: Towards conceptual refinement and operationalization. London: Cambridge University Press. pp. 445–446.

Bart Barthelemy (1997). The Sky Is Not The Limit – Breakthrough Leadership. St. Lucie Press.

Bass, B. M., and Riggio, R. E. (2006). Transformational leadership (2nd ed.). Mahwah, New Jersey: Lawrence Erlbaum Associates Publishers; US.

Bass, B. M.; Avolio, B. J.; Atwater, L. E. (1996). "The transformational and transactional leadership of men and women". Applied Psychology: An International Review. 45: 5–34.

Baumeister, R. F.; Senders, P. S.; Chesner, S. C.; Tice, D. M. (1988). "Who's in charge here? Group leaders do lend help in emergencies". Personality and Social Psychology Bulletin. 14 (1): 17–22.

Baym, Nina *(1992). "Nathaniel Hawthorne and His Mother: A Biographical Speculation".* Feminism and American Literary History: Essays. *Women's studies/American literature. New Brunswick, New Jersey: Rutgers University Press. p. 53.*

Beebe, J. (1987). Rapid appraisal: The evolution of the concept and the definition of issues (p. 47-68). In Proceedings of the 1985 international conference on rapid rural appraisal. Khon Kaen, Thailand: Khon Kaen University.

Benjamin Jowett's translation of Plato's Republic does not use the word "leadership"; Plato discusses primarily a "guardian" class. See Plato *(1892).* The Dialogues of Plato translated into English with Analyses and Introductions by B. Jowett, M.A. *3. Oxford University Press.*

Bennis, W. G. (1975). Where have all the leaders gone? Washington, D.C.: Federal Executive Institute.

Benor, D; Harrison, J. O. and Baxter, M. (1984) Agricultural Extension: The Training and Visit System. The World Bank. Washington. D. C

Berdahl, J. L.; Anderson, C. (2005). "Men, women, and leadership centralization in groups over time". *Group Dynamics: Theory, Research, and Practice. 9: 45–57.*

Berkowitz, L (1953). "Sharing leadership in small, decision-making groups". Journal of Abnormal and Social Psychology. 48(2): 231–238.

Bharati RC, Singh KM, Chandra and Singh AK. (2014). Economic condition of eastern region of India-An statistical evaluation. Journal of Agri Search 1(3): 173-179.

Bird, C. (1940). Social Psychology. New York: Appleton-Century.

Birner R and Anderson J. (2007). How to make agricultural extension demand-driven? The case of India's agricultural extension policy. Discussion Paper 00729.Washington DC, International Food Policy Research Institute. Brewer F.2000. History of Indian Extension.

Birner, R., and N. Palaniswamy (2006).Public administration reform and rural service provision: A comparison of India and China. Washington D.C.: International Food Policy Research Institute.

Blair, Gerard M. (2009). "Presentation Skills." Basic Management Skills. University of Edinburgh, School of Engineering. Web. 06 May.

Bloom, C. (1980). The managerial role (p. 21-46). In L. J. Goodman and R. N. Love (Eds.), Project planning and management: An integrated approach. New York: Pergamon Press.

Boehm, B.W. (1991). "Software risk management: principles and practices". IEEE Software. 8: 32–41.

Bono, J. E.; Ilies, R. (2006). "Charisma, positive emotions and mood contagion". The Leadership Quarterly. 17 (4): 317–334.

Brent D. Ruben, (1984). Communication and Human Behavior, New York, Macmillan Publishing Co.

Brewer, F (2000). History of Indian Extension. (http://web3.canr.msu.edu/vanburen/India/histupd.htm.

Browning, M (2018). "Self-Leadership: Why It Matters"

Brunner, E. and Hsin Pao Yang, E. (1949) Rural America and the Extension Service, Columbia University

Burns, J. M. *(1978).* Leadership. *New York: Harper and Row Publishers Inc.* ISBN 978-0060105884.

Business balls information website: Leadership Theories Page, Integrated Psychological Approach section. Business balls.com. Retrieved 2012-08-15

Business balls management information website – Leadership Theories page, "Integrated Psychological Approach" section: leadership-theories.htm#integrated-psychological-leadership

Cano, PE, Librado F. (2010). Transformation of an Individual Family Community Nation And The World. *Trafford Publishing. p. 134.*

Carson, J. B.; Tesluk, P. E.; Marrone, J. A. (2007). "Shared Leadership in Teams: An Investigation of Antecedent Conditions and Performance". Academy of Management Journal. 50 (5): 1217–1234.

Casley, D. J., and Kumar, K. (1987). Project monitoring and evaluation in agriculture. Baltimore and London: Johns Hopkins University Press (published for the World Bank).

Casley, D. J., and Lury, D. A. (1982). Monitoring and evaluation of agriculture and rural development projects. Baltimore and London: Johns Hopkins University Press (published for the World Bank).

Casley, D. J., and Lury, D. A. (1987). Monitoring and evaluation of agriculture and rural development projects. Baltimore and London: Johns Hopkins University Press (published for the World Bank).

Cecil A Gibb (1970). Leadership (Handbook of Social Psychology). Reading, Massachusetts: Addison-Wesley. pp. 884–89.

Cernea, M. M., and Topping, B. J. (1977). A system for monitoring and evaluating agricultural extension projects. World Bank Staff Working Paper No. 272. Washington, DC: World Bank.

Chan, K.; Drasgow, F. (2001). "Toward a theory of individual differences and leadership: Understanding the motivation to lead". Journal of Applied Psychology. 86 (3): 481–498.

Chand, R and S K Srivastava (2014): "Changes in the rural labour market and their Implication for Agriculture," Economic and Political Weekly, Vol 49, No 10, pp 47-54.

Chand, R., R Saxena and S Rana (2015): "Estimates and Analysis of Farm Income in India, 1983-84 to 2011-12," Economic and Political Weekly, Vol 50, No 22, pp 139-145.

Chandrasekhar, C.P. (1993), "Agrarian Change and Occupational Diversification: Non-agricultural Employment and Rural Development in West Bengal", Journal of Peasant Studies, Vol.20, No.2

Chandre Gowda, M. J., and R. K. Samanta (2002). Dynamics of collaboration in the activities of Krishi Vigyan Kendra, Manage Extension Research Review, 3(1)50-67. DARE (2006) Annual Report, New Delhi.

Chatrerji; Markowitz (2004). "Does the Length of Maternity Leave Affect Maternal Health?" (PDF). *Southern Economic Journal. 72 (1): 16. Chemers, M. (1997). An integrative theory of leadership. Lawrence Erlbaum Associates, Publishers.* ISBN 978-0-8058-2679-1.

Chin, Roger (2015). "Examining teamwork and leadership in the fields of public administration, leadership, and management". Team Performance Management. 21 (3/4): 199–216.

Chitamber, J.B. (1997). Introductory Rural Sociology. Wiley Eastern Limited, New Delhi.

Chomsky, Noam (1999). *"Consent without Consent: Regimenting the Public Mind".* Profit Over People: Neoliberalism and Global Order. *New York: Seven Stories Press (published 2011).*

Chriss Webb (1978) People and Communication, Landon, McMillan Press.

Claudio Zaki Dib (1988). Formal, Non-Formal and Informal Education: Concepts/ Applicability. Cooperative Networks in Physics Education - Conference Proceedings 173", American Institute of Physics, New York, pgs. 300-315or article retrieved from http://techne-dib.com.br/downloads/6.pdf

Clayton and F. Pétry (Eds.), Monitoring systems for agricultural and rural development projects (vol. 1). Rome: FAO.

Clayton, E. (1983). Role, characteristics, and operational features of agricultural monitoring systems (p. 1-9). In E. Clayton and F. Petry (Eds.), Monitoring systems for agricultural and rural development projects (vol. 1). Rome: FAO.

Cole, J.M. (1981). Selecting Extension Teaching Methods. Journal of Extension, 27-32.

Compare: *Vasconcelos e Sá, Jorge (2012).* There is no leadership: only effective management: Lessons from Lee's Perfest Battle, Xenophon's Cyrus the Great and the practice of the best managers in the world. *Porto: Vida Economica Editorial. p. 19.*

Courland l. Bove, and John V.Thill (2000) Business Communication today, New Jersey,Prentice-Hall.

Dahama O.P. and Bhatnagar O.P, (1987) Education and Communication Development, Second Edition, Oxford and IBH Publishing Pvt. Ltd, Calcutta.

Dahama, O.P. and Bhatnagar, O.P. (1980). Education and Communication for Development. Oxford and lBH Publishing Co. Pvt. Ltd., New Delhi.

Daivadeenam, P. (2002). Educational Psychology in Agriculture. Agrotech Publishing Academy, Udaipur. Mangal, S.K. 2000. Educational Psychology. Prakash Brothers, Ludhiana.

Daniel, Jodi G; Patel, Bakul Patel; Quinn, Matthew (2013). "The path toward a risk-based regulatory framework for health IT". Health IT Buzz. Office of the National Coordinator for Health IT (US).

Dansereau, F.; Graen, G.; Haga, W. J. (1975). "A vertical dyad linkage approach to leadership within formal organizations: A longitudinal investigation of the role making process". Organizational Behavior and Human Performance. 13 (1): 46–78.

DARE (2006). Annual Report, New Delhi.

DARE/ICAR (2009-2010). Annual Report, New Delhi. www.icar.org.in/en/node/1107.

Dasborough, M. T. (2006). "Cognitive asymmetry in employee emotional reactions to leadership behaviors". The Leadership Quarterly. 17 (2): 163–178.

De, D. and Jirli, B. (2010). A Handbook of Extension Education. Agrobios (India), Jodhpur.

Deaton, A. and Jean Dreze (2002), "Poverty and Inequality in India: A Re-examination", Economic and Political Weekly.

Deb, U.K (2002) *et al.,* - 'Diversification and Livelihood Options: A Study of Two Villages in Andhra Pradesh, India 1975-2001' Working paper 178, ODI, London

Department of Agriculture and Cooperation (2000). Policy framework for agricultural extension (draft).Extension division, DAC, Ministry of Agriculture, Government of India, New Delhi.

Devito Joseph A. (1978). Communicology, An Introduction to the Study of Communication, New York, Harper and Row.

Donald Markwell (2013) "Instincts to Lead": On Leadership, Peace, and Education, Connor Court: Australia.

Dubey V.K. and Indira Bishnoi (2008) Extension Education and Communication, New Age International Publishers.

Dushi Guari. Notes on the Types of education: formal, informal, non-formal. Retrieved from http://www.preservearticles.com/notes/notes-on-the-types-of-education-formal-informal-nonformal/17995

Economic Survey, Govt. of India (2009-10). Ministry of Finance, Department of Economic Affairs, Economic Division, Feb., 2010. Oxford University Press, New Delhi.

Edgar Dale Audio-Visual Methods in Teaching Revised Edition, Holt, Rinehart and Winston,New York, Chicago.

Elevos, paraphrased from Leaders, Bennis, and Leadership Presence, Halpern and Lubar.

Elgie, Robert *(2015).* Studying Political Leadership: Foundations and Contending Accounts. *Palgrave Studies in Political Leadership. Springer (published 2016).*

Encyclopedia of public relations Archived April 16, 2009, at the Wayback Machine, by Robert L. Heath, 2005.

Essays, U K. (2013). Education System Before ICT Came In To Picture Education Essay.

EU Report on EU/global comparisons in the commercialisation of new technologies

Evenson, R.E. and D. Gollin (1997), "Genetic Resources, International Organizations and Rice Varietal Improvement." Economic Development and Cultural Change, 45(3):471-500.

Extension Education in Community Development (1961). Directorate of Extension, Ministry of Food and Agriculture, Government of India, New Delhi

Extension Methods. (N.D.). Retrieved July 12, 2017 from https://assets.helvetas.org/downloads/e6a_extension_methods.pdf

FAO (1988). Socio-economic indicators for monitoring and evaluating agrarian reform and rural development. WCARRD: Ten years of follow-up (1979-1989). Rome.

FAO (1990). Report of the global consultation on agriculture extension. Rome: Author.

FAO (1992). Sociological analysis in agricultural investment project design. FAO Investment Centre Technical Paper 9. Rome: Author.

Farrington, J., Sulaiman, V.R. and Pal, Suresh (1998). Improving the effectiveness of agricultural research and extension in India: An analysis of institutional and socio-economic issues in rainfed areas. Policy Paper 8, National Centre for Agricultural Economics and Policy research, New Delhi.

Ferguson, S.D. (1999). Communication Planning and Integrated approach. London: Thousand oaks.

Forsyth, D. R. (2009). Group dynamics (5th ed.). Pacific Grove, California: Brooks/Cole.

Forsyth, Donelson *(2010). Group Dynamics. Belmont, California: Wadsworth.*

Forthcoming in "The Handbook for Teaching Leadership", by Werner Erhard, Michael, C. Jensen, and Kari Granger; Scott Snook, Nitin Nohria, Rakesh Khurana.

Foti, R. J.; Hauenstein, N. M. A. (2007). "Pattern and variable approaches in leadership emergence and effectiveness". Journal of Applied Psychology. 92 (2): 347–355.

Freire, P. (1969) Extension & Communicacion, translated by Louise Bigwood and Margaret Marshall and re-printed in 'Education: The Practice of Freedom' (1976), Writers and Readers Publishing Cooperative

Gail E. Myer and Michele T. Myers (1985) The Dynamics of Human Communication: A Laboratory Approach, USA,McGraw-Hill.

Ganesh, R., Mohammad lqbal, 1. and Anandaraja, N. (2003). Reaching the Unreached - Basics of Extension Education. Associated Publishing Company, New Delhi.

Gardner, J. W. (1965). Self-Renewal: The Individual and the Innovative Society. New York: Harper and Row.

George, J. M. (2000). "Emotions and leadership: The role of emotional intelligence". Human Relations. 53 (8): 1027–1055.

George, J. M. (2006). "Leader Positive Mood and Group Performance: The Case of Customer Service". Journal of Applied Social Psychology. 25 (9): 778–794.

Gershenoff, A. G.; Foti, R. J. (2003). "Leader emergence and gender roles in all-female groups: A contextual examination". Small Group Research. 34 (2): 170–196.

Gha, D., Hopkins, M., and McGranahan, D. (1988). Some reflections on human and social indicators for development. Discussion Paper No. 6. Geneva: United Nations Research Institute for Social Development (UNRISD).

Goldsmith Marshall (2016). "Leaders Make Values Visible".

Goleman, D., Boyatzis, R.E., and McKee, A. (2003) The New Leaders: Transforming the art of leadership. London: Sphere.

Goleman, Daniel; Boyatzis, Richard E.; McKee, Annie (2003). New leaders.

Gordon-Till, Jonathan, Ethics (2002): the professional challenge , Business Information Review, 19: 46-54. Print.

Government of India Planning Commission (2001) - Report of the Working Group on Public Distribution System and Food Security for the Tenth Five Year Plan (2002-2007).

Graen, G. B.; Novak, M. A.; Sommerkamp, P. (1982). "The effects of leader-member exchange and job design on productivity and satisfaction: Testing a dual attachment model". Organizational Behavior and Human Performance. 30 (1): 109–131.

Greenleaf, Robert K. *(1977).* Servant Leadership: A Journey into the Nature of Legitimate Power and Greatness. *Mahwah, New Jersey: Paulist Press (published 2002).*

Gregory, Scott (2018). "The Most Common Type of Incompetent Leader". *Harvard Business Review.*

Griffin, Ronald J. Ebert, Ricky W. (2010). Business essentials(8th ed.). Upper Saddle River, New Jersey: Prentice Hall. pp. 135–136.

Grosse, Robert (1996). "international Technology Transfer in Services". Journal of International Business Studies. 27 (4): 782.

Guastello, S. J. (2007). "Nonlinear dynamics and leadership emergence". *Leadership Quarterly. 18 (4): 357–369.*

Guo, Xuezhi (2019). "Traditional Political Thought and Imperial Legacy". The Politics of the Core Leader in China: Culture, Institution, Legitimacy, and Power. *Cambridge: Cambridge University Press. p. 139.*

Gupta, Ashok Kumar (1991). Emerging Pattern of Political Leadership: A Case Study of Punjab. *New Delhi: Mittal Publications. p. 12.*

Hackman, M. and Johnson, C. (2009). Leadership: A communication perspective. Long Grove, Illinois: Waveland Press, Inc.

Hakala, David. "The Top 10 Leadership Qualities". *HR World. Archived from* the original *on 2013-04-10.*

Hanka, S.S. (1999). Entrepreneurship Development. S. Chand and Co., New Delhi.

Hargadon, Andrew. Harvard Business School Working Knowledge for Business Leaders, August 4, 2003.

Headquarters, Department of the Army (2006). "Army Leadership. Competent, Confident, and Agile". FM 6-22. Washington, D.C., 12 October 2006 p. 18. Publication available at Army Knowledge Online (www.us.army.mil) and General Dennis J. Reimer Training and Doctrine Digital Library at (www.train.army.mil).

Heeks, Richard (2016). "From ICT4D to Digital Development?", ICTs for Development. Retrieved 9 September 2016. "ICT in Education", Unesco. Retrieved 10 March 2016.

Hemphill J K (1949) Situational factors in leadership. Ohio State University, Bureau of Educational Research Monograph, 32, xii, 136

Henry P. Knowles; Borje O. Saxberg (1971). Personality and Leadership Behavior. Reading, Massachusetts: Addison-Wesley. pp. 884–89.

Hildebrand, P. E. (1981). Combining disciplines in rapid appraisal: The sondeo approach. Agricultural Administration, 8, 423-432.

Home Science. Higher Secondary First Year. (2005). Tamil Nadu Textbook Corporation.

Hortan, D., Peterson, W., and Ballantyne, P. (1993). M and E principles and concepts (p. 5-19). In D. Hortan *et al.* (Eds.), Monitoring and evaluating agricultural research: A sourcebook. Oxon, UK: CAB International.

Horton, Thomas (1992). New York: The CEO Paradox

Howell, Jon P. (2012). Snapshots of Great Leadership. *London, GBR: Taylor and Francis. pp.* 16–17.

Howell, Jon P. (2012). Snapshots of Great Leadership. *London, GBR: Taylor and Francis. p.* 17.

Howell, Jon P. (2012). Snapshots of Great Leadership. *London, GBR: Taylor and Francis. pp.* 4–6.

Hoyle, John R. Leadership and Futuring: Making Visions Happen. Thousand Oaks, California: Corwin Press, Inc., 1995.

Hoyt, C. L.; Blascovich, J. (2016). "Leadership Efficacy and Women Leaders' Responses to Stereotype Activation". *Group Processes and Intergroup Relations. 10 (4): 595–616.*

http://pdf.usaid.gov/pdf_docs/PNAAW962.pdf

http://pdf.usaid.gov/pdf_docs/PNAAW962.pdf

http://web3.canr.msu.edu/vanburen/India/histupd.htm.India,

http://www.collegeonline.org/library/research-tools/librarian-searchguide.html

http://www.scribd.com/doc/104941836/Classification-of-Extension-Teaching-Methods

http://www.scribd.com/doc/104941836/Classification-of-Extension-Teaching-Methods

ICAR Now and Ahead, New Delhi. ICAR (2005-06). Budget Book, New Delhi.

ICAR, Annual Reports (Various years 1990 onwards), New Delhi. ICAR (1996).

Ilies, Remus; Morgeson, Frederick P.*; Nahrgang, Jennifer D. (2005).* "Authentic leadership and eudaemonic well-being: Understanding leader–follower outcomes". *The Leadership Quarterly. 16 (3): 373–394.*

India, DAC Annual Report (2011-12). Department of Agriculture and Cooperation, Ministry of Agriculture, Government of India. March 2012.

India, Department of Agriculture and Cooperation (2000). Policy framework for agricultural extension (draft).Extension division, DAC, Ministry of Agriculture, Government of India, New Delhi.

Indian Institute of Management, Lucknow (2004a). Impact Assessment Report, on the Innovations in Technology Dissemination (ITD) Component of the National Agricultural Technology Project, Agriculture Management Centre, 2004.

Indian Institute of Management, Lucknow (2004b). Successful Case Studies, Interventions and Innovations in Technology Dissemination, Agriculture Management Centre, 2004.

Ingrid Bens (2006). Facilitating to Lead. Jossey-Bass.

Instructor's Guide for Introduction to Leadership Development (1976): A Course Outline : Leadership Development*. 3. Fort Benning, Georgia: United States Army Infantry School. p. 7.*

International Finance Corporation (2013). Working with Smallholders: A Handbook for Firms Building Sustainable Supply Chains. "Farms 2 Firms". *Archived from* the original *on 2013-12-03.*

International Fund for Agricultural Development (IFAD) (2001). Andhra Pradesh Tribal Development report, Completion evaluation.

Israel, A. (1987). Institutional development: Incentives to performance. Baltimore and London: Johns Hopkins University Press (published for the World Bank).

Issues in Culture, Rights, and Governance Research: 2012 Edition: ScholarlyBrief*. Atlanta, Georgia: Scholarly Editions (published 2013). 10 January 2013.*

ITU (2010). Measuring the Information Society 2010. International Telecommunication Union (ITU): Switzerland.

Jha, D. and Kumar, S. (2006). Research resource allocation in Indian agriculture. Policy Paper No. 23. Delhi, National Centre for Agricultural Economics and Policy Research.

Jones, G.E. and Garforth, C. (1997) The history, development, and future of agricultural extension in Swanson, B. "Improving Agricultural Extension: A Reference Manual (3rd Edition)" FAO

Judge, T. A.; Bono, J. E.; Ilies, R.; Gerhardt, M. W. (2002). "Personality and leadership: A qualitative and quantitative review". Journal of Applied Psychology. 87 (4): 765–780.

Jung, D.; Wu, A.; Chow, C. W. (2008). "Towards understanding the direct and indirect effects of CEOs transformational leadership on firm innovation". The Leadership Quarterly. 19 (5): 582–594.

Jung, Dong I.; Sosik, John J. (2002). "Transformational Leadership in Work Groups: The Role of Empowerment, Cohesiveness, and Collective-Efficacy on Perceived Group Performance"*. Small Group Research. 33 (3): 313–336.*

Kaiser, Robert B.; Hogan, Robert*; Craig, S. Bartholomew (2008).* "Leadership and the fate of organizations"*. American Psychologist. 63 (2): 96–110.*

Kannan, K P and G Raveendran (2012): "Counting and Profiling the Missing Labour Force," Economic and Political Weekly, Vol 47, No 6, pp 77-80.

Kenny, D. A.; Zaccaro, S. J. (1983). "An estimate of variance due to traits in leadership". Journal of Applied Psychology. 68 (4): 678–685.

Kesley, L.D. and Hearne, C.C (1966). Cooperative Extension work. Constock Publishing Associates, Ithaca, New york.

Kickul, J.; Neuman, G. (2000). "Emergence leadership behaviors: The function of personality and cognitive ability in determining teamwork performance and KSAs". Journal of Business and Psychology. 15: 27–51.

Koch, A. J.; D'Mello, S. D.; Sackett, P. R. (2015). "A Meta-Analysis of Gender Stereotypes and Bias in Experimental Simulations of Employment Decision Making". Journal of Applied Psychology. 100 (1): 128–161.

Koenig, A. M.; Eagly, A. H.; Mitchell, A. A.; Ristikari, T. (2011). "Are Leader Stereotypes Masculine? A Meta-Analysis of Three Research Paradigms". Psychological Bulletin. 137 (4): 616–42.

Kumar, K. (Ed.). (1993). An overview of rapid appraisal methods (p. 1-22). In K. Kumar (Ed.), Rapid appraisal methods. Washington, DC: World Bank.

Kumar, P. and Rosegrant, M. W. (1994).Productivity and sources of growth for rice in India, Economic and Political Weekly, Dec. 31, pp 183-188.

Lanjouw, P. and Shariff, A. (2000), "Rural Nonfarm Employment in India: Access, Incomes and Poverty Impact", National Council of Applied Economic Research, New Delhi

Larson, J. R. Jr.; Christensen, C.; Abbot, A. S.; Franz, T. M. (1996). "Diagnosing groups: Charting the flow of information in medical decision-making teams". Journal of Personality and Social Psychology. 71 (2): 315–330.

Law, J. R. (1996). Rising to the occasion: foundations, processes, and outcomes of emergent leadership.

Leadership in Action (1974): A Text for U.S. Marine Corps Junior ROTC. *Quantico, Virginia: Reserve Liaison and Training Branch, Education Center, Marine Corps Development and Education Command. p. 36.*

Lee, J. W. (1990). More than accountability: Evaluating agricultural extension programmes. TRDC Publication No. 172. Amidale, Australia: University of New England, Rural Development Centre.

Leeuwis, C. and van den Ban, A. Communication for Rural Innovation: Rethinking Agricultural Extension (3rd Edition), Blackwell Publishing

Levi-Sanchez, Suzanne (2018). "Civil Society in an Uncivil State". *Journal of International Affairs. 71 (2): 50–72.*

Lewin *et al.* (1939) Patterns of aggressive behavior in experimentally created social climates, Journal of Social Psychology, 10, 271-301

Lewin, K.*; Lippitt, R.; White, R. K. (1939). "Patterns of aggressive behavior in experimentally created social climates". Journal of Social Psychology. 10 (2): 271–301.* doi:10.1080/00224545.1939.9713366.

Lewis, Albert and Gordia, Gautam. ˉFuture of e-learning in India , express computer 14.11.2005.

Lipman-Blumen, J. (2005) The Allure of Toxic Leaders. New York: Oxford. University Press Inc.

Locke *et al.* (1991) Leadership: Do Traits Matter? Executive 5 (2): 48-60

Lord, R. G.; De Vader, C. L.; Alliger, G. M. (1986). "A meta-analysis of the relation between personality traits and leader perceptions: An application of validity generalization procedures". Journal of Applied Psychology. 71 (3): 402–410.

Lorsch J W and John Morse (1974) Organisations and their members: A contingency approach, New York: Harper and Row

Lussier, R. N., and Achua, C. F. (2010). Leadership, Theory, Application, and Skill Development.(4th ed). Mason, Ohio: South-Western Cengage Learning.

Magnusson, D. (1995). "Holistic interactionism: A perspective for research on personality development". In L. A. Pervin and O. P. John (Eds.), Handbook of personality: Theory and research (pp. 219–247). New York: Guilford Press.

Mahendra Dev, S. (1993), "Non-agricultural Employment in Rural India: Evidence at a Disaggregate Level", in Visaria, P. and Basant, R. (eds., 1993), "Non-agricultural Employment in India: Trends and Prospects", Sage publications.

Maheshwari, Shriram. "Rural Development and Bureaucracy in India". *The Indian Journal of Public Administration. XXX (3): 1093–1100.*

Mann, R. D. (1959). "A review of the relationship between personality and performance in small groups". Psychological Bulletin. 56 (4): 241–270.

Marstiller, James K.; Joerding Fickeler, Jennifer (2005). The Power to Innovate: Rewiring the Minds of Individuals and Organizations. *Bloomington, Indiana: Author House. p. 169.*

Matthews, Michael D.; Eid, Jarle; Kelly, Dennis; Bailey, Jennifer K. S.; Peterson, Christopher (2006). "Character strengths and virtues of developing military leaders: An international comparison". Military Psychology. 18 (Suppl): S57–S68.

Maurya, N K and J V Vaishampayan (2012): "Growth and Structural Changes in India s Industrial Sector," International Journal of Economics, Vol 6, No 2, pp 321-331.

Mazumdar, I and N Neetha (2011): "Gender Dimensions: Employment Trends in India, 1993-94 to 2009-10," Economic and Political Weekly, Vol 46, No 43, pp 118-126.

Meena MS, Bhagwat VR, Jain RK and Ilyas SM. (2003). Self-Help Group: An Effective Instrument for Transfer of Technology. CIPHET Extension Bulletin-1.Central Institute of Post-Harvest Engineering and Technology, Ludhiana, India.

Meena MS, Jain Dilip and Meen HR. (2008). Measurement of Attitudes of Rural Women towards Self-Help Groups, The Journal of Agricultural Education and Extension 14(3): 217-229.NSSO (National Sample Survey Organization).2005.

Meena, M. S., Bhagwat, V. R., Jain, R. K. and Ilyas, S. M. (2003). Self-Help Group: An Effective Instrument for Transfer of Technology. CIPHET Extension Bulletin1. Central Institute of Post-Harvest Engineering and Technology, Ludhiana, India.

Meena, M. S., Jain, Dilip and Meena, H. R. (2008). Measurement of Attitudes of Rural Women towards Self-Help Groups, The Journal of Agricultural Education and Extension, 14: 3,217-229.

Meindl, J. R.; Ehrlich, S. B.; Dukerich, J. M. (1985). "The romance of leadership". Administrative Science Quarterly. 30 (1): 78–102.

Melvin., Stogdill, Ralph (1990). Bass and Stogdill's handbook of leadership : theory, research, and managerial applications Bass and Stogdill's handbook of leade. Free Press.

Miltenberger, R. G. (2004). Behavior Modification Principles and Procedures (3rd ed). Belmont, California: Wadsworth/Thomson Learning.

Miron-Shatz, T., Shatz, I., Becker, S., Patel, J., and Eysenbach, G. (2014). "Promoting business and entrepreneurial awareness in health care professionals: lessons from venture capital panels at medicine 2.0 conferences". Journal of Medical Internet Research, 16(8), e184.

Mishra, S. and B.E. Swanson (2009) Extension's Role in Organizing Producer Groups: a Case Study from Orissa, India, Proceedings of the 25thAnnual Conference of the Association for International Agricultural and Extension Education (AIAEE)

Misra, D. C. (1990). New Directions in Extension Training-A Conceptual Framework, Directorate of Extension, Occasional Paper-3, Directorate of Extension, Ministry of Agriculture, New Delhi, India.

Misra, D. C. (1994). Agricultural extension effectiveness in India (mimeo). University of Oxford, Queen Elizabeth House, International Development Centre.

Mobbs, Dean; Hagan, Cindy C.; Dalgleish, Tim; Silston, Brian; PrÃ©vost, Charlotte (2015). "The ecology of human fear: survival optimization and the nervous system". *Frontiers in Neuroscience. 9: 55.*

Mohapatra, S. and Sahoo, B. K (2008). Agricultural Research and Development in India: Trends, Patterns, Determinant and Policy Implication. Presented at the International Conference on Issues in Public Policy and Sustainable Development held in New Delhi.

Mondal, S. (2019). Fundamentals of Agricultural Extension Education. Kalyani publishers, New Delhi.

Mumford, M. D.; Zaccaro, S. J.; Harding, F. D.; Jacobs, T. O.; Fleishman, E. A. (2000). "Leadership skills for a changing world solving complex social problems". The Leadership Quarterly. 11: 11–35.

Murphy, J. (1993). Good enough, soon enough. The Rural Extension Bulletin, 1, 27-33.

Murphy, J., and Marchant, T. J. (1988). Monitoring and evaluation in extension agencies. World Bank Technical Paper No. 79. Washington, DC: World Bank.

NAFES (2005) Consolidating Extension in the Lao PDR, National Agricultural and Forestry Extension Service, Vientiane

Nageshwar, K. (2016). Interpreting Contemporary India *(published 2016).*

National Agriculture and Livestock Extension Programme (NALEP) (2011). A Guide to Effective Extension Methods for Different Situations.

National Institute of Agricultural Extension Management (MANAGE) (2004). Process Change in Agricultural Extension: Experiences under ITD Component of NATP, 2004.

Network, Jules Evans for the Guardian Professional (2012). "What can business leaders learn from ancient philosophers?". *the Guardian.* Retrieved 2018-11-07.

Nichols, P. (1991). Social survey methods: Afield guide for development workers. Development Guidelines No. 6. Oxford: Oxfam.

Note the relative changes in terminology in American English since 1800.

NSSO (National Sample Survey Organization) (2005). Situation assessment survey of farmers: Access to modern technology for farming, 59th round (January–December 2003). Report No. 499(59/33/2). New Delhi: Ministry of Statistics and Programme Implementation.

Ola, Bolarinde (2017). "A Critique of Conceptual Leadership Styles'". In Godbole, Prasad; Burke, Derek; Aylott, Jill (eds.). Why Hospitals Fail: Between Theory and Practice. *Cham (Switzerland): Springer International Publishing. p. 59.*

Pal, S., and A. Singh. (1997). Agricultural research and extension in India: Institutional structure and investments, Policy Paper-7, New Delhi: National Centre for Agricultural Economics and Policy Research.

Palmer, Michael A. (1988). "Lord Nelson: Master of Command". *Naval War College Review. 41 (1): 105–116.*

Papola, T S (2012): "Structural Changes in the Indian Economy: Emerging Patterns and Implications," ISID Working Paper No 12, Institute for Studies in Industrial Development, New Delhi.

Pardey, P. G., and J. Roseboom (1989). ISNAR agricultural research indicator series. Cambridge, UK: Cambridge University Press.

Parvez Ahmad,Mohd Yaseen, Librarian: Web Librarian The Role of the Library and Information Science Professionals As Managers: A Comparative

Pellerin, Cheryl (2014). "DARPA s Plan X Uses New Technologies to „See Cyber Effects", America Forces Press Service, US Department of Defense, Retrieved 21 November.

Pétry, F. (1983). Designing a monitoring and evaluation system (p. 20-49). In E.

Philippe-Joseph Salazar, L'Hyperpolitique (2009). Technologies politiques De La Domination, Paris.

Planning Commission (2008). XIth Five Year Plan (2007–12). Ministry of Finance, Government of India, New Delhi.

Planning Commission (2010). Mid-term Appraisal XI Plan, Planning Commission for Agriculture, Government of India.

Popper, Karl *(1945).* "Preface to the First Edition". *The Open Society and Its Enemies (7 ed.). London: Routledge (published 2012). p. xxxiii.*

Pradhan, K C (2013): "Unacknowledged Urbanization: New Census Towns of India," Economic and Political Weekly, Vol 48, No 36, pp 43-51.

Pratt, B., and Loizos, P. (1992). Choosing research methods: Data collection for development workers. Oxford: Oxfam.

Profit over People: neoliberalism and global order, N. Chomsky (1999). Ch. "Consent without Consent", p. 53

Prokopenko, J. (1987). Productivity management: A practical handbook. Geneva: International Labour Office (ILO).

Publisher, Author removed at request of original (2016), "6.3 Formal Organizations", *Sociology, University of Minnesota Libraries Publishing edition, 2016. This edition adapted from a work originally produced in 2010 by a publisher who has requested that it not receive attribution.*, retrieved 2021-06-21

Raabe K. (2008). Reforming the agricultural extension system in India: What do we know about what works where and why? IFPRI Discussion Paper 00775. Washington, DC.: International Food Policy Research Institute.

Ramasamy, C. and Selvaraj, K. N. (2007). Prioritizing Agricultural Research and Extension, INRM Policy Brief No. 15, Asian Development Bank.

Rangarajan, C, P I Kaul and Seema (2011): "Where is the Missing Labour Force?" Economic and Political Weekly, Vol 46, No 39, pp 68-72.

Rangarajan, C, Seema and E M Vibeesh (2013): "Developments in the Workforce between 2009-10 and 2011-12," Economic and Political Weekly, Vol 49, No 23, pp 117-121.

Rao, CHH (2003), "Reform Agenda for Agriculture", Economic and Political Weekly, Vol.8, No.7, 2003

Rao, K.P.C (1999) - 'Development of Dryland Agriculture: Policy Issues' in Singh, H.P., Ramakrishna, Y.S., Sharma, K.L. and Benkateswarulu, B. (eds.), Fifty years of Dryland Agricultural Research in India, Central Research Institute for Dryland Agriculture, Hyderabad; 565-572

Ravallion, M. (2000), "What is needed for a more pro-poor growth process in India?" Economic and Political Weekly, Vol.35, No.13

Ray G.L. (1996). Extension Communication and Management, Naya Prakash Publications, Calcutta

Ray, G.L. (2006). Extension Communication and Management Naya Prakashan, Kolkata.

Rayuclu, C.S. (1997). Communication. Himalaya Publishing House, Hyderabad.

Reddy MN and Swanson BE. (2006). Strategy for up-scaling the ATMA model in India. In Proceedings of the Association for International Agricultural and Extension Education (AIAEE) 22nd Annual Conference, Clearwater Beach, Florida, U.S.A.

Reddy, A.A (2005). Extension Education. Sri Lakshmi Press, Bapatla.

Reddy, M. N., and Swanson, B.E. (2006). Strategy for up-scaling the ATMA model in India. In Proceedings of the Association for International Agricultural and

Extension Education (AIAEE) 22nd Annual Conference, Clearwater Beach, Florida, U.S.A.

Report of The Grow More Food Enquiry Committee (1952). *Government of India Ministry of Food and Agriculture.*

Report of working group on agriculture research and education for the eleventh for the five year plan (2007-2012). Government of India, Planning Commission.

Richard Wrangham and Dale Peterson (1996). Demonic Males. Apes and the Origins of Human Violence. Mariner Books

Robert Hariman (1995). Political Style, U of Chicago Press.

Rogers, E.M. 2003. Diffusion of Innovations. Free Press, New Delhi.

Roling, N. and Wagemakers, A. Editors.(1998), Facilitating Sustainable Agriculture: Participatory learning and adaptive management in times of environmental uncertainty, Cambridge University Press

Roy G L (2013) Extension Communication and management. New Delhi. Kalayni Publication.

Ruthenberg, I. (1985). Innovation policy for small farmers in the tropics: The economics of technical innovations for agricultural development. Oxford: Oxford University Press.

Sadler, Philip (2003). "The development process". Leadership. *MBA masterclass series (2 ed.). London: Kogan Page Publishers. p. 108.*

Sagar Mondal and Ray, G.L. (2003). Text Book of Entrepreneurship and Rural Development. Kalyani Publishers, Ludhiana.

Saleem, Farida; Zhang, Yingying Zhang; Gopinath, C.; Adeel, Ahmad (2020). "Impact of Servant Leadership on Performance: The Mediating Role of Affective and Cognitive Trust". *SAGE Open. 10 (1).*

Sandhu A. S.(1969) Textbook on Agricultural Communication, Mittal Publications, New Delhi

Sargent, John F., Jr. (2018). "Federal Research and Development Funding: FY2018" (PDF). *Congressional Research Service.*

Saville, A. H. (1965) Extension in Rural Communities: A Manual for Agricultural and Home Extension technician Workers. Oxford University Press.

Saxena Dr., P.K. (2009). Principles of Management, A Modern Approach. *New Delhi: Global India Publications PVT LTD. p. 30.*

Schmid Mast, M (2002). "Female dominance hierarchies: Are they any different from males'?". Personality and Social Psychology Bulletin. 28: 29–39.

Schyns, B. (2006). "The role of implicit leadership theories in the performance appraisals and promotion recommendations of leaders". Equal Opportunities International. 25 (3): 188–199.

Scouller, J. (2011). The Three Levels of Leadership: How to Develop Your Leadership Presence, Knowhow and Skill. Cirencester: Management Books 2000.,

Sells, S. B. (1961). "Abilities and Attributes of Leaders". Military Small Group Performance under Isolation and Stress - An annotated bibliography. *Fort Wainwright, Alaska: Arctic Aeromedical Laboratory (U.S.). p. 23.*

Sen, Abhijit (1998), "Rural Labor Markets and Poverty", in Radhakrishna and A.N. Sharma, (eds.) Empowering Rural Labor in India, Market, State and Mobilization in India, Institute for Human Development, New Delhi

Shaner, W. W., Phillipp, P. F., and Schmehl, W. R. (1982). Farming systems research and development: Guidelines for developing countries. Boulder, CO: Westview Press.

Sharma, A. (2016). Education: Meaning, Definition, types of education and characteristics, Extension education: Meaning, definition and concept.

Sharma, Shailendra D. (1999). Development and Democracy in India. *Boulder, Colorado: Lynne Rienner Publishers, Inc.*

Shaughnessy, Brooke A.; Treadway, Darren C.; Breland, Jacob W.; Perrewé, Pamela L. (2017). "Informal Leadership Status and Individual Performance: The Roles of Political Skill and Political Will". *Journal of Leadership and Organizational Studies. 24 (1): 83–94.*

Simonton, Dean K. (1985). "Intelligence and personal influence in groups: Four nonlinear models". Psychological Review. 92 (4): 532–547.

Singh JP, Swanson BE and Singh KM. (2006). Developing a decentralized, market-driven extension system in India: The ATMA model. In AW van den Ban and RK Samanta, eds. changing roles of agricultural extension in Asian nations pp. 203–223.

Singh, J.P. (2005). From Self-help Groups to Commodity-based Commodity Associations: The Indian Approach to Mobilizing Rural Women, presentation at the Workshop on Building New Partnerships in the Global Food Chain, Chicago, June 29-30, 2005.

Singh, J.P., B.E. Swanson and K.M. Singh (2005). Organizing and Linking Farmers with Markets: Experience of the NATP Project in India, presentation at the 15th Annual World Food and Agribusiness Symposium, Chicago, June 27, 2005.

Singh, J.P., Swanson, B.E and Singh K.M. (2005). Developing a Decentralized, Market Driven Extension System in India: The ATMA Model. Good Practice Paper prepared for the World Bank, Washington DC.

Singh, J.P., Swanson, B.E. and Singh, K.M. (2006). Developing a decentralized, market driven extension system in India: The ATMA model. In A.W. van den Ban and R.K. Samanta, eds. changing roles of agricultural extension in Asian nations, pp. 203–223. Delhi, B.R. Publishing.

Singh, K. M., and B. E. Swanson (2006). Developing a market-driven extension system in India. Proceedings of the 2006 Annual Conference of the Association for International Agricultural and Extension Education (AIAEE) 22: 627–637.

Singh, K. M., Swanson, Burton E., Jha, A. K. and Meena, M. S. (2012). Extension Reforms and Innovations in Technology Dissemination-The ATMA Model in India (October 30, 2012).

Singh, K.M. and Meena, M. S. (2011). Agricultural Innovations in India-Experiences of ATMA Model (Available at SSRN: http://dx.doi.org/10.2139/ssrn.1989823)

Singh, K.M., B.E. Swanson and J.P. Singh (2005). Development of Supply Chains for Medicinal Plants: A Case Study Involving the Production of Vinca Rosa by Small Farmers in the Patna District of Bihar India, paper presented at the Workshop on Building New Partnerships in the Global Food Chain, Chicago.

Singh, K.M; Meena, M.S; and Jha, A.K (2009). Impact assessment of agricultural extension reforms in Bihar, Indian. Res. J. Ext. Edu. 9 (2): 110-114.

Singh, K.M; Meena, M.S; and Jha, A.K (2011). Agricultural Innovations in India Experiences of ATMA Model, Paper presented in International Conference on Innovative Approaches to Agricultural Knowledge Management: Global Extension Experiences from November 9-12, 2011 at NAAS New Delhi organized by International Society of Extension Education (INSEE).

Singh, R.P., Asokan, M. and Walker, T.S. (1982) - 'Size, Composition and Other Aspects of Rural Income in the Semi-arid Tropics of India'. ICRISAT Economics Program Progress Report 33.

Singh, Ravinder; Rani, Ajita (2017). Human Behaviour. *Notion Press.*

Situation assessment survey of farmers: Access to modern technology for farming, 59th round (January–December 2003). Report No. 499(59/33/2). New Delhi: Ministry of Statistics and Programme Implementation.

Slade, R. H., and Feder, G. (1985). Training and visit extension: A manual of instruction (mimeo). Washington, DC: World Bank.

Smith, J. A.; Foti, R. J. (1998). "A pattern approach to the study of leader emergence". The Leadership Quarterly. 9 (2): 147–160.

Sorrentino, Richard M.; Field, Nigel (1986). "Emergent leadership over time: The functional value of positive motivation". Journal of Personality and Social Psychology. 50 (6): 1091–1099.

Soundararaj Francis (2007). Specking and Writing for Effective business communication: 3-7.

Srinivasan, T.N. (1999) 'Poverty and Reforms in India', paper presented on the conference on reforms, organized by NBER and NCAER, December 14-15, 1999, Neemrana Fort Resort, Rajasthan

Srivastava (1998), "Migration of Labor in India", The Indian Journal of Labor Economics, Vol.41, No.4 24

Srivastava, S.K., Chand, R., and Singh, J. (2017): "Changing Crop Production Cost in India: Input Prices, Substitution and Technological Effects", Agricultural Economics Research Review, Conference Issue.

State Extension Leaders Network (SELN) (2006) 'Enabling Change in Rural and Regional Australia: The role of extension in achieving sustainable and productive futures', online resource: www.seln.org.au.

State of The Nation (2008). - Canada's Science, Technology and Innovation System

Stewart, G. L.; Manz, C. C. (1995). "Leadership for self-managing work teams: A typology and integrative model". Human Relations. 48 (7): 747–770.

Stogdill, R. M. (1948). "Personal factors associated with leadership: A survey of the literature". Journal of Psychology. 25: 35–71.

Sulaiman, V.R. and Sadamate V.V. (2000). Privatizing agricultural extension in India. Policy Paper 10, National Centre for Agricultural Economics and Policy Research, New Delhi.

Suman, R.S. (2014). Attitude of farmers towards sustainability of vegetable cultivation. Journal of AgriSearch 1 (1): 1-3.Swanson BE and Mathur PN. 2003. Review of the Agricultural Extension System in India, The World Bank, July 2003.

Sumita Roy, Tej Verma and Pushpa Gupta (2006). Textbook on Family Approach in Extension Programme Management, First edition, Indian council of Agricultural Research, New Delhi.

Sundaram, K. (2001), "Employment-Unemployment Situation in the Nineties: Some results from NSS 55th Round Survey", Economic and Political Weekly, Vol.36, No.11

Swanson, B.E. (2006). Extension Strategies for Poverty Alleviation: Lessons from China and India. Journal of Agricultural Education and Extension, 12(4): 285-299.

Swanson, B.E. (2006). The changing role of agricultural extension in a global economy. Journal of International Agricultural and Extension Education 13(3): 5–17.

Swanson, B.E. (2008). FAO Report on global review of good agricultural extension and advisory service practices (http://www.fao.org/nr/ext/extn_en.htm).

Swanson, B.E. (2008). Global review of good agricultural extension and advisory service practices. Rome: Food and Agriculture Organization of the United Nations.

Swanson, B.E. and Mathur, P. N. (2003). Review of the Agricultural Extension System in India, The World Bank, July 2003.

Swanson, B.E. and R. Rajalahti. (2010). Strengthening Agricultural Extension and Advisory Systems. Washington, D.C. The World Bank.

Sy, T.; Cote, S.; Saavedra, R. (2005). "The contagious leader: Impact of the leader's mood on the mood of group members, group affective tone, and group processes" (PDF). *Journal of Applied Psychology. 90 (2): 295–305.*

Tagger, S.; Hackett, R.; Saha, S. (1999). "Leadership emergence in autonomous work teams: Antecedents and outcomes". Personnel Psychology. *52 (4): 899–926.*

Teaching Methods in Extension. (2009). Retrieved July 11, 2017 from http://www.uppinghamseminars.co.uk/tme.pdf

The Art of War: Complete Texts and Commentaries (2003) 44, translated by Thomas Cleary

The Relationship between Servant Leadership, Follower Trust, Team Commitment and Unit Effectiveness, Zani Dannhauser, Doctoral Thesis, Stellenbosch University 2007

Thomas, J. J. (2012): "India s Labour Market during the 2000s: Surveying the Changes," Economic and Political Weekly, Vol 47, No 51, pp 39-51.

Trager, J. (1996) The Food Chronology, Aurum Press, London

Tsui; Nifadkar; Ou (2007). "Cross-Cultural Organizational Behavior Research: Advances, Gaps, and Recommendations". Journal of Management. 33 (3): 426–478.

Tyagi, Y. and Verma, S. (2004). Economic rate of return of innovations in technology dissemination component of the National Agricultural Technology Project submitted to the MANAGE, Hyderabad, India, 2004.

Uma Joshi (1999). Text Book of Mass Communication and Media, First Edition, Anmol Publications PVT.LTD.

United Nations (2012): "World Urbanization Prospects: The 2011 Revision," ST/ESA/SER.A/322, Department of Economic and Social Affairs, Population Division, New York.

Unni, Jeemol (1996), "Diversification of Economic Activities and Non-agricultural Employment in Gujarat", Economic and Political Weekly, August 17.

Vaidyanathan, A. (1986), "Labor Use in India: A Study of Spatial and Temporal Variations", Economic and Political Weekly, Vol.29, No.50

Van den Ban, A. W. and Hawkins, H.S (1985). Agricultural Extension, Longman. Scientific and Technical. USA, Newyork.

Van der Meij, L.; Schaveling, J.; van Vugt, M. (2016). "Basal testosterone, leadership and dominance: A field study and meta-analysis". *Psychoneuroendocrinology. 72: 72–79.*

Van Vugt, M., and Ahuja, A. (2011). Naturally Selected: the Evolutionary Science of Leadership. Harper Business.

Van Vugt, M.; Hogan, R.; Kaiser, R. (2008). ""Leadership, followership, and evolution": Some lessons from the past"(PDF). *American Psychologist. 63 (3): 182–196.*

Venkata, Ramana, P. (2006). The changing role of librarian in a challenging dynamic web environment. 4th International Convention Caliber-, Gulbarga, 2-4 February, Inflibnet Centre, Ahmedabad. (2006).

Vidyabhushan and Sach Dev, D.R. (1998). An Introduction to Sociology. Kitab Mahal Agencies, Allahabad.

Vinkenberg; Engen; Eagly; Johannesen-Schmidt (2011). "An exploration of stereotypical beliefs about leadership styles: Is transformational leadership a route to womens promotion?"(PDF). *The Leadership Quarterly. 22 (1): 10–21.*

Visaria, P. and Basant, R. (1993), "Non-agricultural Employment in India: Trends and Prospects", Sage Publications, New Delhi

Walker, T.S. and Ryan, J.G. (1990) - Village and Household Economies in India's Semi-Arid Tropics. Baltimore: Johns Hopkins University Press.

Western, Simon (2019). Leadership: A critical text. London: Sage.

Wheatcroft, Andrew; Overy, Richard *(1989).* The Road to War: The Origins of World War II *(Revised ed.). London: Random House (published 2012).*

WHO (World Health Organization) (1989). Evaluation of programmes to ensure food safety. Geneva: Author.

Working Group of Agricultural Extension (2007). Recommendations of working group on agricultural extension for formulation of eleventh five-year plan (2007–12). New Delhi: Planning Commission (23) (PDF) Indian Agricultural Extension Systems and Lessons Learnt.

Working Group on Agricultural Extension (2007). Recommendations of working group on agricultural extension for formulation of eleventh five-year plan (2007–12). New Delhi: Planning Commission.

World Bank (2004). Implementation completion report on a credit and a loan to India for the Diversified Agricultural Support Project. Report No. 29529. Washington, D.C.

World Bank (2005). Implementation completion report on a credit and a loan to India for the National Agricultural.

Yadav; Singh (2015). "Gender and Politics of Culture". Political Science: 77.

Yella Reddy N Audio-Visual aids. EEI Publications, Rajendranagar, Hyderabad

Yella Reddy, N. (1998). Audio-Visual Aids for Teaching, Training and Extension. Haritha Publishing House, Hyderabad

Zaccaro, S. J. (2007). "Trait-based perspectives of leadership". American Psychologist. 62 (1): 6–16.

Zaccaro, S. J., Gulick, L. M. V. and Khare, V. P. (2008). "Personality and leadership". In C. J. Hoyt, G. R. Goethals and D. R. Forsyth (Eds.), Leadership at the crossroads (Vol 1) (pp. 13–29). Westport, Connecticut: Praeger.

Zaccaro, S. J.; and Banks, D. J. (2001). "Leadership, vision, and organizational effectiveness". In S. J. Zaccaro and R. J. Klimoski (Editors), The Nature of Organizational Leadership: Understanding the Performance Imperatives Confronting Today's Leaders. San Francisco: Jossey-Bass.

Zuppo, Colrain M. "Defining ICT in a Boundaryless World: The Development of a Working Hierarchy" (PDF). International Journal of Managing Information Technology (IJMIT). p. 19.

Index

A

Abstract intelligence 231
Adoption process 190
Agribusiness 305
Agricultural credit 307
Agricultural exports 306
Agricultural extension 146, 222
Agricultural extension education 1
Agricultural journalism 180
Agricultural technology information centre 48
Ambivert personality 228
Animal behavior 224
Aristotle's model of communication 174
Atal mission for rejuvenation and urban transformation 73
Atal Pension Yojana 70
Attracting and retaining youth in agriculture 49
Audio-visual aids 154

B

Baroda village reconstruction project 37
Barriers 179
Behaviour 224
Beti Bachao, Beti Padhao Yojana 72
Binet test 232
Booklet 136
Bulletin 135
Business communication 287
Business leadership skills 317

C

Campaign 134
Capacity building 128
Caste leaders 89
Cognitive domain 225
Communication 167, 177, 236, 318
Communication channels 187
Communication models 174
Communication skills 235, 313

Community development 83
Community development programme 77
Community development project 39
Concept 7, 87
Constraints 67
Consumers behavior 225
Curriculum development 120
Cyber extension 52

D

David Berlo's model 176
Deen Dayal Upadhyaya Grameen Kaushalya Yojana 73
Deen Dayal Upadhyaya Gram Jyoti Yojana 73
Definition 5, 22, 86
Differences 4
Digi locker 75
Digital India 72
Downward communication 179
Drone entrepreneurs 295

E

Educational psychology 197, 221
Empowering smallholder 145
Entrepreneurship development 287, 308
Etawah pilot project 38
Evaluate programme planning 32
Evaluation 31, 97, 109
Evolution 1
Exhibition 134
Extension administration 90
Extension evaluation 30
Extension evaluation process 33
Extension learning process 136
Extension organization 143
Extension process 10
Extension programme planning 21
Extension programme planning model 28
Extension system 35
Extension teaching and learning process 130
Extension teaching methods 140
Extension training programmes 119
Extrinsic factors 116

F

Fabian entrepreneurs 295
Farm and home visit 133
Farmer-led extension 55
Farmers FIRST 50
Field tour 134
Financing 328
Firka Bikas Yojana 38
First-line extension system 43
Five assets 67
Flip charts 166
Folder 135
Food grains 305
Food processing 306
Footnotes 259
Formal education 2
Front-line demonstration 128
Function 11, 18, 93
Fundamental objective 17

G

Gold monetisation schemes 74
Group-contact methods 132
Group discussion 134, 278
Grow more food campaign 38
Grow-more-food enquiry committee report 39

Growth 1
Gurgaon project 36

H

High yielding variety programme 42
History 1, 123
Horizontal communication 179
Horizontal technology transfer 102
Human behavior 224

I

ICT applications 144
Imitating entrepreneurs 295
Importance 8
Impromptu presentation 273
Indian village service 38
Individual-contact methods 132
Informal education 2
Innovation decision process 187
Innovative entrepreneurs 295
In-service training 106
Institute village linkage programme 49
Integrated rural development programme 41
Intelligence 231
Intensive agricultural area programmes 41
Intensive agricultural district programme 40
Interpersonal communication 241
Intrinsic factors 116

J

Jawahar Rozgar Yojana 48
Judgmental skills 313

K

Katz and Lazarfeld's model 176
Key village scheme 40
Kisan Vikas Patra 72
Knowledge stage 188
Krishi Vigyan Kendra 44, 122
KVK portal 129

L

Lab to land programme 46
Lasswell's model 175
Leadership skills 312
Levels 15

M

Mahila Udyami Yojana 304
Make in India 71
Market-led extension 56
Marthandam project 37
Mass communication 241
Mass or community-contact methods 132
Meaning 3, 86
Mechanical intelligence 231
Media-mix planning 151
Mera Gaon Mera Gaurav 50
Method demonstration 133
Mission Indradhanush 72
Modern journalism 181
Monitoring and evaluation 95, 313
Monitoring mechanism 126
Motivation 229
Motivation skills 320
Multi-sector support 128

N

Namami Gange Project 77
National agricultural innovation project 48, 49
National agricultural technology project 47, 49

National agriculture extension project 48
National Bal Swachhta Mission 74
National demonstration 43
National extension service 39
National heritage city development and augmentation yojana 74
National innovations on climate resilient agriculture 49
National Institution for Transforming India (NITI AAYOG) 76
National sports talent search scheme 76
Nilokheri experiment 39
Non-formal education 3
Non verbal communication 241

O

Objectives 16, 25
One rank one pension (OROP) scheme 74
On-farm testing 127
Operational research project 44
Operation flood 40
Opinion leaders 90
Oral presentation skill 251
Orientation training 106
Origin 4

P

Page index 258
PAHAL-direct benefits transfer for LPG (DBTL) consumers scheme 76
Pamphlet 135
Pandit Deendayal Upadhyay Shramev Jayate Yojana 73
Perception 228
Personality 226
Philosophy 12
Pilgrimage rejuvenation and spiritual augmentation drive 74
Planning AV aids 163
Political leaders 90
Post-training phase 115
Pradhan Mantri Awas Yojana 70
Pradhan Mantri Fasal Bima Yojana 71
Pradhan Mantri Garib Kalyan Yojanaye 71
Pradhan Mantri Gram Sinchai Yojana 71
Pradhan Mantri Jan Aushadhi Yojana 71
Pradhan Mantri Jan Dhan Yojana 69
Pradhan Mantri Khanij Kshetra Kalyan Yojana 76
Pradhan Mantri Mudra Yojana 69
Pradhan Mantri Suraksha Bima Yojana 70
Pradhan Mantri Ujjwala Yojana 77
Precise writing 263
Pre-service training 105
Pre-training phase 112
Prime minister's employment generation programme (PMEGP) 302
Principles 19, 25, 115
Print media 135
Privatization 50
Priya Darshini Yojana 304
Process 13
Programme-planning process 26
Public speaking 276

R

Radio 135
Radio and television 147
Rashtriya Gokul Mission 76
Rashtriya Gram Swaraj Abhiyan 77

Reading skills 261
Refresher training 106
Religious leaders 90
Responsibilities 127
Result demonstration 134
Rural development 63
Rural entrepreneurship 331
Rural landless employment guarantee programme 48
Rural leadership 85
Rural reconstruction 36
Rural sociology 197, 209

S

Sagarmala project 75
Salesmanship 8
Sansad Adarsh Gram Yojana 70
Scope 10
Setu Bharatam Project 77
Seva-gram 37
Shriniketan project 36
Shyama Prasad Mukherji Urban Mission 75
Simple model 112
Skill development 308
Skill India 72
Smart city mission 74
Social change 219
Social ecology 204
Social entrepreneur 295
Social institutions 215
Social intelligence 231
Social stratification 207
Soil health card 50
Soil health card scheme 72
Speaking 276
Spiral model 113
Staffing pattern 125
Standard curriculum 120
Stand-up India 75
Start-up India 75
Steps 26
Stree Sakthi Package 304
Strengths, weakness, opportunities and threats (SWOT) analysis 297
Strip index 258
Structural communication 236
Subject matter 143
Sukanya Samriddhi Yojana 69
Supply chain management 322
Swachh Bharat Abhiyan 72
Swadesh Darshan Yojana 73

T

Teaching aids 157
Teaching methods 137, 162
Teaching process 138
Technical skills 313
Technology assessment and refinement 47
Television 135
Time-management skills 312
Total quality management 323
Traditional leaders 89
Trainers training centre 46
Training and visit 40
Training for development 104
Training methodology 121
Training process 112
Training strategy 114
Transfer of technology 100
Types 32

U

Udaan scheme 74
UJWAL discom assurance Yojana 76
Upward communication 179

U

Verbal communication 240, 241
Vertical technology transfer 102
Vestibule training 106
Vikalp scheme 76
Visible card index 258
Visual aids 155

W

Watershed development programme 48
Women entrepreneurs 304
Writing skills 248
Written handouts 165